Calving Management and Newborn Calf Care

Your bonus with the purchase of this book

With the purchase of this book, you can use our "SN Flashcards" app to access questions free of charge in order to test your learning and check your understanding of the contents of the book To use the app, please follow the instructions below:

1. Go to **https://flashcards.springernature.com/login**
2. Create a user account by entering your e-mail address, assigning a password and inserting the coupon code here below.

Your personal "SN Flashcards" app code   04B36-084D7-54FA9-A6E6A-EEA83

If the code is missing or does not work, please send an e-mail with the subject "**SN Flashcards**" and the book title to **customerservice@springernature.com**.

João Simões • George Stilwell

Calving Management and Newborn Calf Care

An interactive Textbook for Cattle Medicine and Obstetrics

 Springer

João Simões
Department of Veterinary Sciences
School of Agrarian and Veterinary
Sciences, University of Trás-os-
Montes e alto Douro
Vila Real
Portugal

George Stilwell
Faculty of Veterinary Medicine
University of Lisbon
Lisbon
Portugal

This work contains media enhancements, which are displayed with a "play" icon. Material in the print book can be viewed on a mobile device by downloading the Springer Nature "More Media" app available in the major app stores. The media enhancements in the online version of the work can be accessed directly by authorized users.

ISBN 978-3-030-68170-8   ISBN 978-3-030-68168-5   (eBook)
https://doi.org/10.1007/978-3-030-68168-5

This Springer imprint is published by the registered company Springer Nature Switzerland AG
The registered company address is: Gewerbestrasse 11, 6330 Cham, Switzerland

Drawing on the Book Cover by Maria Stilwell, Instagram Picture Gallery @mariaalapis.

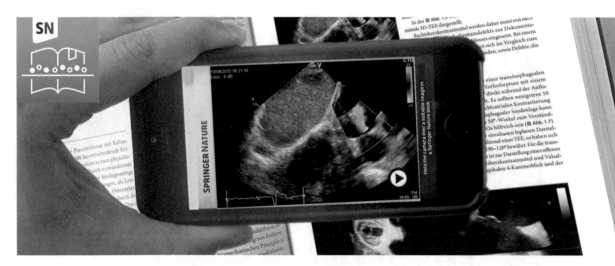

□ Fragments of the Kahun Papyrus. The earliest known writing on veterinary medicine around 1900 BC from Africa. Original from Petrie Museum, University of London. (Adapted from Arey (2014) with permission from Elsevier. Arey BJ. A historical introduction to biased signaling. In: Arey BJ (ed.). *Biased Signaling in Physiology, Pharmacology and Therapeutics*. Academic Press: San Diego, CA; 2014. pp. 1–39. ► https://doi.org/10.1016/B978-0-12-411460-9.00001-X)

"Tell me and I forget. Teach me and I remember. Involve me and I learn."

Uncertain origin. Probable ancient Chinese proverb popularized by Benjamin Franklin (1706–1790)

Foreword 1

As experienced veterinarians and researchers in bovine health management and reproduction we felt immensely proud to be invited to foreword a textbook dealing with this highly relevant and decisive, physiologic (unfortunately, sometimes pathological) phase of calving. We observe very frequently that the "repro-experts" in the field, or even in the Academy, pay extensive attention to the reproductive process: puberty, oestrus cycle, synchronization, insemination, pregnancy diagnosis... but our work and specialty in the bovine reproduction is far from finished after a pregnancy is achieved. However, the next processes and events in the pregnant cows up to the calving are very often underestimated or even ignored.

Parturition is an essential phase in bovine production, both dairy and beef. Besides the relevance of the newborn itself (it constitutes the rearing or replacement animals, that is, the future or our herds), in dairy cattle parturition supposes the entry of the cow into a new lactation. The consequences of a calving have an important impact on the health of the cow, its future and, consequently, the economy of the farm. Therefore, a delivery in good conditions is the basis for the success of the next pregnancy and the entire lactation, and is thus an essential event for dairy production. In beef cattle, calving is even more relevant, since it gives rise to the calf, the single objective of these production systems. The management of this critical phase is therefore crucial for both dam and calf(ves).

We also used the term "phase" when referring to calving, because it is not just the act of the dam delivering its offspring, but also includes many other processes and the related risk factors. Accordingly, the authors include all other events, from breeding to the puerperial cow and the care of the newborn. Consequently, we find in the different chapters the description of the main anatomical issues of the reproductive tract and caudal third of the cow, reproductive physiology, pregnancy and associated problems, description of risk factors, calving prodromes, the calving self and calving assistance (dystocia, vaginal calving, foetotomy and cesarean section), and finally, paying attention to the newborn calf, but, also, including the care of the puerperal mother, which is of the utmost relevance.

The layout of this book is extremely easy to follow and didactic. It includes an accurate selection of images, highlighted relevant conclusions, practical tips, tables, clear, complete and brief contents, and examples of clinical cases and didactic questions-answers which help to efficiently gain the soundest knowledge. Last but not least, this book also contains a recommendation of sound references, visual guides, and teaching textbooks. This text will help to educate professionals to implement a proper obstetrical approach, as well as to design, perform, and evaluate holistic calving protocols in herds, preventing and treating main problems related to the calving cow. It is likely that this is the text the authors would have liked to have and could not find while they were studying, or even when they were addressing a calving challenge in their early daily practice. This book reflects the perfect combination of the veterinarian excellence in both the practical and scientific fields of buiatrics, two qualities that the authors clearly demonstrate. This will become a reference book for many of our residents in training.

We would like to thank the authors for the honor bestowed on us to write the foreword of this book, and we encourage them to continue with this work and teaching philosophy, contributing in this way to instructing new generations of veterinarians, specialists, and animal scientists.

Susana Astiz
ECBHM President,
Oberschleißheim (Munich), Germany

Raphaël Guatteo
ECBHM Past President
Oberschleißheim (Munich), Germany

Foreword 2

For many years, animal health and welfare has been an increasingly important part of animal science and livestock production. One reason for this is that it has a very significant impact on the acceptance of livestock production systems by consumers and the economic success of farms. It is important to note that, most of the time, farmers are confronted first with animal health problems, even before the veterinarians. Normally the 'first aid', such as calving assistance, including moderate dystocias, is performed by people working on farms or by related professionals. Therefore it is essential that these people have the best knowledge to better know the different steps of all related processes, to be able to implement the best management plans (e.g. for calving) and to use the best available practices. Only this can ensure the highest possible level of animal welfare and health. This book by João Simões and George Stilwell is therefore highly recommended for teachers and students of animal science, and for practitioners looking to acquire profound knowledge on the subject of 'healthy calving', i.e. unassisted calving where dam and calf can express their full natural behaviour. As such, I wish the book every success.

Matthias Gauly
President of the EAAP – The European Federation of Animal Science
Rome, Italy

Preface

Towards a Healthy Calving

What is a healthy calving? Perhaps it could be better defined as an unassisted calving where a healthy dam and a healthy calf can express their natural behaviour. This definition involves the ultimate goal of species survival, which is to ensure a viable offspring and a female able to breed in a relatively short period, i.e. the continuity of the species. In a natural environment, calving usually occurs in a quiet and secluded area, away from potential predators.

Dairy and beef industries are based on extensive or intensive production systems, and the breeding management should be adapted to either system. Farmers and stockpersons are primarily responsible for all adopted measures, ensuring animal health and welfare. Veterinary obstetricians should acquire adequate skill and competences to diagnosis, treat and prevent dystocia and other reproductive abnormalities. They are also responsible for designing, taking part in and (re)evaluating the effectiveness of calving management programmes for cattle herds.

High-quality academic and professional training of veterinary students and veterinary technician students, as well animal production students, is a basilar element to achieving the objectives of the aforementioned competences.

Throughout over three decades of academic and professional experience, we have been able to identify the needs and expectations of students, veterinarians, farmers and other stakeholders in the cattle production business. This book aims to contribute to students' education and to help professional people respond to difficulties and doubts related to calving management, bovine obstetrics and newborn calf care.

We wish to thank all the people who, directly or indirectly, contributed to this book. We also wish to thank all farmers, colleagues, students and others who, during these last three decades, taught us so much. We are simply passing on some of this knowledge and experience.

We hope that this book can fulfil its mission: helping people and animals (cows and calves), by minimizing the adverse effects of dystocia or complicated pregnancies and thus improving animal health and welfare, but also farmers' success.

João Simões
Vila Real, Portugal

George Stilwell
Lisbon, Portugal

Votes for this Book from the Community

The role of livestock veterinarians is to ensure that increasing consumer demands for high standards of animal welfare practices are implemented on our farms. Livestock farming requires delivery of a sustainable, climate-change sensitive, yet profitable large ruminant production system that is able to meet the global food security needs of an animal protein-hungry world. Veterinarians involved in servicing these farming businesses, irrespective of whether it is encouraging smallholder farmers on subsistence farms in developing countries, or larger family farming enterprises in developed countries, or even corporate international agricultural trading entities, certainly need to be equipped with the skills in obstetrical and calf management that can meet the challenges outlined and illustrated in this book. Yet they should also be endeavouring to ensure that 'best practice' management practices are adopted that will minimise the need for many of the interventions described in the book. Increasingly, their role should be to assist livestock farming businesses to continually work towards improving the nutrition, genetics, health, and biosecurity and other animal husbandry management inputs that enable our farmed animals to grow and reproduce more efficiently, with fewer remedial interventions required. With this in mind, this book is a very useful manual for those veterinarians and aspiring students involved in provisions of bovine obstetrics and calf management services plus preventive health management advice. As the manual contains an abundance of mostly practical yet very educational information, informed by clinical case situations and excellent illustrations and videos (available online) that address most of the issues arising in cattle breeding management, it is highly recommended.

Peter Windsor | DVSc, PhD, BVSc(Hons)
Grad. Cert Ed. Studies (Higher Ed.), DipECSRHM
Professor, Faculty of Veterinary Science
The University of Sydney
Sydney, NSW, Australia

As the title claims, *Calving Management and Newborn Calf Care* by João Simões and George Stilwell is a textbook that covers the fundamentals of breeding cattle from the beginning of the process all the way through to the care of the dam and calf after calving. It starts with a useful overview of the physiology and anatomy of reproduction and then moves through the problems of mid- and late pregnancy to normal calving and dystocia and chapters on obstetrical manoeuvres, fetotomy and caesarean. The inclusion of a chapter focusing on calf care is a highly welcome addition, as this is an area which is often neglected by textbooks and similar resources.

Throughout the book illustrations are used to emphasise, clarify and elucidate the points of the text. There is even a chapter which is mostly images from obstetrical case studies, which adds considerably to the text presented in earlier chapters. Case studies and tips feature throughout the book, making it more than just another textbook. Similarly, questions on the contents form an integral part of each chapter, culminating in a chapter consisting completely of questions.

These additions and alternative approaches, alongside the authors' clear enthusiasm and expertise, make this a textbook that will be useful for anyone with an interest in managing the calving cow.

Richard Laven | PhD, BVetMed, MRCVS
Professor, School of Veterinary Science
Massey University Press
Auckland, New Zealand

The productivity of dairy and beef cattle largely depends upon their normal reproduction, with calving a high-risk period for both cows and calves. The new book *Calving Management and Newborn Calf Care: An interactive Textbook for Cattle Medicine and Obstetrics* attempts to unravel this complex subject by providing practical guidance and information on all aspects of calving management and calf care. It covers the major diseases and problems and preventative methods, and considers animal welfare issues. Comprehensive, practical, and easy-to-read, this unique book is an excellent resource for veterinary students, veterinary surgeons, and anyone working with cattle. The information is presented concisely and this, together with many images and videos, highlighted relevant conclusions, practical tips, brief synopses, and examples of clinical cases, makes it an ideal immediate source of reference and information. The authors' ability to teach is another plus, with complex issues explained in an understandable and easy-to-remember way. In addition, a didactic questions-answers final chapter makes this book an excellent textbook for veterinary and animal science students. For all these reasons I recommend this book without reservations.

Pascal Oltenacu | DVM, PhD
Professor, Department of Animal Sciences
University of Florida – IFAS
Gainesville, FL, USA

It is with great pleasure that I recommend this excellent new book entitled *Calving Management and Newborn Calf Care: An interactive Textbook for Cattle Medicine and Obstetrics*. This is a great textbook for veterinary students and a great resource for early career cattle veterinarians. I particularly like the 'Important', 'Key Points,' and 'Tips' that are highlighted features within each chapter. The book makes effective use of real cases which provide the reader with very practical application of the information and distinguishes this book from many other veterinary texts as the context helps tremendously with learning, understanding, and application. The figures, pictures, and drawings are quite useful, as are the web links to resources. The chapters on calving management (► Chap. 3) and on vaginal delivery and newborn care (► Chap. 5) are novel and very practical. Finally, the authors' use of picture cases in the last chapter provides a great resource for the young veterinarian as they begin their career.

Todd Duffield | DVM, DVSc
Professor and Chair, Department of Population Medicine
Ontario Veterinary College
University of Guelph
Guelph, ON, Canada

Contents

About the Authors

João Simões
is a professor of Large Animal Medicine and Reproduction at the University of Trás-os-Montes e Alto Douro. As a clinician and academic, he has dedicated 27 years of his professional career to veterinary students and producers, and collaborated with national and European animal production associations. On the basis of this work experience and research, he has written numerous scientific and technical publications. In recent years, he has also edited several books and special issues on animal production and veterinary medicine for scientific journals.

George Stilwell
worked as a practitioner mainly with farm animals for over 15 years before joining the Veterinary Medicine Faculty (FMV) in Lisbon, where he now lectures on Farm Animal Clinics and Deontology and Bioethics. He did his PhD on cattle pain management and is a diplomate at the European College of Bovine Health Management (ECBHM). He leads the Animal Behaviour and Welfare Research Lab (CIISA-FMV). George has been involved in various European projects on ruminant health and welfare (AWIN, Anicare, BovINE) and in several EFSA working groups. He is a board member of the Portuguese Veterinary Council and has published more than 50 peer-reviewed papers and several books on farm animal health and welfare.

Abbreviations and Acronyms

ACTH	Adrenocorticotropic hormone
AKAV	Akabane virus
AV	Aino virus
Bpm	Beats per min
BTV	Blue tongue virus
BVD	Bovine viral diarrhoea
BVDV	Bovine virus diarrhoea virus
BW	Body weight
CAP	Contraction associated proteins
CL	Corpus luteum
Co	Soccygeal vertebrae
CRH	Corticotropin-releasing hormone
C-section	Caesarean section
E_1	Oestrone
E_1S	Oestrone-3-sulphate
E_2	Oestradiol
E_2B	Oestradiol benzoate
EBVs	Estimated breeding values
EPDs	Expected progeny differences
FPT	Failure in passive transfer
FSH	Follicle-stimulating hormone
GnRH	Gonadotropin releasing hormone
HPA axis	hypothalamic–pituitary–adrenal axis
i.m.	Intramuscular
i.v.	Intravenous
IBR	Infectious bovine rhinotracheitis
IFNT	Interferon-tau
Ig	Immunoglobulin
IP	Intraperitoneal
IVF	In vitro fertilization
IVM	In vitro maturation
IU	International units
L	Lumbar vertebrae
LH	Luteinizing hormone
MHC	Major histocompatibility complex
NMS	Neonatal maladjustment syndrome
NSAIDs	Nonsteroidal anti-inflammatory drugs
Mpm	Movements per min

P_4	Progesterone
PCR	Polymerase chain reaction
PGE_2	Prostaglandin E_2
$PGF_{2\alpha}$	Prostaglandin $F_{2\alpha}$
PGHS-II	Prostaglandin H synthase type II
PM	Perinatal mortality
PTA	Predicted transmitting ability
REM	Rapid eye movement
S	Sacral vertebrae
s.c.	Subcutaneous
SBV	Schmallenberg virus
T	Thoracic vertebrae
T_3	Triiodothyronine
T_4	Thyroxine
USP	United States pharmacopeia

Reproductive Anatomy and Physiology of the Nonpregnant and Pregnant Cow

Contents

© Springer Nature Switzerland AG 2021
J. Simões, G. Stilwell, *Calving Management and Newborn Calf Care*,
https://doi.org/10.1007/978-3-030-68168-5_1

1

Learning Objectives

- To describe the anatomy and morphology of the main reproductive structures.
- To identify the main physiological and hormonal mechanisms controlling oestrus cycle and pregnancy.
- To define reproductive cyclicity, pregnancy, uterine involution and ovarian resumption.
- To relate anatomical and physiological findings with clinical approaches to the reproductive tract.

1.1 Introduction

Female fertility can be defined as the ability to successfully conceive, deliver and raise a healthy offspring and be able to reassume, in a short period, a new reproductive cycle. The function of the reproductive system, under hormonal influence, is to ensure these successive tasks. Several other organs and systems (e.g. hormonal organs and musculoskeletal system) contribute to accomplishing this function.

The reproductive tract is composed by the external (vulva and vaginal vestibule) and internal (ovaries, oviducts, uterus and vagina) genitalia. Generally speaking, the ovaries have a folliculogenesis function as well as being responsible for the hormonal support of pregnancy; the uterus ensure the embryonic and foetal development; and the vagina is responsive for the reception of spermatozoids at mating time and transport of the foetus at calving time.

During the prepubertal period, calves grow developing their body systems, including the reproductive tract, which can be defined as postnatal maturation. This prepubertal period varies significantly, from 9 to 24 months, according to breed – puberty in dairy heifers will occur around 6–12 months, while in beef breeds, it can go up to 24 months. A significant development of the reproductive system occurs at puberty time under hormonal influence. Puberty begins when the hypothalamus is able to produce enough gonadotropin-releasing hormone (GnRH) to stimulate ovarian activity and oestrogen production. Nevertheless, heifer's development continues until it reaches full maturity at around 24–26 months of age for dairy breeds or up to 36 months for beef breeds. At puberty, heifers become cyclic presenting an oestrus cycle of 20–21 days which ultimately will lead to mating and to conception. Oestrus cyclicity is interrupted by pregnancy (gestational anoestrus), until postpartum ovarian activity resumption (i.e. postpartum anoestrus, lactational anoestrus or suckling anoestrus). Several conditions, such as persistence of corpus luteum (CL), follicular or luteal ovarian cysts, intense negative energy balance, low body condition (cachexia) and others, can cause pathological anoestrus or interfere in the oestrus cycle.

Outstanding knowledge of the anatomy and physiology of the reproductive system is fundamental for the veterinarian and, especially, for the obstetrician. The adequate recognition of the reproductive organs and related structures, as well as their functioning, allows for proficient reproductive diagnosis (▶ Box 1.1), treatment and prevention of pregnancy disorders, dystocias and their complications [15]. This introductory chapter aims to describe the main anatomical and physiological aspects related to the reproductive system in the female Bovidae.

Box 1.1 Physical Examination and Anomalies of the Reproductive Tract

Identification and anamnesis at clinical examination
- Identification of the cow (ID, breed, age, parity and lactation)
- Reproductive history of the herd (breeding program, pregnancy losses, abortion rates, reproductive indexes, calving difficulty and reproductive diseases' incidence)
- Reproductive history of the cow (oestrus: irregular or not observed; date of last calving and last observed oestrus; increased aggression or nymphomania; inter-oestrus interval, <18 days or >24 days)

General physical examination (register: body condition score; behaviour; overdeveloped neck muscles; signs of virilism; prominence of tail head; relaxed pelvic ligaments; bilateral ventral abdominal enlargement)
   Inspection of the vulva and vaginal vestibule
- Vulva position and conformation (sloping conformation; swollen; lacerated; discharge: colour, quantity and consistency – blood, creamy, haemorrhagic, foul smelling black-brown and foul smelling; placental debris after calving)
- Efficiency of the vulvar seal
- Enlarged clitoris and abundant vulvar tuft of hair

Transrectal examination (manual and ultrasonographic evaluation)

- Position of the genital tract (e.g. retraction of the uterine horns)
- Cervix – position, size and mobility (cervix open/closure; lacerated; purulent discharge; double; absent)
- Uterus/uterine horns – position, size and contents (distended: unilateral or bilateral; contents, fluid, emphysematous, bones-foetal maceration and amorphous mass – foetal mummification; uterine prolapse; mural swelling; abnormal size and shape; signs of pregnancy and foetal life; abortion; stillbirth)
- Ovaries – position, size (ovarian structures: follicles, number, size and consistency; corpora lutea, position, size and age; presence of ovarian cysts – size and possible type)
- Oviducts (enlarged and palpable; thickened)

Vaginal examination (vaginoscopy)
- Vaginal walls (vaginal discharge: purulent, blood creamy, haemorrhagic, mucoid, translucid; urine in anterior vagina, vaginal mucosa erosions, granulation, pustular, necrotic; peri-vaginal swelling; vaginal length; vaginal contents; colour and conformation of the external cervix Os; cervical secretions; vaginal prolapse)

1.2 External Genitalia (Vulva and Vaginal Vestibule)

External genitalia comprise the vulva that includes the labia (*labia majora* and *labia minora*) and (dorsal and ventral) commissures, clitoris, vestibule and vestibular glands. The vulvar labia should appose completely and serve as the first physical barrier for bacterial contamination. Their union forms the dorsal and ventral commissures (◻ Fig. 1.1). The clitoris is a sensitive structure housed in the ventral commissure.

The vestibule is 10–12 cm in length. In its junction with the vagina, the external urethral orifice emerges ventrally. In heif-

1

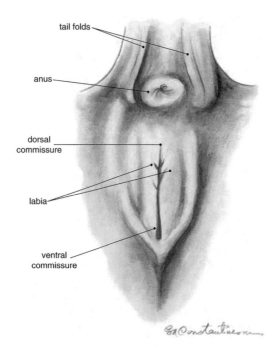

tail folds

anus

dorsal commissure

labia

ventral commissure

☐ **Fig. 1.1** Vulva and perineal region of the cow. (Adapted from Constantinescu [3] with permission from John Wiley and Sons)

ers, the hymen (more or less pronounced) is also observed in this area before mating has occurred. The vulvo-vaginal sphincter muscle prevents urine reflux into the vagina and acts as a second physical barrier for bacterial contamination. Bartholin's glands secrete a viscous fluid that is drained through the Gartner's ducts and tubes, providing lubrication during oestrus.

1.3   Internal Genitalia (Vagina, Uterus, Oviducts, Ovaries and Ligaments)

Vagina   The vagina is a long virtual tube (25–30 cm in length) which increases in size during pregnancy. The vaginal vestibule caudally delimits it. Cranially, the vaginal fornix surrounds the cervix. Sensitive nerve fibres are located in the dorsal region of the vagina which stimulates abdominal muscles contractions during parturition.

Uterus   The uterus is formed by three tubular structures: the cervix, the body and two horns.

In heifers, the uterus is fully located in the pelvic cavity. During pregnancy, its volume increases widely filling the ventral and caudal part of the abdominal cavity. After calving, it takes about 40 days for the uterus to regain its original size (uterine involution). The size of the nonpregnant uterus tends to increase slightly after each successive calving (☐ Fig. 1.2).

(a) *Cervix*

The cervix is a tightly closed tubular sphincter with a length of approximately 10 cm and a diameter of 3 cm. It is composed of dense connective tissue presenting significant collagenous tissue with a small amount of smooth muscle. Usually, it shows two to five internal annular rings or folds and well-delimited internal Os (internal orifice of the uterus in which the uterine body communicates with the cervical canal) and external Os (external orifice of the uterus in which the cervical canal communicates with the anterior vagina). The cervix has no glands. However, several calyciform cells are located between the cervical rings, which secrete thick mucus responsible for selection and transport of spermatozoids. The cervix ensures an aseptic uterine environment during the oestrus cycle, pregnancy (cervical plug) and gives passage to the foetus during parturition. The cervix slightly opens during oestrus and up to 20 cm or more at calving. At this time, foetal pressure on the cervix stimulates its sensitive nerve fibres so that oxytocin is released from the anterior pituitary causing contraction of the myometrium (Ferguson's reflex).

(b) *Body*

The body of the uterus is short measuring approximately 3–4 cm in length. It is the connection between the two uterine horns and the cervix.

(c) *Uterine Horns*

The length of the uterine horns is about 35–40 cm (☐ Fig. 1.3). It is a funnel-shaped structure distally attaching to the oviduct. Ova is fecundated on the oviduct and implanted in one of the uterine horns which progressively increases in size to house the foetus during whole pregnancy.

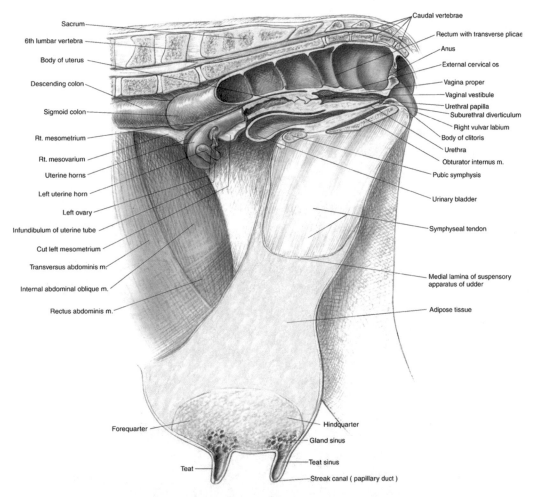

□ **Fig. 1.2** The topography of the female genitalia. Legend: Median section. m muscle, Rt right. (Adapted from Adapted from McCracken et al. [14] with permission from John Wiley and Sons)

The uterine wall has three layers: the endometrium, myometrium and perimetrium. The endometrium is a mucous membrane presenting a stratified columnar epithelium. Tubular glands and 70–140 caruncles are also present. The caruncles are distributed irregularly or in columns on the myometrial surface. Their shape are ovoid, and the size varies from less than 1.5 mm to approximately $12 \times 4 \times 2.5$ cm in the nonpregnant and pregnant uterus, respectively. Under hormonal influence, mainly during proestrus, oestrus and pregnancy, the endometrium vascularity and thickness increase. The endometrium also produces and segregates prostaglandin $F_{2\alpha}$ ($PGF_{2\alpha}$) during the oestrus cycle and at calving.

The myometrium consists of two layers of smooth muscle fibres – the circular and the longitudinal layer. A vascular layer is found between these two muscle sheets. Significant hyperplasia and hypertrophy of the muscle fibres occur during pregnancy. Myometrial contractions facilitate sperm transport from the cervix to the oviduct during oestrus. It also allows for foetus alignment and expulsion during calving. Both sympathetic and parasympathetic nervous systems innervate the myometrium.

Oviducts Each oviduct has approximately a length of 25 cm. It is formed by the utero-tubal junction, isthmus, ampullary-isthmic junction,

1

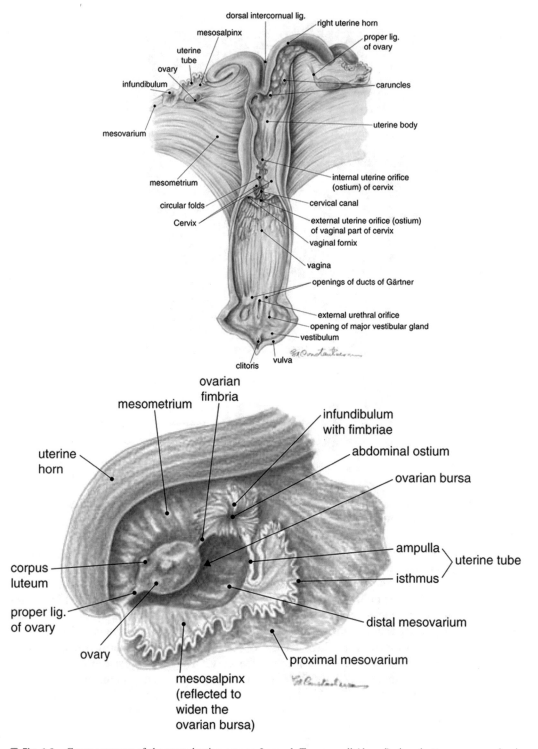

Fig. 1.3 Gross anatomy of the reproductive organs. Legend: Top, overall (dorsal) view; button, ovary and oviduct morphology. (Adapted from Constantinescu [3] with permission from John Wiley and Sons)

ampulla and infundibulum. The fimbriae of the oviduct infundibulum (funnel-shaped) surround the oviduct. The ovarian bursa may facilitate the recovery of the ova by the infundibulum and its fimbria, keeping them from falling into the body cavity. It is within the ampulla that fertilization occurs.

Spermatozoa attach to the isthmus wall, and this is where it is thought that many physiological changes occur to their membranes, which are essential for sperm capacitation.

Ovaries   Each ovary is ovoid shaped, measuring around 3.5 × 2.5 × 1.5 cm. They present a well-vascularized medulla and a cortex where tertiary follicles and corpora lutea emerge. The size of the ovary varies according to the number and size of these superficial structures. The ovaries ensure oogenesis, as well as the hormonal production (namely oestrogen and P_4) during the oestrus cycle and pregnancy.

Uterine and Ovarian Ligaments   The mesovarium is a serous membrane supporting each ovary. It is parietally attached next to the pelvis given the spiral conformation of the nonpregnant uterine horns. Its cranial border is called suspensory ligament of the ovary. Laterally, the mesovarium originates the mesosalpinx which is attached to the oviducts. Caudally, it forms the mesometrium, which is attached to the uterine horns. All three ligaments constitute the broad ligament of the uterus. The mesosalpinx and mesovarium also originate the ovarian bursa. The ovarian bursa is wide and open but closely related with the oviduct.

The intercornual ligament links caudally both uterine horns.

Vascularization   The vascularization of the internal genitalia consists of (1) a pair (left and right) of ovarian arteries which supply the ovaries, oviducts and cranial part of the uterus; (2) a pair of uterine arteries which supply most parts of the uterus; and (3) a pair of vaginal arteries which supply the vagina.

> **Tip**
>
> From the fourth month of pregnancy, the uterine artery ipsilateral to the pregnant uterine horn can be easily detected by transrectal palpation. The fremitus of this uterine artery serves as pregnancy diagnosis, and its calibre will indicate pregnancy time. In case of twins, fremitus will be perceived in both uterine arteries.

Sympathetic Nervous System   The sympathetic nervous fibres emerge from the last thoracic and the first lumbar segments of the spinal cord and reach the myometrium through the hypogastric nerve. Both alpha and beta adrenoreceptors are present in the smooth muscles of the myometrium. The alpha adrenoreceptors are mediated by noradrenaline causing excitation. Inversely, the beta adrenoreceptors are mediated by adrenaline (epinephrine) causing inhibition. If noradrenaline storage granules on postganglionic sympathetic neurons are emptied during the pregnancy, it cannot stimulate the smooth muscles. So, myometrial contractions during labour can be inhibited by adrenalin. This can occur by central nervous stimulation of the adrenal gland, e.g. response to environmental stressful stimuli. As a consequence, the first stage of parturition can be delayed. This inhibition is more evident in mares than in cows.

1.4 Pelvis

1.4.1 Anatomy

The pelvis (pelvic basin or pelvic bowl) assumes a craniocaudal direction and is formed by the hard (bones) and soft (soft tissues) parts of the birth canal (▶ Box 1.2 and ◻ Fig. 1.4). The pelvis girdle is composed by the pelvic symphysis junction of the two hip bones. Each hip bone is the fusion of the ilium, pubis and ischium.

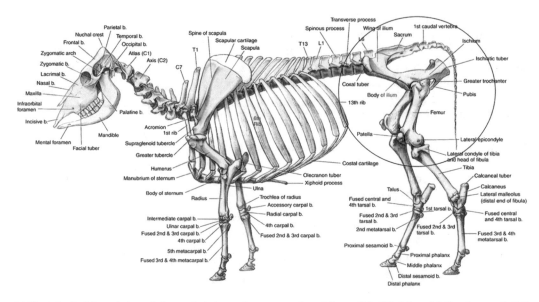

□ Fig. 1.4 Left lateral view of the cow's skeleton. Legend: The pelvis has around 30° inclination craniocaudally. *C* cervical vertebra, *T* thoracic vertebra, *L* lumbar vertebra, *b* bone. (Modified from McCracken et al. [14] with permission from John Wiley and Sons)

In the cranial pelvic opening (*Apertura pelvis cranialis*), the pelvic brim (outer bony edges) defines the circumference of the pelvic inlet. Dorsally, the pelvic brim is delimited by the body of the first sacral vertebra and the inferior part of the sacroiliac joint. Although ossification of the symphyseal cartilage starts at 13–14 months of age [1], the sacroiliac ligament slightly relaxes at calving time. Due to its bony nature, the pelvic brim size represents the most significant limiting factor to the passage of the foetus at calving time. The sacrum, the sacrotuberous ligament and the ischial (ischiatic) arch border the caudal pelvic aperture (*Apertura pelvis caudalis*).

- Blood vessels (supplied by the internal iliac artery).
- Nerves: Sciatic (bilateral superior), pudendal and obturator (bilateral inferior) and the sympathetic pelvic plexus.
- Lymph nodes (deep inguinal and sacral lymph nodes).
- Pelvic diaphragm: Pelvic fascia, deep and superficial perineal fascia and superficial and deep muscles of the perineum.

Pelvic joints
- Intrinsic
  - Sacroiliac.
  - Coccygeal vertebrae (first three joints).
  - Acetabulum.
  - Ischiopubic symphysis.
- Extrinsic
  - Lumbosacral.
  - Coxo-femoral or hip joint.

Box 1.2   Constitution of the Pelvis

Hard part (bones)
- Floor: Pubis (called pelvic floor).
- Lateral wall: Ilium and ischium.
- Roof: Sacrum (five vertebras) + first three coccygeal vertebrae (Co1-Co3).

Soft part
- Ligaments: Sacrotuberous and sacroiliac.
- Cartilages (symphyseal cartilage).

> **Important**
> During labour, both obturator nerves are exposed to trauma (foetal pressure) at its pelvic trajectory at sacroiliac joint level and close to the obturator foramen. The obturator nerve innervates adductor muscles, namely, the external obturator, pectineus and gracilis muscles, which can be desensitized.

1.4.2 Pelvic Shape, Size and Angulation

The pelvis is characterized as dolichopelvic. The pelvic brim is oval and laterally flattened. The sacroiliac (diagonal) diameter is larger than the bi-iliac (horizontal) diameter (□ Fig. 1.5 and □ Table 1.1). This is an important aspect to consider during foetal manipulation and

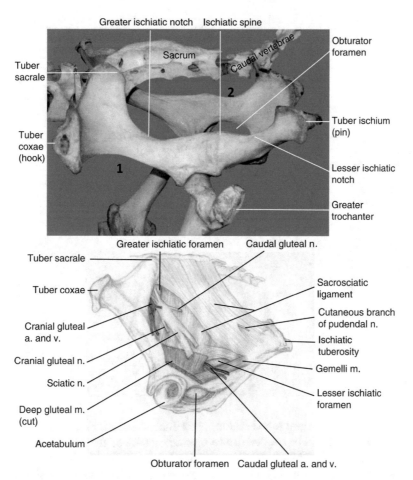

□ **Fig. 1.5** The pelvic structures. Legend: Top, hard part (bones) of the pelvis. (1) Pelvic inlet: the cranial opening of the pelvis delimited by anterior iliac and pubis portions and the first sacral segment (S1). (2) Pelvic outlet: the caudal opening of the pelvis delimited by posterior pubic symphysis portion, ischial tuberosities and the last sacral segment (S5). The posterior sacrotuberous ligaments and pelvic diaphragm (levator ani and coccygeus muscles) complete the boundary; bottom, soft part of the pelvis. Schematic illustration of the sacrosciatic ligament and the structures passing through the greater and lesser ischiatic foramina, such as arteries (a.), veins (v.), muscles (m.) and nerves (n.). Left lateral view. (Modified from Mansour et al. [13] with permission from John Wiley & Sons)

◻ Table 1.1   Estimated internal parameter of the pelvic inlet from 30 German Holstein-Friesian cows evaluated post-mortem using computed tomography

Pelvic measure	x ± SD (cm)	Minimum (cm)	Maximum (cm)	CV (%)
Left sacro-pubic (diagonal) diameter	24.7 ± 1.5[a]	21.5	26.9	6.2
Right sacro-pubic (diagonal) diameter	24.9 ± 1.7[a]	20.4	27.7	7.0
Sacro-pubic (vertical) diameter[1]	25.0 ± 1.9[a]	20.7	28.2	7.6
Bi-iliac (horizontal) diameter[2]	20.3 ± 1.7[b]	16.9	23.8	8.2
Pelvic (inlet) area (cm[2])	444 ± 56	331	537	12.7
Pelvic (inlet) circumference	77.1 ± 4.7	67	84	6.1
Pelvic volume (cm[3])	9617 ± 1096	7242	11,809	11.4

[1]Pelvic height; [2]pelvic width; x ± SD, mean ± standard deviation. [a] vs. [b]: $P < 0.05$
Modified from Tsousis et al. [24] with permission from Elsevier

◻ Table 1.2   Pelvimetry: estimative of pelvic measure parameters from external measurements and age of 30 German Holstein-Friesian cows

Pelvic measure	r^2	Regression equation
Pelvic inlet area (PA)	0.80	PA (cm^2) = −349 + 6.5TcTc + 8.3TcTi − 1916/A
Pelvic inlet circumference (PC)	0.81	PC (cm) = +9.8 + 0.5TcTc + 0.8TcTi − 171/A
Pelvic volume (VOL)	0.81	VOL (cm^3) = −11,693 + 133TcTc + 160TcTi + 32TiTi
Right diagonal diameter (Diar)	0.80	Diar (cm) = −10.5 + 0.3TcTc + 0.3TcTi
Minimum height (Hmin)	0.69	Hmin (cm) = −14.6 + 0.24TcTc + 0.37TcTi

Legend: r^2 Regression coefficient, $P < 0.001$, *A* age (in months), *TcTc* hip width (most lateral point of the two tuber coxae), *TcTi* hip length (most cranial point of the *tuber coxae* until the most caudal point of the ipsilateral *tuber ischiadicum*), *TiTi* pin bones width (most lateral point of the two *tuber ischiadicum*)
Modified from Tsousis et al. [24] with permission from Elsevier

extraction at calving. The largest foetal region, i.e. shoulders and hip, should be rotated (up to 45°) when entering the pelvic inlet.

The pelvic diameters, area and volume, can directly or indirectly be measured or estimated (pelvimetry), and these evaluations can be used in genetic improvement programs [9]. For example, they can be applied to breeding programs to exclude heifers with small pelvic area and so reduce dystocia. However, its utility to predict dystocia at calving is limited and should only be used as an indicative esteem in practice. The main pelvic parameters can be estimated, measuring the external distance between the ileal (*tuber coxae*) and ischial (*tuber ischiadicum*) tuberosities

(◻ Table 1.2). Heifers should be mated only after reaching 60% of the adult weight of the respective breed. Although the cow's body weight is closely correlated with pelvic size, about 90% of the pelvic area enlargement can be accomplished without large increase in cow size [16].

Although pelvis size varies according to breed and parity, pelvic height is commonly between 15 and 28 cm, while the pelvic width is approximately 20% smaller [5, 12, 24, 25].

In cows, the pelvis has a craniocaudal inclination of approximately 30°. The pelvic angulation or tilt is represented by a sigmoidal line, mainly due to the pelvic floor conformation. This aspect also should be to take in consideration when conducting a calving. Its passage through the birth canal will be easier if the first half the foetus is pulled dorsally.

> **Tip**
>
> Pelvic diameters can be directly measured to estimate pelvic area, using a pelvimeter (e.g. Rice pelvimeter) by transrectal or vaginal routes, at calving.

The prepubic tendon is composed by several tendons originated from the abdominal muscles (pectineus, rectus and obliquus abdominis and gracilis muscles) as well as from the pelvic attachments of the linea alba and the yellow abdominal tunic. It is attached to the iliopubic eminences and ventral pubic tubercle of the pubis, and the symphyseal tendon. The prepubic tendon is a ventrocaudal support to the pregnant uterus. Its rupture is rare in cows.

Case Study 1.1 Selection of Beef Heifers for a Breeding Program Using Pelvimetry

A beef farm wants to start a breeding program to reduce dystocia due to foetopelvic disproportion. The aim is to select Hereford heifers to go on to the natural mating season due to start in 3 weeks. The threshold for the cull rate was defined as 20%. The age of the heifers varied between 350- and 410-old-days. Each heifer was restrained, and a Rice pelvimeter was transrectal used to measure the pelvic height (distance between the pubic symphysis and the sacral vertebrae) and pelvic width (distance between the widest shafts of the ilium). Concomitantly, a transrectal evaluation of ovaries and uterus was made to evaluate the ovarian cyclicity and detect congenital anomalies (e.g. freemartins and agenesis or hypoplasia of the tubular genitalia). The pelvic height ranged between 12 and 17 cm. The pelvic width ranged between 9 and 14 cm. The pelvic area was calculated by multiplying the pelvic height and the pelvic width. The following formula obtained the prediction of the 365-day pelvic area: 365-day pelvic area = actual pelvic area (cm^2) + [0.27 × (365 – age in days)] for measurements taken between 320 and 410 days. It was also decided that only heifers weighing >600 kg (approximately 65% of the mature body weight) would be included in the breeding program. According to the 20% threshold, heifers presenting a pelvic area up to 130 cm^2 were candidates to culling, since they will be more likely to have calving difficulty. In this case, the heifers' selection to breeding was made according to the following order: (1) non-congenital anomalies of the reproductive tract; (2) weight >600 kg; (3) pelvic area >130 cm^2; and (4) cycling heifers.

1.5 Hypothalamic-Pituitary-Gonadal Axis

Reproduction is regulated by the hypothalamic-pituitary-gonadal axis (◻ Fig. 1.6). GnRH is produced and secreted by GnRH-expressing neurons located in the hypothalamus. It is responsible for the release of follicle-stimulating hormone (FSH) and luteinizing hormone (LH) from the anterior pituitary. Both these hormones act at ovarian level stimulating oogenesis and luteogenesis during different phases. FSH stimulates the emergence of follicular waves and development of FSH-dependent antral follicles. The antral follicles progressively secrete oestrogen and inhibin, which have negative feedback on FSH secretion. During the follicular deviation, the dominant follicle acquires LH receptors and become LH-dependent. Atresia occurs in the remaining follicles in the same follicular growing wave. In the presence of an active CL, the regression of the dominant follicle also occurs, while a new follicular wave emerges. Usually, two to three follicular waves are observed during the oestrus cycle. The last follicular wave is called the ovulatory wave: when luteolysis or regression of the CL occurs, the dominant follicle increases in size up to approximately 18–20 mm and oocyte matura-

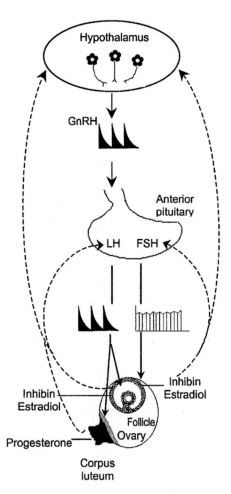

◻ Fig. 1.6 Schematic representation of the hypothalamic-pituitary-gonadal axis. Legend: Solid arrows indicate stimulatory effects; dashed arrows indicate inhibitory effects. *GnRH* gonadotropin-releasing hormone, *FSH* follicle-stimulating hormone, *LH* luteinizing hormone. (Adapted from Hafez et al. [8] with permission from John Wiley and Sons)

tion occurs; high levels of oestradiol mainly produced by the preovulatory follicle(s) will have positive feedback on GnRH neurons of the hypothalamus; finally, a LH peak promotes ovulation and luteogenesis.

Despite the central role of the hypothalamic-pituitary-gonadal axis, several other hormones are involved in the different steps of the reproductive process (◻ Table 1.3).

1.6   Oestrous Cycle

In cattle, the length of the oestrous cycle is 20–21 days but can vary between 17 and 24 days. It is composed of four successive phases under hormonal control: oestrus, metaoestrus, dioestrus and proestrus (◻ Fig. 1.7). The oestrus has a duration of approximately 18 h, and ovulation occurs 6 h after heat signs end. During oestrus, the cow shows specific behaviour accepting to be mounted by the male (or other cows). The oestradiol mainly produced (around 90%) by the preovulatory follicle induces this oestrus behaviour.

After ovulation, the ruptured follicle initiates luteogenesis, and the metestrus starts. Firstly, it develops a *corpus hemmoragicum* refractory to $PGF_{2\alpha}$ during the first 3 days. LH stimulates the development of small and large luteal cells and angiogenesis to reach its maturity as CL in the following days. The plasmatic level of progesterone (P_4) quickly exceeds 1 ng/mL (active CL) starting the dioestrus phase. In the absence of conception, around the 16–17th day of the oestrus cycle, the endometrium (luminal epithelial cells) releases $PGF_{2\alpha}$ which quickly causes functional luteolysis of the CL. Under the influence of increasing concentrations of oestradiol, oxytocin stimulates this release of $PGF_{2\alpha}$ from the endometrial epithelial cells. This hormonal action also increases the number of myometrial oxytocin receptors increasing uterine contractibility during the next oestrus phase. Similar hormonal mechanism occurs at calving. Prostaglandin F2α is transported from the uterine vein to the ovarian artery, bypassing the systemic blood circulation. This mechanism prevents the degradation of prostaglandin in the lungs. At this time, plasmatic P_4 level drops to less than 1 ng/mL in 4–6 h, allowing the development of the next dominant follicle to the preovulatory stage. In this pre-oestrus phase, several alterations of the reproductive tract occur to prepare the cow for copulation and conception.

■ Table 1.3 Main functions of the reproductive hormones in bovine females

Hormone	Production and release	Main functions
CRH	Hypothalamus.	Stimulates ACTH (stress response).
GnRH	Hypothalamus.	Stimulates of FSH and LH release.
Oxytocin	Hypothalamus (production); anterior pituitary (release). Corpus luteum. Luteal cells of placenta.	Induces the uterine contractions. Stimulates the milk let-down. Increases the gamete transport. Stimulates the synthesis of prostaglandins.
ACTH	Anterior pituitary.	Stimulates cortisol synthesis and secretion.
Prolactin	Anterior pituitary.	Simulate the lactopoiesis. Contributes to the development, differentiation and function of mammary tissue.
FSH	Anterior pituitary.	Stimulates the FSH-dependent follicles growth. Stimulates follicular maturation.
LH	Anterior pituitary.	Stimulates the dominant follicle growth. Induces the ovulation of follicle (preovulatory surge). Luteinization of granulosa cells to form the *corpus hemmoragicum*.
Growth hormone (somatotropin)	Anterior pituitary.	Simulates the lactopoiesis. Simulates the hepatic gluconeogenesis and amino acid uptake.
Progesterone	Corpus luteum. Placenta (trophoblast giant cell of placentome).	Regulation of the GnRH. Stimulates the growth of the endometrial gland. Stimulates the secretory activity of the oviducts and endometrial glands before embryo's implantation. Myometrial contractility prevention during pregnancy.
Oestrogen	Granulosa cells (ovarian follicle) under FSH-regulated aromatase from androgen. Theca cells produce androstenedione, under LH stimulus.	Induces sexual behaviour and receptivity. Increases the oviduct's secretory activity. Stimulates of the growth of the endometrial gland. Stimulates the mammary duct growth. Increases the enzymatic activity of the mammary gland. Regulates the GnRH and prostaglandins hormones.
Relaxin	Corpus luteum (late gestation).	Cervix dilatation. Relation of the pelvic ligaments.
Androgens (androstenedione, testosterone, dehydroepiandrosterone)	Theca cells (ovarian follicle). Placenta.	Precursors of oestrogens. Regulate the ovarian follicle growth. Regulate the hypothalamic-pituitary-gonadal axis. Stimulate the prolactin secretion.

(continued)

Table 1.3 (continued)

Hormone	Production and release	Main functions
Cortisol	Adrenal cortex.	Increases the number of oxytocin receptors. Releases prostaglandins. Contributes to the conversion of progesterone to oestrogens and stimulates oestrogen production.
Thyroid hormones (T_3 and T_4)	Thyroid gland.	Simulate the lactopoiesis.
Insulin	Beta cells of the pancreatic islets.	Stimulates milk production.
$PGF_{2\alpha}$	Endometrium.	Luteolysis of the corpus luteum. Uterine contractions.
Placental lactogen	Placenta (binuclear giant cells).	Regulates the ovarian steroidogenesis, mammogenesis and lactogenesis luteotrophic action. Regulates the nutrients supply of the foetus.
Pregnancy-associated glycoproteins	Mono- and binuclear giant cells of the placenta; remain detectable in plasma up to 90 days postpartum.	Maternal recognition of pregnancy. Trophoblast adhesion and implantation. Luteotropic and luteopreventive action. Induces the release of prostaglandin E_2 in ovaries and placenta.

CRH corticotropin-releasing hormone, *ACTH* adrenocorticotropic hormone, *GnRH* gonadotropin-releasing hormone, *FSH* follicle-stimulating hormone, *LH* luteinizing hormone, T_3 triiodothyronine, T_4 thyroxine, $PGF_{2\alpha}$ prostaglandin $F_{2\alpha}$

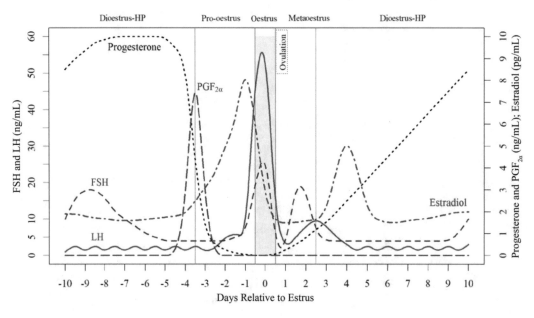

Fig. 1.7 Hormonal pattern during the oestrus cycle in cattle. Legend: Dioestrus-high progesterone (dioestrus-HP) ranging from −10 to −4 d ($n = 27{,}574$); proestrus ranging from −3 to −1 d ($n = 12{,}302$); oestrus at d 0 ($n = 4144$); metaoestrus from 1 to 2 d ($n = 8275$); and dioestrus increasing progesterone (dioestrus-HP) ranging from 3 to 10 d ($n = 33{,}052$). (Adapted from Toledo-Alvarado et al. [23] with permission from Elsevier)

During pregnancy, the oestrus cycle is suspended occurring what is known as gestational anoestrus. Even if follicular activity may still be present, the high levels of P_4 prevent ovulation. However, sometimes an erratic oestrus and ovulation may be observed. In postpartum, the anoestrus persists and is designated postpartum anoestrus. Its length varies according to the effect of several inhibitory stimuli. Typically, a transient FSH surge occurs approximately between three and five days postpartum, and an ovulation may be detected about 15 days after calving in dairy cows and about 30 days in beef cows (suckling-induced anoestrus; ◻ Fig. 1.8). A significant delay in the ovulation time is observed in high-yielding dairy cow (>60 days) and

in beef cow presenting poor body condition (approximately 70–100 days) [4].

1.7 Pregnancy

At conception, a disruption of the pulsatile release of $PGF_{2\alpha}$ from the endometrium occurs. During maternal recognition of pregnancy, the interferon tau (IFNT) secreted by the embryo's mononuclear cells of trophectoderm maintains the CL. The interferon tau is produced between the 16th and 24th days after conception. It inhibits the endometrial production of oxytocin receptors. In consequence, the oxytocin segregated by the CL and anterior pituitary cannot stimulate the

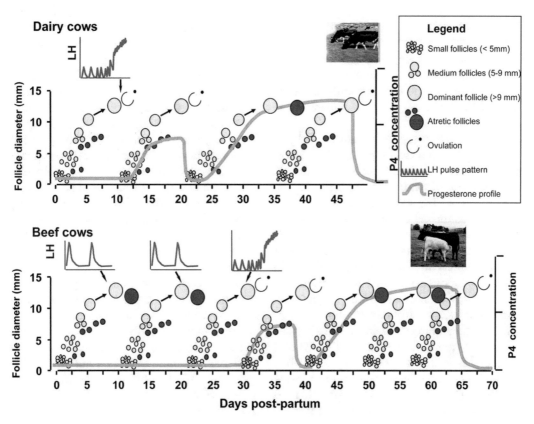

◻ **Fig. 1.8** Resumption of dominant follicles and ovarian cycles during the postpartum period in dairy and beef suckler cows. Legend: The prolonged suppression luteinizing hormone (LH) pulses originate a delay of the ovulation in absence of progesterone (P_4). In high-yielding Holstein-type cows, the irregular pattern of LH is mainly caused by severe negative energy balance, dystocia, retained placental membranes and uterine infections. In

beef cows, the suckling also interferes with hypothalamic release of Gonadotropin-releasing *hormone*. Overall, the first ovulation is silent (sub-oestrus; no behavioural oestrus) and followed (>70%) by a short cycle in >70% of the ovulations. LH pulse frequency is that occurring during an 8-h window where cows are blood sampled every 15 min. [18]. (Reprint from Crowe et al. [4] with permission from Cambridge University Press)

1

endometrial production of $PGF_{2\alpha}$. This mechanism only prevails during the first month of pregnancy [26].

The placentation starts to develop on the 21st day. Consecutively, endometrial implementation of the blastocyst occurs on the uterine wall of uterine horn ipsilateral to the ovary which ovulated.

In cows, the CL is responsive for pregnancy maintenance. Nevertheless, the placenta also produces progesterone in middle pregnancy, approximately between 150th and 240th day of gestation [20]. During this period, pregnancy can be maintained by progesterone produced by the placenta, even if luteolysis of the CL occurs. However, the cow usually aborts at the end of that period.

1.7.1 Placenta

The placenta is composed of four membranes, i.e. chorion, allantois, amnion and vitelline membranes, and the umbilical cord (◘ Fig. 1.9) to ensure the development of the foetus.

The amniotic sac is constituted by a double membrane surrounding the foetus, approximately 30 days after conception. It derivates from the extra-foetal portion of the somatopleure. The external amniotic membrane forms the primitive chorion. Ventrally, the amniotic membrane takes part in the formation of the umbilical cord. Whitish and firm amniotic plaques (<1 cm of length) cover the inner surface of the amnion. The amniotic cavity contents a clear semitransparent fluid derived from the amniotic epithelium and foetal lungs, unkeratinized skin and urine. The amniotic plaques and fluids and the foetal hair and skin become yellow when stressed foetuses defecate (meconium) inside the uterus [19]. The volume of amniotic fluid varies between 5 and 8 litres at mid-pregnancy when it usually reaches the maximum quantity.

> **Tip**
>
> During the initial gross examination of the placenta, the detection of the amniotic plaques can be helpful to identify the amniotic membrane.

The allantoid membrane initially evaginates from the hindgut and develops between the amnion and the primitive chorion. The allantois develops and blends with the primitive chorion originating the chorioallantoic membrane. This membrane attaches to maternal caruncles at around the same time than the amniotic sac is formed. It surrounds the amniotic membrane and forms the allantoid cavity. The external part of the amniotic membrane is attached to the internal part of the allantoid membrane. The interplacentomal (intercotyledonary) space of the chorioallantoic membrane is also denominated as "smooth chorioallantois". Ventrally, it also takes part in the formation of the umbilical cord. Other than nutritional functions, the chorioallantoic membrane also serves to produce several hormones and growth factors. The allantoid fluid is clear and derivates from the secretory activity of the chorioallantoic membrane and foetal urine drained through the urachus. The volume of allantoic fluid is 10–15 L at term pregnancy.

The vitelline or yellow sac plays an important role in early embryonic nutrition and development, e.g. vasculogenic and haematopoietic function. In cows, the epithelium seem to be functional until 50th day and persist at until 70th, remaining vestigial and finally disappearing. The involution of the vitelline sac is required to establish chorioallantoic placentation [6].

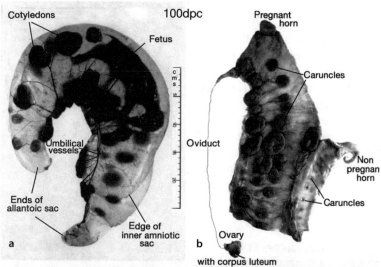

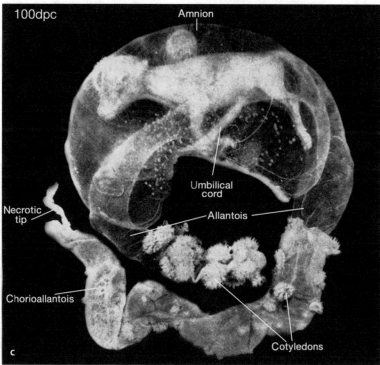

◘ **Fig. 1.9** Amnion and allantois at day 100 of pregnancy. Legend: **a** opened out flat to expose caruncles in both pregnant and nonpregnant uterine horns. A degeneration and coagulative necrosis usually occurs in the ends of the chorioallantois sac (3–5 cm in length) and is called "necrotic tips". **b** A day 100 foetus with chorion dissected away over the central region to reveal amnion and allantois; **c** villi on foetal cotyledons on the nondissected part are well demonstrated. (Original from Dr. Wooding. Adapted from Peter [17] with permission from Elsevier)

In cows, the umbilical cord, with a length up to 50 cm, presents two arteria and two veins. Both veins merge at the umbilical ring (◘ Fig. 1.10). The allantoid duct is the continuation of the urachus which is attached to the apex of the balder transporting foetal urine until the allantoid cavity. The mucous connective tissue (Wharton jelly) surrounds the umbilical blood vessels preventing their collapse during pregnancy. At calving, Wharton jelly helps to collapse these vessels, stimulated by quick contrast to a low environmental temperature.

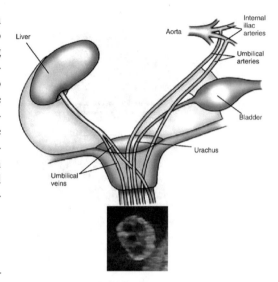

1.7.2 Foetomaternal Interface

The cow has a cotyledonary placenta, i.e. presents focal microvillous aggregations. The placentome is the anatomical and functional unit to transport nutrients from the mother to concept and interchange gases (O_2 and CO_2). It is formed by the attachment between maternal caruncles and foetal cotyledons (◘ Fig. 1.11). The number of placentomes

◘ **Fig. 1.10** Umbilical structures of the foetus. Legend: The umbilical veins of the umbilical cord merge after reaching the umbilical stalk. (Modified from Jackson and Cockcroft [11] with permission from Elsevier). A sonogram (original from João Simões) of the umbilical cord was added. The sonogram (B-mode; 7.5 MHz) represents a transversal section of the umbilical cord with approximately 2.5 cm in diameter. The four vessels are evident (pregnancy of 130 days of a Holstein-Friesian cow)

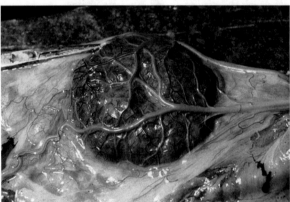

◘ **Fig. 1.11** Macroscopic aspect of the placentome. Legend: Left, placentome from a term pregnancy with umbilical arteries injected with coloured latex. The branches of the umbilical arteries and veins penetrate or leave the placentome forming an intricate vascular pattern. Right: the cotyledonary villi are partially separated from the caruncular crypts. This photograph was taken under water. (Adapted from Schlafer et al. [19] with permission from Elsevier)

progressively increases up to approximately the 70th day of pregnancy. The largest placentomes can be observed closer to the middle of the pregnant uterine horn and to the attachment of the uterine artery.

The bovine placenta is currently classified as synepitheliochorial. The surface of the chorioallantois membrane becomes irregular over the maternal caruncles. The fingerlike villous projections of the trophoblast (chorionic epithelium) are connected with caruncle epithelium of the maternal crypts. The fusion between the trophoblast giant cells (binucleate cells) and uterine epithelial cells originates trinucleate foetomaternal hybrid cells. Trophoblast giant cells produce steroids hormones such as progesterone (P_4), prostaglandin I_2 and prostaglandin E_2 as well protein hormones such as placental lactogen and pregnancy-associated glycoproteins.

Other than P_4 and other steroids, placenta also produces oestrogens, mainly oestrone-3-sulphate (E_1S). Oestrone sulphate can be detected in early pregnancy, from the second month, and reach its maximum levels (15–30 nmol/L) after the 265th day [10]. At this time, oestrone (E_1) levels quickly and significantly increases (up to ×15). Cows carrying twins have higher oestrone levels than cows carrying singletons [22]. Oestrogens play a significant role in the preparation for calving. They stimulate placental maturation, myometrial activity (increasing the number of myometrial oxytocin receptors) and softening of the birth canal. It was suggested by Hoffmann and Schuler [10] that placental P_4 and oestro-gens may play a significant role in regulating caruncular growth, differentiation and function during pregnancy.

1.8 Uterine Involution

After calving, the enlarged uterus involutes to approximately its previous size and will be fully located inside the pelvic cavity, except in older cows. This uterine restauration process is denominated uterine involution and is normally completed by 40th–42nd day after calving, the timeframe that defines the puerperal period. During this period a regeneration of the endometrium occurs with tissue loss and repair. Lochia, a reddish blend of tissue, blood and mucous, are expulsed by uterine contraction. Uterine size and uterine wall thickness can be evaluated by manual transrectal palpation (► Box 1.3) as well as by ultrasonographic measurement. During the first week postpartum, the diameter of the pregnant uterine horn (>13 cm) and even the nonpregnant uterine horn (>8 cm) is too large to obtain an entire cross-sectional area. The symmetry of both uterine horns (difference of <1 cm in diameter) is reached in 18th–20th day after calving. Ecbolic drugs such as the exogenous repeated administration of oxytocin (40 IU) or $PGF_{2\alpha}$ (twice a day and during 7 days) do not improve uterine involution [21]. Several factors, e.g. dystocia, uterine adherences, clinical metritis, hypocalcaemia and ketosis, can delay the morphological uterine involution (◘ Fig. 1.12).

Box 1.3   Scoring Uterus Size by Transrectal Palpation According to Grunert [7]
- Score 1: The uterine horns are ≤2 cm in diameter and can be retracted.
- Score 2: The uterine horns are 3–5 cm and can be retracted.
- Score 3: The uterine horns are 6–8 cm and can be retracted.
- Score 4: The uterine horns are 9–20 cm and can be delimited by the hand.
- Score 5: The margins of the uterus can be only partially palpable.
- Score 6: The margins of the uterus cannot be delineated by hand.

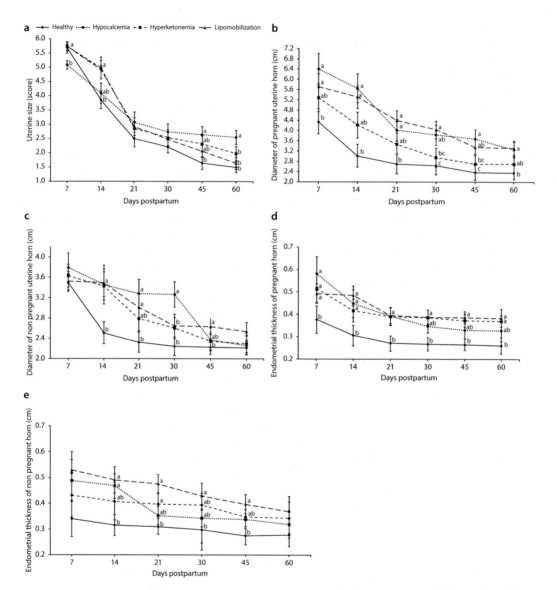

◘ Fig. 1.12 Morphological uterine involution in multiparous Holstein dairy cows. Pattern of uterine involution in healthy (solid line; $n = 14$) cows or suffering hypocalcaemia ($n = 11$; serum calcium levels <8.0 mg/dL from 3 to 8 h after delivery), hyperketonaemia ($n = 11$; β-hydroxybutyric acid serum levels >1200 μmol/L in at least one of three samples at the points: parturition, +1 and + 7 days postpartum) and lipomobilization ($n = 14$; serum of non-esterified fatty acids levels >0.4 mmol/L in at least one of three samples at the points: −7, −4 and −2 two days before calving). **a** Uterine size; **b** diameter of pregnant uterine horn; **c** diameter of nonpregnant horn; **d** endometrial thickness of pregnant horn; **e** endometrial thickness of nonpregnant horn (least squares means ± SEM) in relation to days postpartum. [a–c] Different superscript letters differ at $P < 0.05$. (Adapted from Braga Paiano et al. [2])

Key Points

- Sexual maturity is completed several months after puberty.
- The reproductive tract involves external and internal genitalia, each one with specific functions.
- The pelvis consists of a hard part (bones) and a soft tissue part, housing the ovaries, uterus and vagina in non-pregnant cow and forming the birth canal at calving.
- The hypothalamic-pituitary-gonadal axis regulates the reproductive cycle and pregnancy, but the endometrium and placenta also assume hormonal functions.
- The 21-day oestrus cycle presents four successive phases, occurring gestational anoestrus and the resumption of ovarian activity during the postpartum.
- The embryo is implanted in one uterine horn where the foetus develops until pregnancy term.
- The ruminants have a cotyledonary and synepitheliochorial placenta forming a placental barrier to macromolecules including immunoglobulins.
- The puerperal period is defined as the 40–42 days postpartum, during which the uterine involution is completed.

❓ Questions

1. What are the more relevant aspects of the pelvic structures in relation to calving and its assistance?
2. Describe the ovulatory activity in adult cows.
3. Describe the main features of the bovine placenta.

✔ Answers

1. The pelvis consists of a bone structure (hard part) connected by symphyseal cartilage, ligaments and joints, and involving soft tissue (soft part) such as blood vessels, nerves, lymph nodes and pelvic diaphragm. The pubis, ilium and ischium together with the sacrum and the first three coccygeal vertebrae delimit the pelvic cavity, where the genitalia are housed and forming the birth canal at calving time. The pelvic brim delimits the cranial pelvic aperture presenting an oval shape. The sacroiliac ligament only slightly relaxes at calving time, so that the pelvic inlet area represents the most relevant limitation to foetal passage through the birth canal. The foetus should be slightly rotated at elbows and hip levels during traction adapting its shape to the largest (diagonal) diameter of the pelvis. Also, the sigmoid angulation and 30° craniocaudal inclination imply that the first half of the foetus should be pulled in a parallel direction. The obturator and *sciatic* nerve run along the medial surface of ilium and the ventral surface of the sacrum, respectively, and are at risk of foetal pressure during the passage of the foetus through the birth canal. Damage to these nerves may cause maternal obstetrical paralysis leading to the *Downer cow syndrome.*

2. The ovulatory activity starts at puberty, and one or two follicles ovulate each 20–21 days, on average. This activity is under hypothalamic-pituitary-gonadal axis promoting an oestrus cycle (proestrus, oestrus, metaoestrus and dioestrus phases) involving mainly GnRH, FSH, LH, P_4 and oestrogen hormones. During the oestrus phase, a preovulatory LH peak induces ovulation approximately 20 h from its start and 6 h after the end of the oestrus phase. If fertilization occurs, maternal recognition of the pregnancy starts from the 16th after ovulation, where the interferon tau produced by the embryonic trophectoderm is the main sign to indirectly inhibit endometrial $PGF_{2\alpha}$ production. CL persists during whole pregnancy inhibiting additional ovulatory activity (gestational anoestrus). In the postpartum, in the absence of progesterone, LH pulses from the anterior pituitary are suppressed, and the resumption of ovulatory activity is delayed but usually occurs within the first 3 weeks after calving (postpartum anoestrus). Usually, the cow

presents a sub-oestrus (silent ovulation, in which heat signs are inapparent). The resumption of ovulation can be delayed due to several factors (e.g. poor nutrition, negative energy balance or illness). Milking (lactational) or suckler cows (suckling-induced) anoestrus are normal, probably due to the negative influence of prolactin hormone on LH production through GnRH inhibition.

3. Approximately 30 days after conception, the extra-foetal portion of the somatopleure originates the amnios, a double amniotic membrane and an external primitive chorion that will originate the chorion. The allantoid membrane develops approximately 1 week later between the amnion and the primitive chorion. Externally, the fusion between the allantoid membrane and the primitive chorion originates the chorioallantoic membrane. Internally, the allantoid membrane attaches to the amniotic membrane. These membranes also shape the umbilical cord. Both amniotic and chorioallantoic membranes delimit the amniotic and allantoid sac, respectively. Moreover, these membranes segregate fluids, contributing to the production of amniotic and allantoid fluids. The vitelline membrane is also present in early pregnancy but disappears by the 70th day of gestation. Meanwhile, placentomes increase in number, progressively up to 140. The synepitheliochorial junction between maternal caruncles and foetal cotyledons ensures foetal nutrition, gas exchange and serve as a placental barrier, constituting the bovine foetomaternal unit. At term, the bovine placenta weighs approximately 5 kg.

References

1. Bassett EG. The comparative anatomy of the pelvic and perineal regions of the cow, goat and sow. N Z Vet J. 1971;19(12):277–90.

2. Braga Paiano R, Becker Birgel D, Harry Birgel E Jr. Uterine involution and reproductive performance in dairy cows with metabolic diseases. Animals (Basel). 2019;9(3):93. https://doi.org/10.3390/ani9030093.

3. Constantinescu GM. The Genital apparatus in the ruminant. In: Schatten H, Constantinescu GM, editors. Comparative reproductive biology. Iowa, USA: Blackwell Publishing; 2007. p. 33–8. https://doi.org/10.1002/9780470390290.ch2e.

4. Crowe MA, Diskin MG, Williams EJ. Parturition to resumption of ovarian cyclicity: comparative aspects of beef and dairy cows. Animal. 2014;8(Suppl 1):40–53. https://doi.org/10.1017/S1751731114000251.

5. de Munck F. Pelvimetry: the repeatability and reproducibility of the Rice pelvimeter in South-African beef cattle. Master thesis. Faculty of Veterinary Medicine, Utrecht University; 2019.

6. Galdos-Riveros AC, Favaron PO, Will SE, Miglino MA, Maria DA. Bovine yolk sac: from morphology to metabolomic and proteomic profiles. Genet Mol Res. 2015;14(2):6223–38. https://doi.org/10.4238/2015.June.9.8.

7. Grunert E. Female genital system. In: Rosenberger G, Dirksen G, Gründer HD, Grunert E, Krause D, Stöber M, editors. Clinical examination of cattle. 1st ed. Berlin/Hamburg, Germany: Paul Parey; 1979. p. 323–50.

8. Hafez ESE, Jainudeen MR, Rosnina Y. Hormones, growth factors, and reproduction. In: Hafez B, ESE H, editors. Reproduction in farm animals. 7th ed. Philadelphia, USA; 2000. p. 31–54. https://doi.org/10.1002/9781119265306.ch3.

9. Hiew MW, Constable PD. The usage of pelvimetry to predict dystocia in cattle. J Vet Malaysia. 2015;27(2):1–4.

10. Hoffmann B, Schuler G. The bovine placenta; a source and target of steroid hormones: observations during the second half of gestation. Domest Anim Endocrinol. 2002;23(1–2):309–20. https://doi.org/10.1016/s0739-7240(02)00166-2.

11. Jackson PG, Cockcroft PD. Clinical examination of the gastrointestinal system. In: Jackson PG, Cockcroft PD, editors. Clinical examination of farm animals. Oxford, UK: Blackwel; 2002. p. 81–112. https://doi.org/10.1002/9780470752425.ch8.

12. Kolkman I, Hoflackb G, Aerts S, Murray RD, Opsomer G, Lips D. Evaluation of the Rice pelvimeter for measuring pelvic area in double muscled Belgian Blue cows. Livest Sci. 2009;121(2–3):259–66. https://doi.org/10.1016/j.livsci.2008.06.022.

13. Mansour M, Wilhite R, Rowe J. Chapter 4: The pelvis and reproductive organs. In: Guide to ruminant anatomy: dissection and clinical aspects. New Jersey, USA: Wiley; 2018. p. 139–71. https://doi.org/10.1002/9781119379157.ch4.

14. McCracken TO, Kainer RA, Spurgeon TL. Spurgeon's color atlas of large animal anatomy: the essentials. Oxford, UK: Wiley-Blackwell; 2011.

15. Millward S, Mueller K, Smith R, Higgins HM. A post-mortem survey of bovine female reproductive tracts in the UK. Front Vet Sci. 2019;6:451. https://doi.org/10.3389/fvets.2019.00451.

16. Morrison DG, Williamson WD, Humes PE. Estimates of heritabilities and correlations of traits associated with pelvic area in beef cattle. J Anim Sci. 1986;63(2):432–7.

17. Peter AT. Bovine placenta: a review on morphology, components, and defects from terminology and clinical perspectives. Theriogenology. 2013;80(7):693–705. https://doi.org/10.1016/j.theriogenology.2013.06.004.

18. Roche JF, Crowe MA, Boland MP. Postpartum anoestrus in dairy and beef cows. Anim Reprod Sci. 1992;28:371–8. https://doi.org/10.1016/0378-4320(92)90123-U.

19. Schlafer DH, Fisher PJ, Davies CJ. The bovine placenta before and after birth: placental development and function in health and disease. Anim Reprod Sci. 2000;60-61:145–60.

20. Schuler G, Greven H, Kowalewski MP, Döring B, Ozalp GR, Hoffmann B. Placental steroids in cattle: hormones, placental growth factors or byproducts of trophoblast giant cell differentiation? Exp Clin Endocrinol Diabetes. 2008;116(7):429–36. https://doi.org/10.1055/s-2008-1042408.

21. Stephen CP, Johnson WH, Leblanc SJ, Foster RA, Chenier TS. The impact of ecbolic therapy in the early postpartum period on uterine involution and reproductive health in dairy cows. J Vet Med Sci. 2019;81(3):491–8. https://doi.org/10.1292/jvms.18-0617.

22. Takahashi T, Hirako M, Takahashi H, Patel OV, Takenouchi N, Domeki I. Maternal plasma estrone sulfate profile during pregnancy in the cow; comparison between singleton and twin pregnancies. J Vet Med Sci. 1997;59(4):287–8. https://doi.org/10.1292/jvms.59.287.

23. Toledo-Alvarado H, Vazquez AI, de Los CG, Tempelman RJ, Gabai G, Cecchinato A, Bittante G. Changes in milk characteristics and fatty acid profile during the estrous cycle in dairy cows. J Dairy Sci. 2018;101(10):9135–53. https://doi.org/10.3168/jds.2018-14480.

24. Tsousis G, Heun C, Becker M, Bollwein H. Application of computed tomography for the evaluation of obstetrically relevant pelvic parameters in German Holstein-Friesian cows. Theriogenology. 2010;73(3):309–15. https://doi.org/10.1016/j.theriogenology.2009.09.014.

25. Weiher O, Hoffmann G, Sass D. The relationships between the internal and external pelvic measurements of Schwartzbunt cows. Dtsch Tierarztl Wochenschr. 1992;99(11):452–4.

26. Wiltbank MC, Mezera MA, Toledo MZ, Drum JN, Baez GM, García-Guerra A, Sartori R. Physiological mechanisms involved in maintaining the corpus luteum during the first two months of pregnancy. Anim Reprod. 2018;15(S1):805–21. https://doi.org/10.21451/1984-3143-AR2018-0045.

Online Visual Guide

Drost M, Samper J, Larkin PM, Gwen CD. Female Reproductive System. Visual guides of animal reproduction. Florida, USA: UF College of Veterinary Medicine; 2019. From: https://visgar.vetmed.ufl.edu/en_bovrep/uterus/uterus.html. Accessed on 17 Sept 2020.

Suggested Reading

Forde N, Lonergan P. Interferon-tau and fertility in ruminants. Reproduction. 2017;154(5):F33–43. https://doi.org/10.1530/REP-17-0432.

Haeger JD, Hambruch N, Pfarrer C. The bovine placenta in vivo and in vitro. Theriogenology. 2016;86(1):306–12. https://doi.org/10.1016/j.theriogenology.2016.04.043.

Lefebvre RC, Gnemmi G. Anatomy of the reproductive tract of the cow. In: DesCôteaux L, Gnemmi G, Colloton J, editors. Practical atlas of ruminant and camelid reproductive ultrasonography. Howa, USA: Wiley-Blackwell; 2010. p. 27–34. https://doi.org/10.1002/9781119265818.ch3.

Thatcher WW. A 100-year review: historical development of female reproductive physiology in dairy cattle. J Dairy Sci. 2017;100(12):10272–91. https://doi.org/10.3168/jds.2017-13399.

Problems and Complications Occurring in Mid and Late Pregnancy

Contents

The original version of this chapter was revised. The correction to this chapter can be found at https://doi.org/10.1007/978-3-030-68168-5_12

© Springer Nature Switzerland AG 2021, corrected publication 2022
J. Simões, G. Stilwell, *Calving Management and Newborn Calf Care*,
https://doi.org/10.1007/978-3-030-68168-5_2

2

Learning Objectives

- To identify the causes and predisposing factors of the main complications occurring during pregnancy.
- To set a clinical approach decision, regarding the diagnosis and treatment of pregnancy complications.
- To determine the relationships between physical findings detected at transrectal and vaginal palpation and prognosis, in uterine torsion and vaginal prolapse.
- To describe the advantages and disadvantages of using different conservative or surgical treatments.
- To recognize procedures that may prevent these main complications occurring during pregnancy.

2.1 Introduction

Pregnancy length in cows is, on average, 278 days but can vary in some days according to breed and age. It is accepted that the term "pregnancy" should be used for a 260 days gestational period, after which a viable offspring, even if premature, is delivered. Several problems or complications can occur between conception and pregnancy term. In the first stages, the most prevalent problem is embryo loss occurring within the first 40 days (embryonic period) of pregnancy [6]. It can be caused by a multitude of factors, namely, poor follicle or ova quality, hormonal imbalance, failure in maternal recognition or embryo implantation, embryo genetic defects, uterine inflammation and infection, poor nutrition and heat or other environmental caused stress. However, this book does not aim to address in detail this very early period.

Also, congenital anomalies (teratogenic defects) can develop during the embryonic period, when the organogenesis process is significant. The leading causes of developmental defects are of genetic, toxic, environmental or infectious origin. Some relevant congenital defects are reported in other chapters of this book.

During the embryonic or in the early foetal period (after the 40th day of gestation), the dead concept can be expulsed or reabsorbed. Once ossification is well developed, from the 70th day of pregnancy, the foetus is expulsed (abortion) or, in some few cases, foetal maceration or mummification occurs.

Towards pregnancy term, several non-infectious complications may occur and should be diagnosed and treated promptly. Some of them, such as uterine torsion or vaginal prolapse, are periparturient diseases and can cause dystocia. Also, in late pregnancy, the development of hydrops is an impressive pregnancy abnormality leading to foetal death and impacting cow's welfare and performance.

The main objective of this chapter is to address the most representative problems occurring during mid and late pregnancy, with particular focus on pre- and periparturient period.

2.2 Uterine Torsion

Uterine torsion is defined as a rotation of the uterus across its long axis. Both foetus and foetal membranes also rotate alongside with the uterus. It is one of the most significant causes of dystocia at calving time [9]. Usually, uterine torsion occurs just before or during Stage I of labour, when the cervix is already partially dilated. However, a slight twist, usually up to approximately 180° (partial torsion), can be detected during the second half of pregnancy. In most cases neither foetal nor uterine blood supply is affected, and pregnancy continues until its term. In Stage I, successive lying down and rising of the dam and increased movement of large foetuses are probable mechanical predisposing factors that induce uterine rotation.

Although uterine torsion should be considered an emergency, according to our experience, mild cases can be successfully solved for more than 24 h after the onset of labour, followed by an uneventful vaginal delivery of a live foetus without complications to the dam.

2.2.1 **Etiopathogenesis**

Several anatomical (▶ Box 2.1), physiological, behavioural and mechanical factors, acting sole or in combination, may predispose the cow to uterine torsion during the last trimester of pregnancy. The pregnant uterus is like a long tube which can rotate across its longitudinal axis. During pregnancy, the myometrium is quiescent, originating a poor muscle tone. Also, foetal development occurs in one uterine horn destabilizing the uterus, which means that an empty horn will move dorsally much more easily. The presence of a large foetus is also a significant risk factor for uterine torsion – foetal weight is above the mean in 89% of the uterine torsion cases [10]. Additionally, movement of large foetuses increase the chance of a rotational drive. Cow movements when getting up, rising first its hind part allowing for a momentary free suspension of the uterus, are also a predisposing element.

Box 2.1   Anatomical Instability of Uterine Structures in the Cow
- The broad ligament is loose and relatively large.
- The broad ligament has a sub-iliac attachment.
- Some cows present thin broad ligament with poorly developed smooth muscle.
- The greater curvature remains free due to the attachment of the broad ligaments along the lesser curvature of the uterus.
- During pregnancy, the pregnant uterine horn extends massively beyond the area of attachment.
- The cow has a deep abdomen, which is spacious especially if the rumen is partially empty, which happens frequently in dairy cows at the end of gestation.
- As a result, the pregnant uterine horn can rotate freely without significant resistance.

An additional event that can trigger uterine rotation is moving the cow in late pregnancy, such as in truck transport along bouncy roads.

The occurrence of uterine torsion in cows carrying twins is negligible. In bicornuate pregnancies, both foetuses stabilize the uterus, impeding its rotation.

Uterine torsion is more prevalent in multiparous cows than in primiparous cows, as they present a larger abdominal cavity, greater pelvic ligaments pliability and more movable and extendable broad ligaments. Also, the uterine wall usually shows a feebler muscle tone resulting in a much looser organ.

Uterine torsions occur more frequently to the left (counter-clockwise) than to the right direction, and this can be related to the higher proportion of pregnancies in the right than in the left uterine horn. Usually, the pregnant horn rotates over the nonpregnant horn [24].

The torsion pivot can be pre-cervical (cranial to the cervix), intra-cervical or post-cervical (vaginal). The post-cervical torsion is more common than the other ones. Concentric and parallel spiral folds of the vaginal wall are evident by palpation as they are dragged in the same direction by the uterine torsion. These vaginal folds are more distinct in the post-cervical torsion, since the vaginal wall is directly involved, but they can also be marked in high rotation degrees in the intra-cervical torsion.

The degree of uterine torsion can go from 45° to more than 360°, but most of them vary between 180° and 270°. In complete uterine torsions, the birth canal is completely occluded, and the foetus and placenta cannot be touched by the obstetrician's hand.

Uterine torsion is considered partial up to 180°. In these cases, the birth canal is only partially occluded. Both the foetus and the placenta are easily accessible up to 90°, but the birth canal gradually closes from 90° to 180°, increasing the difficulty for the obstetrician's hand to reach the interior of the uterus.

Oedema of the broad ligament and uterus, thromboses of the uterine vessels, uterine haemorrhages and necrosis and (bloody) asci-

2

tes are common sequels and are due to vascular and circulatory disturbances. Depending on the degree of torsion, uterine blood vessels may become progressively occluded, decreasing blood flow to the uterus resulting in oedema and friability of the uterus and, ultimately, foetal death and uterus wall necrosis. Clinical signs severity and the impact on the dam and foetus are related to the degree of torsion.

> **Tip**
>
> The main differential diagnosis for partial intra-cervical uterine torsion is insufficient cervical dilatation. However, the walls in partial torsion are less tense than in the case of cervical stenosis.

2.2.2 Clinical Signs and Diagnosis

The diagnosis of uterine torsion is mainly achieved by behaviour observation and palpation. The clinical signs vary according to the degree of the torsion. Signs of discomfort and abdominal pain occur due to stretching of the broad ligament and include straining, kicking the abdomen, lying and raising repeatedly, lateral decubitus and bruxism. The cow will also show loss of appetite, rumen stasis and obstipation. An increase in respiratory movements (tachypnoea), heartbeats (tachycardia) and blood pulsation may also be observed during clinical examination. Usually, vaginal discharge (e.g. cervical mucous plug, foetal fluids) is absent but can be occasionally observed (i.e. mucoid, blood, fetid or serosanguinous discharge) [10].

Due to vagina rotating and stretching, the vulva sinks and seems to be slightly twisted. Some tension on the perineum can also be observed.

In less severe cases, these behaviour and general clinical signs may go undetectable. Typically suspicion of uterine torsion comes from having a cow apparently on Stage I of parturition (vulvar and udder oedema, milk leaking, gelatinous vaginal discharge and some restlessness), but not showing any signs

of going into Stage II after several hours (no contractions and no signs of allantois or foetus). Some cows may even reassume normal dry-cow behaviour, with farmers completely missing all these subtle signs.

Transrectal and vaginal palpations are essential to unmistakably diagnose partial or total uterine torsion, its direction and the degree of rotation. Pre-cervical torsion is only detected by transrectal palpation. Foetal viability, uterine friability and concomitant lesions should be evaluated at this time (▸ Box 2.2).

> **Box 2.2 Complications or Sequelae Associated with Uterine Torsion**
> - Uterine friability.
> - Uterine necrosis.
> - Uterine perforation or rupture.
> - Uterine adhesions with surrounding viscera.
> - Ovarian vein rupture.
> - Haemoperitoneum.
> - Intestinal obstruction.

■ Transrectal Palpation

The displacement of both broad ligaments is detected by transrectal palpation. The broad ligament contralateral to the side of torsion is tense and stretched over the uterus in a diagonal trajectory. The ipsilateral broad ligament is pulled ventrally under the uterus (◘ Fig. 2.1).

In case of doubt, the route of the uterine arteries can aid in the final diagnosis – one artery will be tense and crossing the caudal abdominal cavity over the uterus, while the other artery will not be evident or will be palpated disappearing under the uterus.

The evaluation of the elasticity of the uterine wall is indicative of uterine friability. The increase thickness and gradual loss of elasticity of the uterine wall corresponds to increasing oedema and eventually to uterine necrosis. As a result, the uterine wall feels spongy to touch. Elasticity evaluation can serve as an indicator of uterine rotation degree and uterine damage/viability and, thus, of the dam's survival chances and future fertility [22].

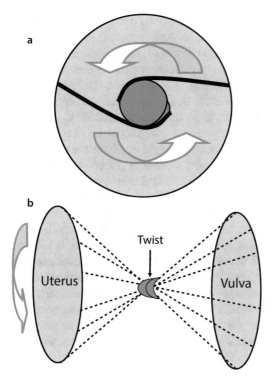

□ **Fig. 2.1** Left uterine torsion. Legend: **a** Both broad ligaments are pulled around the uterine horns and remain under strong tension. **b** The mucosa of soft tissues caudal to the place of rotation is dragged concentrically **b** forming spiral folds

Uterine adhesions to other abdominal structures should also be checked.

■ **Vaginal Palpation**

When introducing the hand into the birth canal, this will be found partially or completely occluded. In partial torsion, the birth canal narrows to less than 10 cm in diameter in the case of intra-cervical or post-cervical torsions. A similar finding is observed in the pre-cervical torsion but cranial to the cervix. In all cases, and contrary to complete torsion, the placenta and the foetus are accessible. The position of the foetus is normally lateral (~90° torsion) or ventral or dorso-pubic (>180° torsion). In complete post-cervical torsion, the cervix cannot be touched, and the vagina will be felt as a *cul-de-sac*.

Particularly in the post-cervical torsion, the spiral vaginal folds are very evident and allows for the identification of the torsion's direction – if the arm rotates counter-clockwise, when entering the vagina, a left

torsion is present; a right torsion will cause the arm to turn clockwise

■ **Complementary Exams**

Transrectal ultrasonography can directly and objectively evaluate the thickness of the uterine wall and tissue damage. Murakami et al. [16] observed that cross-sectional thickness of the uterine wall, between 15 and 25 mm, and the presence of multiple hypoechogenic areas corresponded to macroscopic vascular disrupts in the uterus.

Dam blood lactate concentrations can also be useful to evaluate the degree of uterine necrosis and, consequently, the therapeutic approach and prognosis. Lactate increase indicates low tissue perfusion and peripheral vascular impairment. Murakami et al. [15] proposed a cut-off of >5.0 and >6.5 mmol/L for detection of uterine necrosis and poor prognosis in dams, respectively.

Also, prognosis is poor when torsion is accompanied by a sizeable uterine rupture, bleeding and necrosis. Uterine necrosis is characterized by a dark purple uterine wall, vessel thrombosis and thickening of uterine myometrium. Through paracentesis on the right ventro-caudal abdomen, abundant reddish fluid may be collected, confirming ascites and hemoperitoneum.

> **Tip**
>
> At calving, uterine torsion should be suspected in those cows whose abdominal contractions have diminished in frequency and intensity for several hours, placental membranes are not observed outside the vulva, and urination is absent.

> **Tip**
>
> In most uterine torsion cases, the obstetrician's hand rotates slightly in the same direction of the rotation when inserted into the vagina. This is due to the oblique trajectory of the concentric vaginal folds. It is a very important sign allowing for the diagnosis of the torsion direction.

2.2.3 Treatment and Prognosis

Reduction of the uterine torsion should be performed as soon as possible. Regarding 144 cases of uterine torsion in field conditions, Klaus-Halla et al. [11] observed a lower foetal survival when the uterine torsion lasted for more than 12 h (34.8%; $p < 0.001$) compared with 85.7% and 92.2%, when it was corrected at 6 h or 6–12 h, respectively.

Manual intravaginal reduction, using a plank with the cow standing, rolling the cow using a plank to fix the uterus (Schaffer's method) or caesarean section (C-section) are the treatment options. Conservative correction should always be the first treatment option. C-section should be performed in chronic (i.e. occurring during pregnancy), marked torsion, torsions irreducible by other methods or predicted insufficient cervical dilatation after reduction [23].

Intravaginal manual reduction is used mainly in partial torsions, in which the foetal legs are accessible. The method consists in swinging the foetus and the uterus, like a pendulous, in the opposite direction of the torsion, using the head or limbs as a force hinge. The rotations forces can be directly applied on the legs or head by the obstetrician's hands or using ropes attached to a rod or a Camerer's distortion fork. The objective is to mutate the body of the foetus on to a dorsal position or, at least, to less than 45° torsion.

The best technique is as follows: the left arm should be used to swing the head/forelimbs from a 3–6 h position (using a clock display as orientation) to a 7–9 h; when this is achieved, the right arm should be used for the final swing towards a 11–12 h position that will allow for a normal vaginal delivery. This method permits using the obstetrician's arm maximum strength is a much more logic way and has allowed for the correction of many (if not all) under -270° torsions (Stilwell, personal communication).

Some care is needed during these procedures to prevent injuries to the dam. The use of large amounts of obstetrical gel is recommended to avoid trauma to the birth canal, due to friction of the mucosa with the obstetrician's hands and material. All this procedure should be performed under a low epidural anaesthesia (5–7 mL procaine or lidocaine).

Rolling the cow according to Schaffer's method is a popular and useful obstetrical technique (◨ Fig. 2.2). However, the torsion direction needs to be accurately defined before rolling the dam to prevent iatrogenic increase in the torsion degree. The cow should be lying on same lateral decubitus (right or left) as the uterine torsion direction; a plank is placed on the upward flank to transabdominally press and fix the uterus; the dam is rolled 180° in the same direction of the uterine torsion to the opposite lateral decubitus. The foetal limbs should be fixed through the vagina while the dam is rotated. What is expected is that the abdominal wall and viscera, except for the uterus and its content, are moved – so that the cow is rolled to "meet" the uterus. A second or third rolling can be necessary for complete reversion in uterine rotations over 270°. At the end of each rolling, the dam is raised, and the uterus and foetal positions are evaluated. This method is particularly suitable for partial torsions. In complete uterine torsion, it is not adequate to roll the cow without a plank, as only very slight reduction is expected.

The disadvantages of this method are as follows: it can be very stressful to the cow; uterine rupture can occurs if the uterus is very full and tense; and it is physically demanding for the obstetrician that has to lie down behind the cow to hold the limbs.

> **Important**
>
> In the presence of a friable and devitalized uterus, or significant uterine rupture, an ovariohysterectomy can be performed to save the dam. The most frequent conditions predisposing for a friable uterus are uterine rupture, uterine torsion, autolytic or emphysematous foetus.

> **Tip**
>
> In the Schaffer's method, the obstetrician should have one arm/hand in the birth canal holding the foetus' limbs simultaneously to the rolling, to help fix its position but also to monitor the uterus repositioning effectiveness.

◘ Fig. 2.2 Schaffer's method. Legend: The cow is lying down in lateral decubitus ipsilateral to the side of the uterine torsion. A plank with approximately 3.5 meters length and 25 cm width is placed in the flank. A slight sliding of the plank occurs during the procedure, so its superior end should be projected upward, up to 1 meter outside the dorsum. A person, up to approximately 90 kg, fixes the plank, and the cow is simultaneously rotated to the contralateral side. The abdominal wall and its content rotate up to 180°, while the uterus and foetus are fixed due to the transabdominal pressure. The head of the cow should remain close to the floor during the procedure, preventing the cow from swinging to get up. Original from João Simões. Button- Schematic representation (the Reuff's method, to force the cow down, is represented in this illustration). Courtesy and original by Gabriela Silva (University of Coimbra)

Some signs are usually apparent when the uterus reaches its normal position: the cervix borders become more obvious; discomfort is relieved; placental membranes appear in the birth canal or outside the vulva; and urination occurs immediately or within the next few minutes, once the urethra is no longer stretched or compressed due to bladder's dislocation.

In severe torsions, when the uterine wall is severely affected, the reposition of the uterus

2

can cause fatal uterine rupture or uterine and ovarian arteries' rupture.

Regarding the C-section option, the ventrolateral or ventral laparotomy approach (see ▶ Chap. 8) allows for a better accessibility to the uterus. After opening the abdominal cavity, the torsion can be corrected by just lifting and rotating the uterus – this is a good approach for obstetricians with short arms and lower physical strength. If uterine reposition is very difficult due to its size, trocarization of the uterus can be made in order to remove the allantoid and/or amniotic fluid, such as is described in the hydrops conditions (see below). This procedure improves foetal manipulation through the uterine wall and facilitates uterine rotation and repositioning. After repositioning, an evaluation of the cervix by vaginal palpation is recommended. If cervix dilatation is considered unlikely, the C-section should be continued and concluded. On the other hand, if the cervix shows signs of dilatation, vaginal delivery will probably be possible and so opening the uterus will not be necessary.

Sometimes, insufficient dilatation of the cervical canal after a successful reposition of the uterus is detected. This insufficient cervical dilatation is probably due to physical pressure on the cervix, when it was not fully dilated at the onset of the uterine torsion. Extraction can be delayed while waiting for adequate cervical dilatation. After uterus reposition, in most of the cases, the dam is relieved and resumes abdominal contractions. Immediate extraction is only an option when the cervix shows enough dilation or responds well and promptly to mechanical stimulation. It should be remembered that the injury caused to the dam increases proportionally to the duration of foetal manipulation.

Ovariohysterectomy should be performed if the uterine wall is significantly friable, vascular system is compromised [22], a sizeable uterine rupture is observed, or the viability of the uterine wall is questionable.

Overall, prognosis depends on torsion degree, foetus condition (e.g. alive, emphysematous) and uterine friability/viability due to the vascular compromise. In moderate and quickly resolved cases, prognosis is positive, but in severe, advanced or when obstetrical intervention is prolonged, prognosis is reserved.

Case Study 2.1 Partial Uterine Torsion

A 3-year-old pregnant Holstein cow at full term weighing approximately 700 kg was presented to the veterinarian due to straining for more than 3 h. According to the farmer, the cow showed normal pre-calving behaviour for the 24 h prior to starting these abdominal contractions. During general physical examination, the rectal temperature was found to be normal (38.8 °C), but respiratory (43 breaths per minute) and cardiac (95 beats per minute) frequencies were slightly increased, justified by the prolonged straining. The cow was alert without evidence of dehydration or hypovolemic shock (capillary repletion time <3 sec.). During transrectal palpation, tension on both broad ligaments was registered. The right ligament covered the upper part of the pregnant uterine horn continuing down vertically to the left side. A slight twist, towards the left side, was detected at cervical level. By vaginal palpation, the intact foetal membranes, as well as two foetal pinching-responsive forelegs, were palpated in the pelvic inlet. Also at this level, a strong circular tension involving approximately 1 cm thick cervical wall was detected. The cervical opening was about 7 cm. Vaginal folds were not evident. Both forelimbs were obliquely positioned, with hoof sole facing upwards. A diagnosis of partial uterine torsion, approximately 140°, to the left side, was proposed. Both foetal hoof circumferences were evaluated using a Calfscale® tape at the coronary band suggesting a large foetus (average of 19.5 cm: estimated body weight of 45–46 kg). The pelvic area was estimated as being large enough due to a 29.5 cm inter-ischial measure. This data was indicative of a low-risk foetopelvic disproportion. Due to the presence of a live large foetus, it was decided to roll the cow, to achieve uterine torsion reduction. The dam was forced on to a left lateral decubitus.

The chorioallantoic sac was intentionally ruptured to improve foetal accessibility, and an anterior presentation and a left lateral position of the foetus were confirmed. The obstetrician's hands fixed both foetal legs, and the dam was rolled towards the right side until the foetus was in dorsal position. At this moment, the birth canal was again evaluated. The circular tension at the twist point disappeared, but the diameter of the cervical canal remained at approximately 12 cm (insufficient cervical dilatation). The cow was compelled to stand, after which a new cycle of abdominal contractions, as well as urination, was observed. Because the amniotic broke during the obstetrical manoeuvres, a manual mechanical cervical stimulation was initiated so that full dilatation was accomplished within 10 min. Moderate forced traction was necessary to deliver a live and healthy male calf. Calf vitality was considered normal according to the Apgar scoring parameters. Oxytocin (40 IU; i.m.) and 70 mL of 100 mg/mL oxytetracycline were administered i.v. (and continued for 3 days).

2.3 Vaginal and Cervicovaginal Prolapses

2.3.1 Etiopathogenesis and Diagnosis

Vaginal prolapse is an eversion and exteriorization of the vaginal wall through the vulva. Usually, it occurs during the last two weeks of pregnancy (acute vaginal prolapse), but it can also be observed during the last trimester of pregnancy, in the early postpartum or during oestrus. It can occur between calvings and repeatedly during a short or a medium period of time. High oestrogen levels, relaxation of the pelvic ligaments, very wide vulva, very large gravid uterus, successive lying downs bouts and cranial-caudal slope of the bed all predispose to vaginal prolapse. Excessive body condition, parity (multiparous cows), the presence of multiple foetuses and previous trauma to the perineal area, among others (▶ Box 2.3), are also predisposing factors for vaginal prolapse.

It is usually a recurrent condition, being more common when the cow is lying down but disappearing when the animal gets up. However, increase in the volume prolapsed due to trauma, dryness of the mucosae and oedema will eventually lead to a vicious circle and a permanent prolapse (chronic prolapse).

> **Box 2.3 Predisposing Factors of Vaginal Prolapse**
> — Increasing levels of oestrogens.
> — Phytoestrogens (e.g. present in *Trifolium subterraneum*).
> — Mycoestrogen (zearalenone).
> — Relaxation of the sacro-tuberous ligaments.
> — Enlarged rumen (e.g. poor coarse forage, chronic ruminal acidosis increasing ruminal osmolarity, rumen impaction).
> — Excessive body condition (e.g. excessive deposition of abdominal and peri-vaginal fat).
> — Multiparous cows.
> — Multiple foetuses.
> — Breed predisposition (Brahman, Brahman crossbreeds and Hereford breeds).
> — Genetic (inherited chronic vaginal prolapse).
> — Oversize gravid uterus (increase in intra-abdominal pressure).
> — Excessive traction on the foetus.
> — Perineal and peri-vaginal injuries causing tenesmus.

Generally, the determining factors inducing spontaneous vaginal prolapses are unknown. Typically, a fold of the vaginal floor starts evaginating just cranially to the vestibulo-vaginal junction. Consequently, oedema and

2

Grade	Description	Relevance	Treatment
I	Intermittent prolapse of the vagina; most commonly when lying down.	Likely to progress to Grade II if not treated.	Temporary retaining suture; cull after calving or perform permanent fixation technique if embryo flush cow.
II	Continuous prolapse of the vagina ± urinary bladder retroflexed.	Urinary bladder involvement (common) can obstruct urination or cause persistent straining.	Temporary retaining suture; cull after calving or perform permanent fixation technique if embryo flush cow.
III	Continuous prolapse of the vagina, urinary bladder and cervix (external Os visible).	Can compromise urine outflow and ureters. Should be treated quickly to prevent life-threatening injury.	Perform permanent fixation technique if embryo flush cow. Induce parturition or perform elective C-section in commercial cows.
IV	Grade II or III with trauma, infection or necrosis of vaginal wall. (a) Sub-acute such that replacement into vaginal vault is possible. (b) Chronic with fibrosis such that the vagina cannot be replaced.	Grade IV a: repair laceration, debride wounds, treat infection and replace into vaginal vault. Grade IV b: Requires elective C-section or vaginal resection.	Perform permanent fixation technique if embryo flush cow. Induce parturition or perform elective C-section if commercial cow.

◻ **Table 2.1**   Clinical grading scale for vaginal prolapse according to Wolfe and Carson [26]

vaginal mucosa irritation will cause discomfort and straining, promoting the eversion of the vaginal wall through the vulva. Wolfe and Carson [26] propose a classification for vaginal prolapse in cows, involving four grades according to clinical expression and lesions' severity (◻ Table 2.1; ◻ Fig. 2.3). Above Grade II, the prolapse of the bladder may also occur, and so the case should be considered an emergency condition. The bladder may be filled due to the occlusion of the urethra. In more severe cases (Grade IV), the cervical external Os also evaginates (cervical prolapse). Local vascular alterations occur quickly, developing inflammation of the vaginal mucosa and oedema. The everted mucosa becomes more susceptible to trauma and to contamination, resulting in further inflammation, infection, bleeding and a more friable tissue. Once these events start to develop, spontaneous reduction is improbable, and a chronic vaginal prolapse is established. If no intervention is applied, infection, necrosis, adhesions, toxaemia, dystocia and even death can occur.

❯ **Important**

Tissue injuries subsequent to vaginal prolapse can be minimized with timely and appropriate intervention.

2.3.2 Treatment and Prognosis

Treatment aims at repositioning the vagina, cervix and bladder into the pelvic/abdominal cavity and to prevent recurrence using a retention suture. In Grade I, the elevation of the cow's hindquarters can be enough and should be tried before making a retention suture.

A low epidural anaesthesia with procaine or lidocaine (5–10 mL) is performed. The prolapsed vagina is cleaned and washed with cold water lower than 15 °C to reduce vasodilation. Disinfection using diluted iodopovidone, chlorohexidine or potassium permanganate solution (1:1000) is recommended. Large tears should be sutured, and, if included and full, the bladder should be drained before attempting reposition. The bladder should be punctured with a large needle [14–16 G × 2.5″

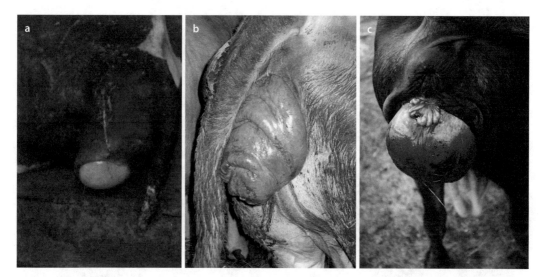

Fig. 2.3 Vaginal Prolapse. Legend: **a** Vaginal prolapse Grade I of a Holstein-Friesian cow. Firstly, the vaginal wall is seen outside the vulva only when the cow is lying down and especially when the bed is sloping cau- dally. **b**, **c** Vaginal prolapse Grade IV of beef breeds. The external Os is located at end of the prolapsed mass **b** and on top **c**. Originals from George Stilwell **a**, António Carlos Ribeiro **b** and João Simões **c**

(1.6–2 mm × 50–60 mm)]. However, usually, iatrogenic elevation of the vaginal mass alleviates the tension on the urethra, allowing urination ad emptying of the bladder.

Epidural anaesthesia and lubrication of the prolapsed mass, using sterile vaseline or obstetric gel, is always required to ease reduction and to prevent trauma.

> **Tip**
>
> Besides lubrication activity, glycerine can also be applied on to the prolapsed vaginal surface to reduce oedema and congestion during a few minutes (osmotic action).

Once reposition of the prolapsed structures is completed, a retention suture is usually made. The Bühner suture is a temporary but popular retention suture. It is easy to perform, is effective and has no major complications. However, it is traumatic and can cause pain and discomfort for a few days. The umbilical tape is passed round the vulva, deeply at the junction of the vulvar lips with the skin, entering dorsal and ventral to the vulvar commissures using a Bühner needle (**Fig. 2.4**). Sometimes, the stitch may tear the dorsal vulva or

vestibular wall. This situation can be mitigated using a modified Bühner suture incorporating a horizontal mattress-like suture [18]. This suture is placed 2 cm above and laterally to each side of the dorsal vulvar commissure. It allows to redirect the suture tension towards thicker and more resistant tissues.

Other temporary sutures, such as deep vertical and horizontal mattress techniques, can also be used. Three to four sutures are placed deep at the junction of the vulvar lips with the skin, passing through them (see clinical case in ▶ Chap. 10). The deep vulvar insertion is required to prevent vulvar tears especially if the dam is straining. In the bootlace technique, an umbilical tape is used to make up five small eyelets on either side of the vulva – this technique is less traumatic and allows for the farmer to easily remove the suture in case of calving.

In all these methods, a gap for urination should be left close to the ventral vulvar commissure.

The main limitation of all these techniques is the potential occlusion of the birth canal, if the cow is pregnant. So, the retention sutures need to be removed before the onset of Stage II of labour to allow for foetal delivery and to prevent vulvar tears during contractions.

2

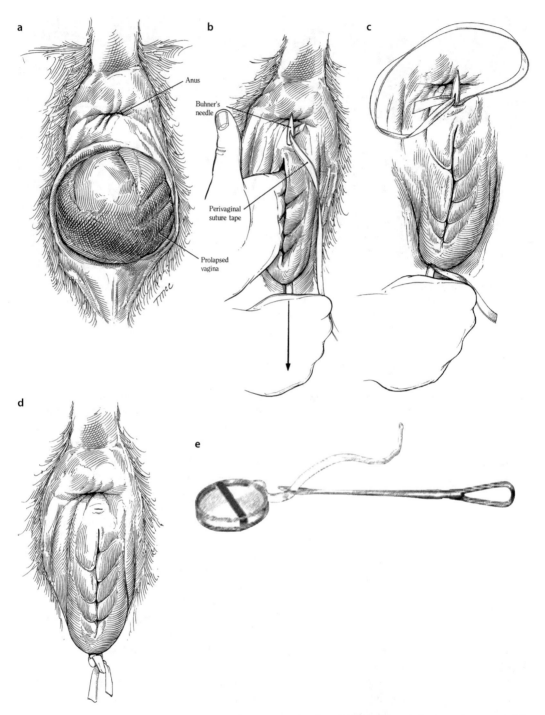

Fig. 2.4 Schematic representation of the Bühner's suture method. Legend: **a** After the reduction of the prolapsed mass, a Bühner needle is introduced 2 cm ventrally to the ventral vulvar commissure and pushed through the junction of the vulvar lips with the skin of the rump. **b** Two centimetres above the dorsal commissure, the needle exits, and a treaded umbilical tape is pulled back to the ventral commissure. **c** The same procedure is performed on the other side. **d** At the end, the suture is appropriately tight to prevent prolapse recurrence, and a reinforced (surgical) knot is made. **e** Bühner's needle, 12″ (30.5 cm long) and umbilical tape, 1/8″, 1/4″ or 3/8″ × 20″ yds (0.3 cm, 0.6 cm or 1.0 cm large × 18.3 m long). (**a–d**: Adapted from Baird [3] with permission from Wiley Blackwell. **e** Modified (FotoSketcher 3.60) from Jorgensen Laboratories, Inc. (USA) with permission)

The Minchev method is more appropriate for periparturient cows when the floor of the vagina is prolapsed (◘ Fig. 2.5a). Even though it is considered a traumatic technique, it has the advantage of keeping the birth canal unblocked. The development of peri-vaginal adhesions prevents the recurrence of the vaginal prolapse, so it is considered a permanent suture. Inadvertent sciatic nerve damage during the procedure or in case of abscessation (bacterial contamination) is the major complication described. Also, the puncture of the internal pudendal artery can occur and should be carefully avoided. A large needle is threaded with an umbilical tape and passed through the dorsolateral vaginal wall. A gauze, plaque or button (inert material) is attached to end of the umbilical tape to anchor the suture. The needle is vertically pushed through the sacroiliac ligament in the dorsal area of the lesser sciatic foramen to avoid the sciatic nerve and through the gluteal musculature. In the gluteal area, the needle is pulled out through the skin. The suture line is pulled up, and another gauze, plaque or button is placed over the rump, without excessive tension. This suture can also be applied using a Bühner needle from the gluteal area down towards the dorsolateral vaginal wall.

The modified Minchev method is similar to the previous one (◘ Fig. 2.5b), but the suture is more cranial – the suture is placed anterior

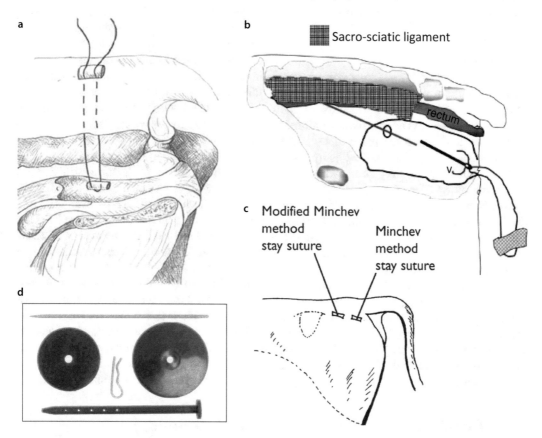

◘ **Fig. 2.5** Schematic representation of the Minchev method. Legend: **a** Vaginopexy (colpopexy) according to the Minchev method. (Adapted from Peter [17] with permission from John Wiley and Sons). **b** Modified Minchev method. Cranial insertion of the needle from the dorsolateral vaginal wall through the sacrotuberous ligament and gluteal musculature; *v* vaginal vault. (Adapted from Miesner and Anderson [14] with permission from Elsevier). **c** Local of external (gluteal) fixation of the suture. (Adapted from Ames [2], with permission from John Wiley and Sons). **d** Trochar and two large plastic buttons. The steel needle is inserted in the plastic rod to make the puncture. (Modified (FotoSketcher 3.60) from Jorgensen Laboratories, Inc. (USA) with permission)

to the lesser sciatic foramen. The use of the button and pin (Johnson button) technique is similar to the Minchev technique. A pin is used to replace the needle and the umbilical tape, and two plastic buttons are inserted in each pin end after puncture. Postoperative tenesmus is expected as vaginal tissue innervated by sacral nerves S3, S4 and S5 may be affected.

To prevent the risk of peritonitis or septicaemia in severe vaginal prolapse, broad-spectrum antimicrobials, e.g. oxytetracycline (10 mg/kg body weight; BW) or 7.0 mg amoxicillin (7.0 mg/kg BW) plus clavulanic acid (1.75 mg/kg BW), should be parenterally administered for 3–4 days. Anti-inflammatory drugs, e.g. meloxicam (0.2 mg/kg BW) or flunixin meglumine (2.2 mg/kg BW), are also recommended for the first 2–3 days.

2.4 Hydrops

Dropsical conditions or hydrops are defined as excessive accumulation of fluids in the uterus, due to placenta and/or foetus malfunctioning. Depending on the place of excessive fluid storage, hydrops is classified as hydroallantois (allantoid cavity) or hydramnios (amniotic cavity). The occurrence of hydroallantois is sporadic but represents 85–95% of the hydrops cases [8]. Hydramnios (*hydrops amnii*) is a much rarer condition. Usually, hydroallantois is associated with placental abnormality and hydramnios with congenital foetal abnormalities. Under normal circumstances, fluids are produced by placental membrane interchanges and by the foetus (e.g. urine) (see ▶ Chap. 1). Excessive placental transudate and reduced fluid reabsorption, as well as excessive foetal fluid production, can originate hydrops.

2.4.1 Hydrallantois

The development of hydrallantois is quick and progressive and usually evolves in less than one month within the last two months of pregnancy. A high volume of fluid, up to 250 L, can be stored inside the allantoid cavity. The reduction in the number of placentomes, less

than 70, and an increase in underdeveloped primitive villous (adventitious placenta) cause placental dysfunction. Some placentomes may increase in size to compensate for this dysfunction. Some assisted reproductive technologies such as somatic cell nuclear transfer (foetal overgrowth and decrease in number of placentomes [5]), in vitro fertilization (renal failure [13, 25]) and somatic nuclear transferred embryos [5] can predispose to hydrallantois. Sasaki et al. [21] observed that a genetic mutation (SLC12A1) caused foetal polyuria and hydronephrosis, leading to this condition (☐ Fig. 2.6). Also, Reis et al. [20] observed an association between *Sida carpinifolia* (swainsonine alkaloid) poisoning and hydrallantois.

The abdominal silhouette is seen as an apple-shaped bilateral abdominal distension (☐ Fig. 2.7). The uterine volume provokes significant pressure on the rumen and lungs. The cow is anorectic, without rumen activity, and, most times, obstipation and respiratory difficulty are also present. Placentomes and the foetus are difficult to detect by transrectal palpation due to the severe distention of the uterine wall.

If salvage slaughter is not an option, the treatment consists in parturition induction – a luteolytic dose of prostaglandin $F_{2\alpha}$ ($PGF_{2\alpha}$) in combination with 20–40 mg of dexamethasone will induce abortion or calving within 32–72 h. A C-section can also be performed as an alternative. Allantocentesis or amniocentesis can be used to reduce the tension on the uterine wall before performing the C-section. A cannula (at least 4 mm in diameter), attached to a tube, is inserted into the allantoic or amniotic cavity after laparotomy. The fluid should be slowly removed to prevent hypovolemic shock. The cannula puncture should be closed with a cruciate suture after enough fluid is drained.

2.4.2 Hydramnios

The amniotic fluid increases gradually during several months. In hydramnios, around 7th or 8th month of pregnancy, the abdominal outline will display a pear-shaped appearance due to excessive storage of amniotic fluid (☐ Fig. 2.8). Renal agenesis or dysgen-

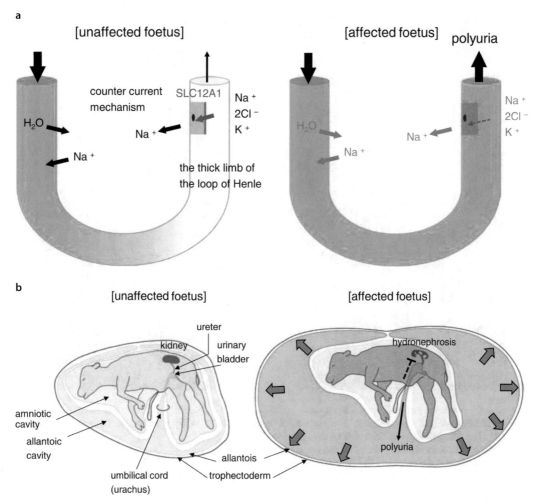

□ **Fig. 2.6** Pathophysiology of hydrallantois due to genetic mutation (SLC12A1). Legend: This model was created to explain the occurrence of hydroallantois in 33 Japanese Black cows conceiving from the same sire. A missense mutation in solute carrier family 12, member 1 (SLC12A1), was observed. SLC12A1 is a key regulator, controlling the urine concentration via a counter current mechanism in the kidneys (left panel; **a**). SLC12A1 regulates the reabsorption of Na+-K+-2Cl– molecule of in the tubules. In the affected foetus kidneys (right panel, **a**), SLC12A1 dislocates from the apical membrane to cytosol, in turn, impairing reabsorption of Na+-K+-2Cl– in the thick limb of the loop of Henle. Consequently, the affected foetuses exhibit polyuria. At mid-gestation, excessive volumes of allantoic fluid obstruct the outflow of foetal urine into the allantoid cavity (right panel **b**, dotted line) due to the limited space in the uterus, resulting in hydronephrosis. The maternal abdomen becomes progressively distended and increases the pressure on the internal organs and foetus in the uterus, causing foetal death. (Modified from Sasaki et al. [21])

esis of the foetus seems to be the most significant cause for this type of hydrops. Excessive pressure on the foetus also induces foetal hydrops and, eventually, death. The medical treatment of hydramnios is similar to hydrallantois (▶ Box. 2.4). The differential diagnosis between hydrallantois and hydramnios is listed in □ Table 2.2.

In the surgical treatment, amniocentesis can be used to reduce the uterine volume during the hysterotomy.

The prognosis for dam's survival and future fertility is good, although premature calving and parturition complications (e.g. metritis) will negatively affect initial and overall lactation.

2

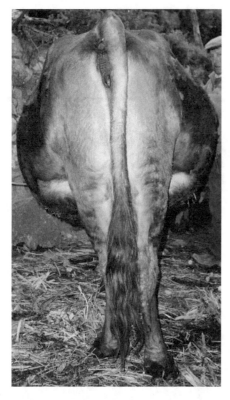

■ **Fig. 2.7** Apple-shaped abdomen in a case of hydroallantois in a beef cow. Legend: The abdominal silhouette has an apple-like contour due to the distension of the allantoic cavity and posteriorly confirmed by caesarean section (at the onset of the 8th month of pregnancy). The uterine wall was found very tense. The placentomes could not be detected by transrectal palpation, as well as the foetus body by abdominal ballottement. In this case, dyspnoea (diaphragm compression) and anorexia (rumen compression) occurred. The ruminal contractions were well identified by transrectal palpation. However, no evidence of ruminal contractions was detected by left flank auscultation, probably due to cranial dislocation of the dorsal ruminal sac. (Original from João Simões)

■ **Fig. 2.8** Pear-shaped abdomen in a case of hydramnios in an 8-month pregnant cow. (Original from George Stilwell)

2.5 Abortion

We can define abortion as the premature expulsion of a dead or inviable foetus, i.e. before the 260th day of pregnancy. Abortion can be due to different reasons: infectious (virus, bacteria, fungi and protozoa; ■ Table 2.3) or non-infectious (e.g. physical, nutritional, toxicological and genetic). When it occurs in late pregnancy, it can cause dystocia.

Abortion can be sporadic with an annual (and acceptable) incidence of less than 3–5%. However, when this threshold is exceeded, or in epidemic abortions outbreaks, the definitive cause of foetal death should be investigated. A thorough physical examination, a complete clinical history and hormonal evaluation should be made when foetal death is suspected and the cervix is closed (▶ Box 2.5).

Box 2.4 Protocol for Parturition or Abortion Induction

- Cloprostenol 500 μg i.m.
- Dexamethasone 20 mg i.m. or i.v.
- Dexamethasone + cloprostenol (20 mg + 500 μg).
- Betamethasone 20 mg i.m.
- Dinoprost 25 mg i.m.

Characteristic	Hydrallantois	Hydramnios
Prevalence	85–95%	5–15%
Rate of development	Rapid (within 1 month)	Slow over several months
Shape of the abdomen	Round and tense	Piriform, not tense
Transrectal detection of placentomes and foetus	Nonpalpable (tight uterus)	Palpable
Gross characteristics of fluid	Watery, clear, amber-coloured transudate	Viscid, may contain meconium
Foetus	Small, normal	Malformed
Placenta	Adventitious	Normal
Refilling after trocharization	Rapid	Does not occur
Occurrence of complications	Common	Uncommon
Outcome	Abortion or maternal death common	Parturition at approximately

Adapted from Drost [8] with permission of Elsevier

After foetal expulsion, a gross examination of the placenta should be made to identify inflammatory lesions such as necrosis or haemorrhage of the cotyledons, intercotyledonary thickening or presence of indicative lesions (e.g. white plaques and leather-like placenta are pathognomonic of fungal infection). Samples of both structures should be collected for bacteria and fungi cultures (fresh samples) or for histopathology analysis (placed in a 10% neutral buffered formalin solution). Also, the entire foetus or foetal parts (gastric content, eye, brain, lung, heart, liver, thymus, thyroid, lymph nodes, tongue, diaphragm, kidney and spleen) should be sent to the laboratory.

Box 2.5 Evidence of Intrauterine Foetal Death, Before the Expulsion of the Foetus
- Transrectal palpation: abnormal volume of foetal fluids, size of pregnant uterine horn or size of foetal body parts.
- Absence of foetal fluids in the pregnant uterine horn.
- Vaginoscopy: sanguineous or purulent discharge.
- Foul smell (emphysematous calf).
- Absence of uterine artery fremitus (from 4 months to the end of pregnancy) Beware: artery pulsation may be felt if the foetus is dead, but not fremitus.

Aetiological diagnosis remains a challenge for most cases of abortion (◘ Table 2.3) as the determining cause remains unclear in more than 50% of the foetal necropsies submitted to the laboratory. Any conclusion should be established on clinical, epidemiological and laboratory findings (◘ Table 2.3; ◘ Fig. 2.9). Laboratory methodologies are based on histopathology, bacterial and fungi cultures, polymerase chain reaction (PCR) and dam's serology tests. Cow's serology evaluation can indirectly provide evidence of exposure, e.g. bovine viral diarrhoea (BVD), infectious bovine rhinotracheitis (IBR) and Leptospira, but paired samples with at least 2 weeks interval are required to detect an increase in antibody titres after abortion. In many cases, abortions are sporadic, and clinical signs or lesions are not evident, and only environmental bacteria are isolated.

When the exact cause is identified, it is perceived that approximately half of all abortions are caused by infectious agents. Although any systemic infection can provoke foetal death and abortion, some agents are particularly or exclusively abortive. Laboratory identification of the specific agent, from foetal tissues or placenta, is crucial for diagnosis and to implement adequate control measures.

2

□ Table 2.3 Main infectious causes of bovine abortion

Abortifacient	Time of abortion
Bacteria	
Brucella abortus	Third trimester
Listeria monocytogenes and *L. ivanovii*	
Chlamydia psittaci	
Salmonella Dublin and *S.* Brandenburg	
Campylobacter fetus subspecies *venerealis* and *C. fetus* subspecies *fetus*	Early embryonic losses Second to third trimester
Leptospira interrogans serovars *pomona, hardjo, canicola, grippotyphosa* and *icterohaemorrhagiae*	Third trimester Last trimester
Bacillus cereus and *B. licheniformis*	Third trimester
Ureaplasma diversum	
Mycoplasma bovis	
Trueperella pyogenes	All trimesters
Viruses	
Bovine viral diarrhoea virus	Early embryonic loss; all trimesters
Bovine herpesvirus type I	All trimesters; usually 4 months to term
Infectious bovine rhinotracheitis virus	4 months to term
Bluetongue virus	Second to third trimester
Aino/Akabane virus	
Protozoa	
Neospora caninum	3–8 months; usually 5 months
Tritrichomonas foetus	Early embryonic losses; first half of gestation
Babesia bovis and *B. bigemina*	All trimesters

□ Table 2.3 (continued)

Abortifacient	Time of abortion
Fungi	
Aspergillus fumigatus	4 months to term
Mortierella wolfii	
Zygomycetes (*Mucor* spp.; *Absidia* spp., *Rhizopus* spp.)	

Modified from Reichel et al. [19]
Epizootic bovine abortion (pathogen not defined); vector: *Ornithodoros coriaceus* (pajaroello tick); last trimester
Other emerging diseases causing abortion in cattle: Epizootic haemorrhagic disease virus (culicoides-transmitted virus); Rift Valley fever virus (mosquitoes-transmitted virus); Akabane virus; Lumpy skin disease; Wesselsbron virus; Schmallenberg virus (culicoides-transmitted virus)

❯ Important

Foetuses are immunocompetent from approximately the 110th day of pregnancy. Before this date, infections agents crossing the placental barrier usually cause embryonic or foetal death. Several viruses and toxins may induce the development of congenital anomalies during organogenesis, but not necessarily leading to abortion (□ Table 2.4).

2.6 Foetal Maceration

Maceration is a biotic cadaveric phenomenon that occurs from the middle 3rd month of pregnancy, after foetal ossification (70 d gestation) is well established. In contrast to abortion, instead of being expelled, the dead foetus is retained inside the uterus. Environmental bacteria may invade the uterus through the vagina so that the uterine wall initiates an inflammatory process and shows evident increase in thickness. Only the skeleton resists

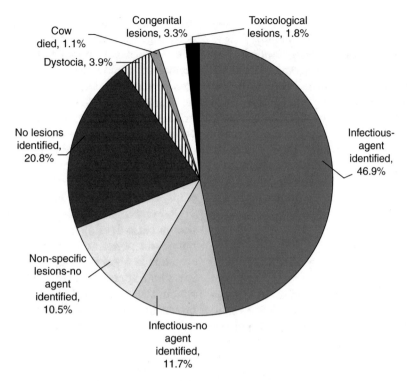

Fig. 2.9 Etiologic classes of bovine abortion identified by laboratory exams. A total of 665 bovine abortion cases were examined at California Animal Health and Food Safety Laboratory System from 2007 to 2013. (Adapted from Clothier and Anderson [4] with permission from Elsevier)

Table 2.4 Associations between virus infection and central nervous muscle and musculoskeletal congenital lesions

Lesion	Definition	BVDV	SBV	BTV	AKAV/AV
Hydranencephaly	Extensive loss of cerebral tissue with replacement by clear fluid.	×	×	×	×
Porencephaly	Cystic fluid filled cavities in the brain tissue.	×	×	×	×
Hydrocephalus	Dilation of the lateral ventricles by cerebrospinal fluid.	×	×	×	
Microencephaly	Reduced size of the cerebrum.	×	×	×	×
Cerebellar hypoplasia	Reduced size of the cerebellum.	×	×	×	
Kyphosis	Dorsal vertebral column curvature.		×		
Lordosis	Ventral vertebral column curvature.		×		
Scoliosis	Lateral vertebral column curvature.		×		
Torticollis	Twisted cervical vertebral column curvature.		×		
Arthrogryposis	Joint contraction of the limbs.		×		×

Adapted from Agerholm et al. [1]

BVDV Bovine virus diarrhoea virus, *SBV* Schmallenberg virus, *BTV* bluetongue virus, *AKAV* Akabane virus, *AV* Aino virus

2

to maceration. The greyish-red vaginal and putrid discharge is the predominant clinical sign. Intermittent or transient anorexia and mild fever may also be observed. The diagnosis is made by transrectal palpation and ultrasonography detecting the foetal parts and increase thickness of the uterine wall. Clearing of the uterine content can be made by giving a luteolytic dose of $PGF_{2\alpha}$. In some cases, this procedure is unsuccessful, and a C-section is required to remove all content [7]. The prognosis for the cow's future fertility is poor.

2.7   Foetal Mummification

Foetal mummification occurs between the 3rd and 8th months of pregnancy (after foetal ossification). The development of this pathology requires foetal death in an aseptic and oxygen-deprived environment (cervix tightly closed), followed by gradual absorption of foetal fluids [8]. It is associated with a persistent corpus luteum which maintains pregnancy. An involution of the caruncles occurs quite quickly. Both foetus and placental membranes are progressively dehydrated, resulting in a compacted structure. Several weeks are required for the completion of the mummification process. At the end, the placental membranes and uterine wall adhere to the foetus body (◻ Fig. 2.10).

The cause for this pathology remains unknown. However, mechanical factors inducing hypoxia, e.g. obliteration of the umbilical cord, defective placentation or early uterine torsion, and genetic defects or hormonal anomalies have all been proposed. Some infectious diseases such as BVD, leptospirosis, trichomo-

niasis, fungi (mould) and neosporosis have also been suggested as possible causes. Neospora infection in a naïve population caused an "outbreak" of mummifications (G. Stilwell, personal communication).

During transrectal palpation, the uterine horns are found empty of fluid, and the walls are completely adhered to the surface of a rigid dry foetus. After removal, foetus and placenta have a brownish-black leathery appearance with the foetus skull showing empty orbital cavities. The mass is odourless.

Papyraceous and haematic mummification can also develop in cows. In papyraceous mummification, a stiff and dry foetoplacental structure is observed with papyrus characteristics and appearance. In hematic mummification, the mummified foetus is covered with brown and viscous adhesive material originated from placentomes' haemorrhages.

> **Tip**
>
> Foetal mummification can be suspected when the udder development is not observed towards the expected term pregnancy, or no pre-calving signs are evident up to the 10th month after matting.

A single $PGF_{2\alpha}$ dose causes luteolysis of the corpus luteum, similarly to parturition induction. Two days after treatment, the cervix dilatates and the mummified foetus will pass into the vagina. Most times the foetus has to be removed by hand from the vagina, as the vulva will not dilate, and no contraction will be present. Very rarely a uterine lavage is needed after removal of the foetus. If the single $PGF_{2\alpha}$ treatment is unsuccessful,

◨ **Fig. 2.10** Foetal mummification. Legend: mummification process. Top: A mummified foetus and placenta that were removed through the vagina 2 days after $PGF_{2\alpha}$ injection. Courtesy of João Fagundes. Bottom: an early stage of foetal mummification from a cow presenting endocarditis. (Original from George Stilwell)

2

repeated administrations or a combination with Oestradiol-17β and oxytocin, if the cervix is opened, can be tried [12]. Prostaglandin E$_2$ (dinoprostone) to dilate the cervix can also be used.

When these procedures are not successful or when the foetus is too large, C-section is the only solution. An ipsilateral flank approach to the pregnant uterine horn is recommended to ensure adequate exteriorization of the uterus.

Except in complicated cases, fertility prognosis is fair, and the cow may conceive in the following oestrus. However, because preparation for lactation does not occur, if mummification is only diagnosed at term (over 9 months of pregnancy), culling/slaughter of the dry cow is usually the best course of action. Having a mummified foetus will not prevent approval of the carcass at slaughter.

Key Points
- Uterine torsion mainly occurs during Stage I of labour and causes dystocia due to the partial/total occlusion of the birth canal.
- Uterine torsion up to 270° can be resolved through the vagina by inducing a pendulous movement to the foetus-uterus.
- Schäffer method is also an easy and successful procedure to solve uterine torsion >180°.
- In embryonic and early foetal death (<70th day of pregnancy), the foetus can be completely reabsorbed.
- Congenital malformations develop mainly during the embryonic period and are related to organogenesis. Abortion, stillbirth or dystocia can occur.
- Vaginal prolapse may be seen as a reproductive emergence, especially the more severe grades and when close to end of pregnancy.
- The Bühner and Minchev techniques are the most appropriate vaginal retention sutures.
- Contrary to hydramnios, hydrallantois is a recurrent condition, and fertility prognosis is poor even after treatment.

- Diagnosis of abortion involves the interpretation of individual and epidemiological data, clinical examination of the dam and laboratory results.
- Contrary to foetal maceration, foetal mummification is an aseptic condition with good fertility prognosis after the induction of parturition with a single luteolytic dose of PGF$_{2\alpha}$.

Recurrence of foetal mummification has been reported.

? Questions
1. Describe the clinical approach to remove the foetus after uterine torsion repositioning.
2. Is suture retention required to prevent vaginal prolapse recurrence at pregnancy term?
3. Describe a field protocol for bovine abortion investigation including what to collect for laboratory identification of bacteria, fungal or parasite agents.

✓ Answers
1. Usually, a C-section is not required to solve dystocia caused by uterine torsion. Foetal manipulation, rotating the foetus and uterus or rolling the cow is normally sufficient to treat this condition. However, insufficient cervical dilatation is commonly observed after reduction, due to the physical occlusion of the soft birth canal (intra-cervical or post-cervical twists) or uterus (pre-cervical twist). There are three suitable options to solve insufficient cervical dilatation after uterus reposition: (1) wait for cervical dilatation caused by the foetus pressure during normal labour; (2) manual mechanical dilatation of the cervix; or (3) perform a C-section.

   While the amniotic sac remains intact, prolongation of labour can occur without causing too much foetal stress or death. This timeline allows for normal cervical dilatation and foetal progression after uterine torsion reduction. The C-section option should be used when signs of significant vascular involvement (e.g. uterine haemorrhages or hypovolemic shock) or loss of uterus elasticity are evident. If the foetus is not delivered within 2–3 h, reassessment is required. Manual mechanical dilatation of the cervix, lasting up to 15 min., is a common practice that is usually successful, and so forced traction of the foetus can be made almost immediately. This choice is recommended when the amniotic sac is ruptured increasing the risk of foetal death. Finally, in more complicated cases, such as those presenting uterine haemorrhages, uterine rupture or hypovolemic shock or when the cervix remains insufficiently dilatated, a C-section is recommended. Previous stabilization of the dam may be required.

2. The aetiology of vaginal prolapse is multifactorial. Nevertheless, at term pregnancy, some hormonal and anatomical changes of the reproductive tract can increase the risk of vaginal prolapse: the rise in oestrogens level and the relaxation of the sacrotuberous ligaments facilitate vaginal eversion. A foreign-body feeling will lead to tenesmus that will cause the exteriorization of part of the vaginal wall. Initially, prolapses are only evident when the cow is lying as abdominal viscera will push out the vagina fold. The vagina will re-enter spontaneously when the animal stands. In time, oedema and trauma will increase the volume and tenesmus, leading to permanent prolapse (◘ Fig. 2.11). All these factors usually persist after the replacement of the vagina, so that suture retention is required to prevent relapses. Bühner, vertical or horizontal mattress techniques are simple and temporarily effective allowing for a gradual reduction in volume. These sutures should be removed before Stage II of labour starts to allow for foetus delivery and to prevent vulvovaginal tears. As a more elaborated

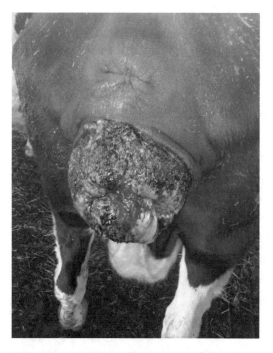

◘ **Fig. 2.11** Trauma consecutive to vaginal prolapse. Legend: When primary prolapse is not resolved, trauma and oedema will cause an important increase in volume so that the vagina will not re-enter even when the cow gets up. (Original from George Stilwell)

and permanent alternative, a vaginopexy (Minchev method or Minchev modified method) can be made. In these cases, the vulva can fully dilate at calving time.

3. Causes for abortion in cattle include a large variety of different factors of infectious and non-infectious origin. Some can be determinants and others only predisposing factors. A preliminary field investigation regarding affected animals, farm management and the environment is required to reduce the list of potential biotic or abiotic abortive agents. The number, age, pregnancy length and vaccination protocol of aborted cows as well as any epidemiological data (e.g. reproductive diseases prevalence's in the farm and region; animal movements and other factors with biosecurity impact; nutritional or environmental hazards such as psychological or heat stress or chemical or plant toxins) should be recorded and

2

checked. Both fresh foetus and placenta should be quickly sent to the laboratory under chilled ambience, for necropsy and sample collection. As an alternative, or when the foetus is too large, small parts should be sampled after a careful necropsy: (1) foetal abomasal contents, lung, liver, kidney, adrenal gland, thyroid, spleen, diaphragm, lymph node, tongue and skeletal muscle brain should be aseptically collected and refrigerated. Conventional freeze (−20 °C) should be used for PCR testing, especially for agents from abomasal contents (e.g. bovine viral diarrhoea virus, *Ureaplasma diversum*, *Neospora caninum*, *Listeria monocytogenes*, *L. ivanovii*, *Bacillus licheniformis*, *Aspergillus fumigatus*, *Mortierella wolfii* and *Leptospira* spp.). Also, 0.5–1 cm sized fixed tissues (e.g. 10:1 formalin:tissue) are required for histopathological analyses. Refrigerated and fixed samples of cotyledons should also be collected. Foetal thoracic fluid or foetal heart blood can also be useful as well as dam's serum for seroconversion testing (e.g. bovine viral diarrhoea virus, bovine herpesvirus, *Leptospira* spp., *N. caninum*, etc.). The results should be interpreted aggregating epidemiological data, farm management and clinical records and necropsy, histopathology and agents' identification results.

References

1. Agerholm JS, Hewicker-Trautwein M, Peperkamp K, Windsor PA. Virus-induced congenital malformations in cattle. Acta Vet Scand. 2015;57(1):54. https://doi.org/10.1186/s13028-015-0145-8.
2. Ames NK. Noordsy's food animal surgery. 5th ed. Iowa: Wiley; 2014.
3. Baird AN. Bovine urogenital surgery (Chapter 14). In: Hendrickson DA, Baird AN, editors. Turner and McIlwraith's techniques in large animal surgery. 4th ed. Iowa: Wiley Blackwell; 2013. p. 235–71. https://doi.org/10.1016/j.theriogenology.2007.04.023.
4. Clothier K, Anderson M. Evaluation of bovine abortion cases and tissue suitability for identification of infectious agents in California diagnostic laboratory cases from 2007 to 2012. Theriogenology. 2016;85(5):933–8. https://doi.org/10.1016/j.theriogenology.2015.11.001.
5. Constant F, Guillomot M, Heyman Y, Vignon X, Laigre P, Servely JL, Renard JP, Chavatte-Palmer P. Large offspring or large placenta syndrome? Morphometric analysis of late gestation bovine placentomes from somatic nuclear transfer pregnancies complicated by hydrallantois. Biol Reprod. 2006;75(1):122–30. https://doi.org/10.1095/biolreprod.106.051581.
6. Diskin MG, Parr MH, Morris DG. Embryo death in cattle: an update. Reprod Fertil Dev. 2011;24(1):244–51. https://doi.org/10.1071/RD11914.
7. Dolník M, KaDaši M, Pošivák J, TóTh M, MuDroň P. Hysterotomy of a dairy cow with long-time macerated foetus. Vet Arhiv. 2019;89:903–12.
8. Drost M. Complications during gestation in the cow. Theriogenology. 2007;68(3):487–91. https://doi.org/10.1016/j.theriogenology.2007.04.023.
9. Faria N, Simões J. Incidence of uterine torsion during veterinary-assisted dystocia and singleton live births after vaginal delivery in Holstein-Friesian cows at pasture. Asian Pac J Reprod. 2015;4(4):309–12. https://doi.org/10.1016/j.apjr.2015.07.009.
10. Frazer GS, Perkins NR, Constable PD. Bovine uterine torsion: 164 hospital referral cases. Theriogenology. 1996;46:739–58.
11. Klaus-Halla D, Mair B, Sauter-Louis C, Zerbe H. Uterine torsion in cattle: treatment, risk of injury for the cow and prognosis for the calf. Tierarztl Prax Ausg G Grosstiere Nutztiere. 2018;46(3):143–9. https://doi.org/10.15653/TPG-170680.
12. Lefebvre RC, Saint-Hilaire E, Morin I, Couto GB, Francoz D, Babkine M. Retrospective case study of foetal mummification in cows that did not respond to prostaglandin F2alpha treatment. Can Vet J. 2009;50(1):71–6.
13. MacDonald R. Hydrops in a heifer as a result of in-vitro fertilisation. Can Vet J. 2011;52(7):791–3.
14. Miesner MD, Anderson DE. Vaginal and uterine prolapse (Chapter 80). In: Anderson DE, Rings DM, editors. Current veterinary therapy: food animal practice, vol. 5th. St. Louis: Saunders, Elsevier; 2009. p. 382–91. https://doi.org/10.1016/B978-141603591-6.10080-6.
15. Murakami T, Nakao S, Sato Y, et al. Blood lactate concentration as diagnostic predictors of uterine necrosis and its outcome in dairy cows with uterine torsion. J Vet Med Sci. 2017;79(3):513–6. https://doi.org/10.1292/jvms.16-0203.
16. Murakami T, Sato Y, Sato A, Mukai S, Kobayashi M, Yamada Y, Kawakami E. Transrectal ultrasonography and blood lactate measurement: a combined diagnostic approach for severe uterine torsion in dairy cattle. J Vet Med Sci. 2019;81(9):1385–8. https://doi.org/10.1292/jvms.18-0500.

17. Peter AT. Vaginal, cervical, and uterine prolapse (Chapter 43). In: Hopper RM, editor. Bovine reproduction. Iowa: Willey Backwell; 2015. p. 383–95. https://doi.org/10.1002/9781118833971.ch43.

18. Pittman T. Practice tips. A retention stitch technique for vaginal prolapse repair in cattle. Can Vet J. 2010;51(12):1347–8.

19. Reichel MP, Wahl LC, Hill FI. Review of diagnostic procedures and approaches to infectious causes of reproductive failures of cattle in Australia and New Zealand. Front Vet Sci. 2018;5:222. https://doi.org/10.3389/fvets.2018.00222.

20. Reis MO, Cruz RAS, Oliveira LGS, Bassuino DM, Schwertz CI, Bianchi MV, Sonne L, Pavarini SP, Driemeier D. Hydrallantois in cows naturally poisoned by Sida carpinifolia in Brazil. J Vet Diagn Investig. 2019;31(4):581–4. https://doi.org/10.1177/1040638719850610.

21. Sasaki S, Hasegawa K, Higashi T, et al. A missense mutation in solute carrier family 12, member 1 (SLC12A1) causes hydrallantois in Japanese Black cattle. BMC Genomics. 2016;17(1):724. https://doi.org/10.1186/s12864-016-3035-1.

22. Schönfelder A, Richter A, Sobiraj A. Prognostic indicators for conservatively incorrectable uterine torsion in the cow. Tierarztl Umschau. 2003;58:512–7.

23. Schönfelder AM, Sobiraj A. Cesarean section and ovariohysterectomy after severe uterine torsion in four cows. Vet Surg. 2006;35(2):206–10. https://doi.org/10.1111/j.1532-950X.2006.00133.x.

24. Sloss V, Dufty JH. Dystocia. Displacement of the gravid uterus. In: Sloss V, Dufty JH, editors. Handbook of bovine obstetrics. Baltimore: Williams & Wilkins; 1980. p. 108–83.

25. van Wagtendonk-de Leeuw AM, Aerts BJ, den Daas JH. Abnormal offspring following in vitro production of bovine preimplantation embryos: a field study. Theriogenology. 1998;49(5):883–94. https://doi.org/10.1016/s0093-691x(98)00038-7.

26. Wolfe DF, Carson RL. Surgery of the vestibule, vagina, and cervix: cattle, sheep and goats. In: Wolfde DF, Moll HD, editors. Large animal urogenital surgery. 1st ed. Baltimore: Williams and Wilkins; 1999. p. 421–4.

Online Guides

Animal Health Diagnostic Center. Bovine diagnostic plans and panels. New York: Cornell University College of Veterinary Medicine; 2020. From: https://www.vet.cornell.edu/animal-health-diagnostic-center/testing/diagnostic-plans-and-panels/bovine. Accessed on 19 July 2020.

Drost M, Samper J, Larkin PM, Gwen Cornwell D. Female reproductive system. In: Visual guides of animal reproduction. Florida: UF College of Veterinary Medicine; 2019. Accidents of Gestation (VISGAR). From: https://visgar.vetmed.ufl.edu/en_bovrep/uterine-torsion/uterine-torsion.html. Accessed on 17 Sept 2020.

Suggested Reading

Ali H, Ali AA, Atta MS, Cepica A. Common, emerging, vector-borne and infrequent abortogenic virus infections of cattle. Transbound Emerg Dis. 2012;59(1):11–25. https://doi.org/10.1111/j.1865-1682.2011.01240.x.

Miesner MD, Anderson DE. Management of uterine and vaginal prolapse in the bovine. Vet Clin North Am Food Anim Pract. 2008;24(2):409–19. https://doi.org/10.1016/j.cvfa.2008.02.008.

Momont H. Bovine reproductive emergencies. Vet Clin North Am Food Anim Pract. 2005;21(3):711–27, vii. https://doi.org/10.1016/j.cvfa.2005.07.004.

Prado TM, Schumacher J, Dawson LJ. Surgical procedures of the genital organs of cows. Vet Clin North Am Food Anim Pract. 2016;32(3):727–52. https://doi.org/10.1016/j.cvfa.2016.05.016.

Normal Birth (Eutocia)

Contents

Electronic Supplementary Material The online version of this chapter (https://doi.org/10.1007/978-3-030-68168-5_3) contains supplementary material, which is available to authorized users. The videos can be accessed by scanning the related images with the SN More Media App.

3

- To list the main hormonal events regarding both dam and foetus and their role in initiating and supporting parturition.
- To describe the physiological changes timeline pattern involved in the preparatory period and along the three stages of parturition.
- To identify the normal foetal disposition and its progression until vaginal delivery.
- To define the main respiratory and cardiovascular changes in the newborn calf.

3.1 Introduction

Eutocia derivates from the Greek words *Eu* (normal) + *tokos* (parturition) + *ia* and is defined as a spontaneous calving (synonymous: parturition) after a normal duration gestation, i.e. the ability of a dam to naturally expel the foetus(es). Several alternative definitions have been presented based on physiological or clinical points-of-view. Some will include only the expulsion phase of parturition, while others will include the first two phases, all three phases or even all the normal pregnancy. These differences are important to define the "normal duration" of calving. Moreover, significant individual characteristics may also influence calving duration. Classically, a variation between 30 min. and 4–6 h has been considered for the foetal expulsion phase, which is the more obvious event related to parturition. Currently, these values have been re-evaluated [29] to a shorter duration, up 70–80 min., in order to support recommendations for the moment for external calving assistance.

Generally, eutocia means a parturition without any need for assistance, i.e. an absolute ability for the dam to calve. However, a lever of calving assistance has recently been proposed for the dystocia definition (see ▶ Chap. 4). By opposition to a difficult parturition, a low degree of assistance by one person to solve a "slight problem" (e.g. unilateral carpal flexion or ventral deviation of the head) without mechanical traction assistance [18, 25] may be included in the clinical definition of normal calving [20].

Independently of the different interpretative notions of eutocia, the calving process should be considered as part of the fertility concept. A fertile female needs to conceive and deliver a viable calf and be ready for new conception in an appropriate time. Consequently, a normal parturition needs to ensure that the dam is able to deliver a foetus and foetal annexes in due time, without significant postpartum complications which can affect her future fertility as well as the viability of the offspring, so as to preserve the species continuity.

This chapter allows to understand the main anatomic and physiologic features, in both dam and foetus, related to all preparatory and labour stages of normal calving. This knowledge is the basis to define when and at what level the obstetrician or other professionals should intervene, as well as to design and manage a herd calving plan.

3.2 Parturition Induction and Hormonal Changes

Towards the end of gestation, several functional and morphologic changes are necessary to prepare and complete the calving, from which should result a live calf and a healthy and successful lactation. These changes are driven by hormones acting on myometrial contractility, cervical dilation, lactogenesis and specific behaviour of the dam that will allow extra-uterine survival of the foetus.

The parturition is induced by several maternal and foetal factors. Some occur gradually towards the end of pregnancy accompanying foetal and placental maturation, while others develop suddenly just before parturition. Although some events are still poorly understood, an overall model has been suggested and is currently widely accepted for cows.

The activation of the foetal hypothalamic-pituitary-adrenal axis is fundamental for foetal cortisol release, which triggers the production of other hormones by the placenta, finally leading to an inversion in the progesterone (P_4)/oestradiol ratio (■ Fig. 3.1).

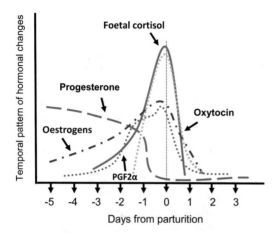

■ **Fig. 3.1** Relative hormonal changes during the peripartum period

It is believed that hormonal, biochemical and physical stimuli act as stressors, thus activating these events. The rapid foetal growth during the last two weeks of gestation plays a significant role in uterine distention increasing its irritability. At the same time, fatty degeneration and infarcts develop in the placenta, which affects blood gases and nutrients exchange at placentomes' level. This concept agrees with what happens in multiple pregnancies in cattle, or in polytocous species (e.g. swine and dogs), in which pregnancy length and individual foetal size are inversely related to the number of foetuses, although some significant inductive differences occur between species. Cows carrying twins present higher plasma cortisol levels at parturition time than those carrying singletons.

At foetal hypothalamic-pituitary-adrenal axis level, corticotropin-releasing hormone (CR) is secreted by the paraventricular nucleus of the hypothalamus, probably in response to stress. This hormone induces the synthesis of pituitary adrenocorticotropic hormone (ACTH) which increases the production of foetal cortisol by the suprarenal glands and its release into the foetus and placental circulation. Maternal cortisol also increases at this time, due to peripartum stress, and is transferred from the maternal to the placenta circulation. The foetal cortisol probably plays a key role in maternal alterations regarding uterine contraction elicitation and contributing for the quick increase of other hormones, e.g. oestrogens, prostaglandin $F_{2\alpha}$ ($PGF_{2\alpha}$) by the endometrium and prostaglandin E_2 (PGE_2) by trophoblastic cells, and of hormonal receptors such as endometrial oxytocin receptors (▶ Box. 3.1).

Box 3.1   Effects of Some of the Main Maternal Hormones at Parturition

— *Oestradiol (E_2)* (1) contributes to the formation of gap junctions in the myometrium, which allow the passage of ions and small molecules between neighboring cells; (2) stimulates the synthesis of $PGF_{2\alpha}$ by the endometrium and the increase in the number of oxytocin receptors in the myometrium; (3) increases myometrial contractions; (4) softening of the cervix; (5) increase in mucus segregation to dissolve the cervical seal; (6) increases vascular permeability in the mammary gland.

— *Progesterone (P_4)* (1) indirect source of oestrogen precursors via $C_{17,20}$ lyase enzyme; (2) the P_4 fall allows for the increase in myometrial activity.

— *$PGF_{2\alpha}$* (1) induces luteolysis of the CL; (2) induces greater myometrial activity and smooth muscle contraction and prevents P_4 block; (3) disrupts the feto-maternal contacts; (4) and stimulates the synthesis of relaxin.

— *PGE_2* stimulates the final functional changes in the cervix, including greater water content and changes in the content or composition of proteoglycans, shortly before or even during parturition.

— *Relaxin* increases collagenase activity and relaxes the pubic symphysis, sacrosciatic ligaments (facilitates pelvic expansion) and the cervix (softening the cervix).

— *Oxytocin* induces (intense) contractions of the myometrium smooth muscles, mainly mediated by Fergusson's reflex, and indirectly increases the pressure on the cervix by the uterine content.

At placenta level, the 17α-hydroxylase/ C17,20-lyase and aromatase enzyme activities are stimulated by cortisol, which promotes the hydroxylation and aromatization of P_4 to free oestrogens, mainly oestradiol (E_2) by induction of the cytochrome P450$_{c17}$, oestrone sulphate (E_1S; represents approximately 80% of total oestrogens) and oestrone (E_1) [12]. Oestrone sulphate is also produced from the androstenedione precursor at placentome level, starting at mid-pregnancy and being an indicator of foetoplacental function and placental viability. One day antepartum, maternal circulating concentrations of oestrone sulphate and Oestradiol-17β increase to their maximal levels (approximately 28 and 1 ng/mL, respectively) and then decrease significantly until 1-day postpartum (about to 10 and 0.4 ng/mL, respectively) in Holstein dairy cows [30]. Inadequate maternal E_1S may reduce myometrial activity and endometrial synthesis of $PGF_{2\alpha}$, resulting in dystocia due to weak uterine contractions [38]. $PGF_{2\alpha}$ induces a prompt functional luteolysis of the corpus luteum (CL), which causes an abrupt progesterone (P_4) fall, concomitantly to the P_4 aromatization by enzyme activity (◘ Fig. 3.2). The major $PGF_{2\alpha}$ metabolite is 15-keto-13,14-dihydro $PGF_{2\alpha}$ (PGFM), which can be easily measured in maternal plasma.

It is well-known that P_4 is the main hormone responsible for the maintenance of pregnancy. In cows, P_4 is mainly produced by the CL and, to a lesser scale, by the maternal adrenals. Additionally, some significant production also occurs at placenta level, mainly during mid-gestation. The P_4 hormone, as well as relaxin, prostacyclin (PGI_2) and catecholamines, is responsible for keeping the uterus quiescent throughout gestation. This uterine quiescence, low intensity and low myometrial activity, occurs because of the hyperpolarization of myometrial cells that show few gap junctions (intercellular connections which improve electrical coupling).

Case Study 3.1 Therapeutic Induction of Parturition

A debilitated Holstein cow at preterm pregnancy (273 days) with a body score of 2 (scale 1–5) is presented to the veterinarian for calving management decision. The cow, weighing about 500 kg, is recovering from a severe chronic illness, but no other abnormal clinical signs are currently evident. Despite a small udder enlargement, with pre-colostrum present, no prodromal external signs are yet observed. At vaginal palpation, the cervix is felt still closed. Foetal vitality is evaluated by the immediate response to hoof pinching and by the presence of fremitus of the uterine artery. A clinical decision to induce parturition by iatrogenic hormone injection is made to prevent further foetal growth and to ensure convenient monitorization of the dam. Three alternative treatments are discussed: (1) single administratzion of a corticosteroid: 20–35 mg dexamethasone i.m. or 8–10 mg flumethasone i.m.; (2) a single administration of 500 μg cloprostenol i.m. or 25 mg dinoprost i.m., a synthetic prostaglandin analogue; and (3) the concomitant administration of both drugs. Due to the cow's weakness state, a combination of both drugs was selected and administered early in the morning. This option can avoid a 10–20% probability of induction failure and can reduce the time to parturition from approximately 44 h when only one of the drugs is administered. Normal parturition occurred 32 h after the administration of the hormones. Calving was conveniently monitored, and a live calf was delivered. Close monitorization was continued due to retained placenta.

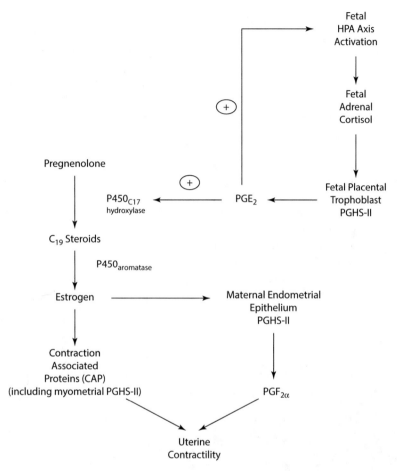

◻ Fig. 3.2 Hypothetical cascade of parturition induction. (Adapted from Whittle et al. [37] with permission of Oxford University Press). Legend: The foetal cortisol induces an increase of the placental trophoblast expression and activity of prostaglandin H synthase type II (PGHS-II), leading to prostaglandin E_2 (PGE$_2$) production. PGE$_2$ stimulates the placental expression and activity of the microsomal enzyme p450$_{c17}$ and sustains foetal hypothalamic-pituitary-adrenal (HPA) axis activation. The enzyme p450$_{c17}$ hydroxyase catalyses the 17α-hydroxylation of pregnenolone and progesterone (lesser extent) followed by the cleavage of the C_{17-20} carbon bond to C_{19} steroids (androstenedione). In the next steroidogenesis step, the enzyme p450 aromatase converts the androgens to estrone and the 17ß-hydroxysteroid dehydrogenase estrone to estradiol-17β. These oestrogens induce the expression of contraction associated proteins (CAP, e.g. prostaglandin and oxytocin receptors and Ca^{2+}-channels) and increase PGF$_{2\alpha}$ production, which also promotes the luteolysis of the corpus luteum. These hormonal events ensure the myometrial contractility and consequent labour. This model was adapted from ewes

A gradual decrease in plasmatic P_4 levels starts between one and two months before parturition. Two to three days before parturition, endogenous PGF$_{2\alpha}$ is released by the endometrium and peak on the day of calving (day 0). This release is time-related with the abrupt decrease of P_4 on day −1 as well as in increase in oestradiol-17β and foetal and maternal cortisol, which start to significantly increase on day −5, peaking on day −1. At 12–24 h before parturition, a P_4 plasmatic level of <1.2 ng/mL is usually found and is a highly accurate and sensitive way to predict calving time [34]. Both decrease of P_4 and increase of E_2 are essential to promote uterine mobility, which is further aided by the PGF$_{2\alpha}$ increase. The uterine mobility triggers foetal movement with the extension of feet and head towards the birth canal, during the first stage of labour.

3.3   Normal Foetal Orientation and Assessment

The foetal orientation is a crucial factor that need be assessed at parturition time. During the late pregnancy, the foetus is largely accommodated in the pregnant uterine horn in a characteristic dorsal (superior) position with limbs and neck slightly flexed (◘ Fig. 3.3). This U-shaped conformation can also fill the proximal part of contralateral uterine horn in the last week of pregnancy. At the onset of labour (see further down the description of parturition), the foetus acquires a typical orientation to enter the birth canal. The normal foetal orientation at this time is shown in ▶ Box 3.2 and ◘ Fig. 3.4. Abnormal (faulty) foetal dispositions at the onset of the first stage of labour

are significant causes of foetal-origin dystocia, hampering the entrance of the foetus or its progression along the birth canal, and, consequently, natural delivery.

The identification of normal or abnormal foetal disposition (see ▶ Chap. 4) is achieved during physical (obstetrical) exam by vaginal palpation, at full cervical dilation time, considering the relative topography of the foetus. The foetal disposition can also be assessed, at least partially, by transrectal palpation. It should be recalled that in both situations, only the caudal part of the uterus can be evaluated.

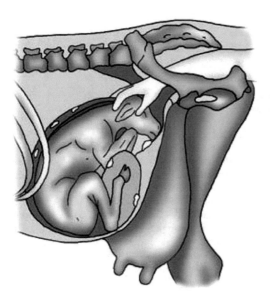

◘ **Fig. 3.3** Normal foetal position during the last half of the pregnancy. Legend: After the 6th month of pregnancy, the foetus can be easily touched by transrectal palpation, emerging dorsally due to its growth. Usually, the dorsum, head and forelimbs (or hindlimbs in posterior presentation) are the most accessible foetal parts. When twining pregnancy occurs, usually the foetuses adopt a complementary orientation according to their longitudinal axes opposing heads (anterior vs. posterior presentation). The foetus is responsive to the hand tactile stimulation, being a sign of foetal viability. (Adapted from Jackson and Cockcroft [11] with permission of John Wiley and Sons)

> **Box 3.2   Normal Foetal Disposition at Parturition Time**
> — Presentation: anterior (prevails in up to 95% in singletons) or posterior longitudinal; in twin pregnancies, it is common to have one of the foetus (usually the first entering the birth canal) in a posterior longitudinal presentation.
> — Position: dorsal (dorso-sacral).
> — Posture: complete forelimbs extension and head extension over the carpus; for posterior presentation: complete hindlimb extension.

The identification of the different foetal anatomical areas is made by palpation considering their shape and conformation: body (thoracic, abdominal, dorsum and lumbar areas), head and neck, rump/tail and limbs. Because external morphologic differences between forelimbs and hindlimbs may be obvious, the evaluation, via vaginal palpation, of the relative direction of each articulation flexion serves as a valuable tool in the forelimbs, the pastern joint (proximal interphalangeal articulation), fetlock joint (metacarpophalangeal articulation) and carpal joints (carpal articulation) flex all in the same direction and all in opposite direction to the elbow joint (composite joints between the humerus, radius and ulna bones); while in the hindlimbs, the pastern and fetlock joints (metatarsophalangeal

Fig. 3.4 Normal foetal orientation at the onset of Stage I of labour (anterior longitudinal presentation, dorsal position and extension of forelimbs and head). (Original from Jacqueline Zurowski)

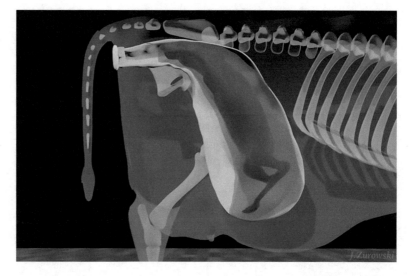

articulation) are flex in the opposite direction of the tarsal (hock) joint (composite joints between the tibia, fibula, tarsal and metatarsals bones). The hooves' position (vertical, horizontal or oblique) can help in the differentiation and in the determination of the foetal position, or even in the pre-diagnosis of a uterine torsion (see ▶ Chap. 2).

3.4 Preparatory Period and Prodromal External Signs of Parturition

Several external signs serve as indication of the proximity of parturition. This preparatory period to calving occurs up to 4 days antepartum and results from complex hormonal changes. During this period, the distension of the uterine walls and the uterine pressure on other internal organs also contribute to behavioural changes in the periparturient cow. These behaviours are mainly changes in physical and feeding activities and are more evident on the calving day, in particular in last 2–4 h before the onset of the parturition (▶ Box 3.3). When getting close to parturition, the cow tends to isolate from the herd, and abdominal contractions may start up to 8 h before parturition, peaking in the last 2 h.

Box 3.3   Preparatory Behavioural Changes on the Calving Day [28]

Physical activity

— Increase steps, stamping and restlessness with head turning towards the abdomen and lateral lying, mainly in last 2 h before the onset of parturition.

— Increase number of lying/standing transitions in the last 4–6 h (peak in the last 2 h with increasing lying time).

— Increased time with tail raised, mainly in the last 2–4 h.

Food activity

— Reduce dry matter intake in last 24 h and more accentuated in the last 6 h; feeding time decrease in the last 2–6 h.

— Reduce rumination time in last 24 h and more accentuated in the last 4–6 h.

External phenotypic changes also occur, namely, in the udder (mammary gland and teats), the vulva and the rump (pelvic ligaments and tail base).

The mammogenesis (lobuloalveolar growth of the mammary gland) is regulated throughout pregnancy by several hormones,

3

□ **Table 3.1**   Normal prodromal external signs in primiparous ($n = 8$) after hormonal induction of parturition with $PGF_{2\alpha}$

Trait	Showing signs (%)	Hours before parturition (Mean ± SEM)	Detection of signs 12 hours before parturition (%)
Udder distension			
Slight	100	82 ± 5	0
Moderate	100	45 ± 4	0
Prominent	100	32 ± 2	100
Udder oedema			
Slight	50	36 ± 4	50
Prominent	37.5	8 ± 1	37.5
Leaking of colostrum	12.5	15 ± 2	12.5
Vulva swelling			
Slight	100	60 ± 6	0
Moderate	100	37 ± 4	0
Prominent	100	26 ± 2	100
Sighting of vaginal discharge	100	35 ± 3	100
Relaxation of the pelvic ligaments			
Slight	100	37 ± 4	37.5
Prominent	87.5	25 ± 3	87.5

Adapted from Kornmatitsuk et al. [14] with permission of John Wiley and Sons
Legend: Age of dam at calving, 26.3 ± 0.1 months (±SEM); pregnancy length at first $PGF_{2\alpha}$ administration, 270 ± 0.5 days; time from the first $PGF_{2\alpha}$ administration to parturition, 59 ± 7 hours; calf alive, 100%; mean calf weight, 32 ± 2 1.0; two doses of $PGF_{2\alpha}$ (25 mg) were administered 24 hours apart

i.e. growth hormone, P_4, prolactin, oestrogens and probably placental lactogen produced by the binucleate cells of the placenta. An exception occurs in lactating cows during the onset of the dry period, when approximately half of the mammary epithelial cells suffer apoptosis (programmed cell death) under local lysosomal enzymes activity and are replaced by adipocytes leading to mammary involution. Towards the end of pregnancy, lactogenesis (induction of milk synthesis) starts as colostrum, being regulated by prolactin, insulin and glucocorticoids, mainly when P_4 is suppressed. When approaching calving, natural udder inflammation is progressively visible (□ Table 3.1). The udder enlarges and becomes tense due to the internal pressure caused by glandular tissue development and colostrum accumulation. The leakage of colostrum, due to internal pressure and also as a response to oxytocin release, may be seen hours before parturition in some cows. Sometimes, close to parturition, excessive inflammation can cause rupture of blood vessels, causing a change in colostrum colour – from yellowish to reddish or brown (oxidative haemoglobin). Some blood clots may also be detected at first milking's. Udder oedema, due to increased capillary pressure and obstruction of lymphatic drainage, can expand subcutaneously onto the ventral abdomen, reaching the brisket in very severe cases.

Rupture of the udder ligaments may occur in some cases. High intakes of potassium and/or sodium and lack of exercise during late pregnancy predispose cows to udder oedema by increasing fluid retention. Very severe oedema, which may be seen in all parities but with special incidence in primiparous, may be treated by starting milking the cow before parturition or by inducing parturition. After calving, the use of emollients or even diuretics (furosemide) and corticosteroids may be recommended for severe cases in order to avoid mastitis and reduce discomfort.

A few (1–3) days before parturition, oestrogens (E_1S and E_2; ◻ Fig. 3.5) will cause vulva swelling that will widen and increase the dorsal to ventral commissure axis. The elasticity of the vulva will also increase reducing the possibility of tissue tearing. The cervical mucus plug (cervical seal), an antibacterial gelatinous (semi)transparent mucus filling and sealing the cervical canal during pregnancy, liquefies and usually is seen hanging from the vulva and/or tail at this time. This happens because of the gradual ripening and softening of the cervix in last days of pregnancy, which can be evaluated either transrectally or by vaginal palpation evaluating the consistence of the external Os or even introducing one finger through it. This mucus is thicker and more abundant than the small amounts of vaginal discharge that sometimes is seen hanging from the vulva in the last two months of pregnancy.

The rump conformation also changes mainly due to the sacrosciatic ligaments' slackening, resulting in an apparently raised head tail and gluteal muscles sinking. This relaxation starts approximately 1–2 day before parturition and is often used to predict parturition onset.

The body temperature of the dam also starts to decrease about 48 h before parturition, dropping 1 °C at approximately 16 h and remaining low until parturition time ([14]; ◻ Fig. 3.6). This decrease in body temperature is often used to decide for elective ceaserean section in cows with a high probability for dystocia [13].

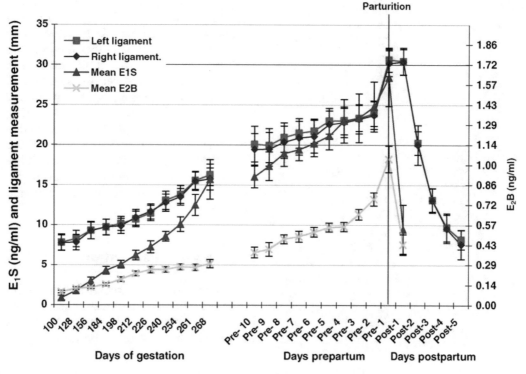

◻ **Fig. 3.5** Timeline pattern of maternal plasma oestrone-3-sulphate (E_1S) and Oestradiol benzoate (E_2B) levels and relaxation of the sacrosciatic ligaments in Holstein-Friesian cows ($n = 20$). (Reproduced from Shah et al. [30] with permission of Elsevier)

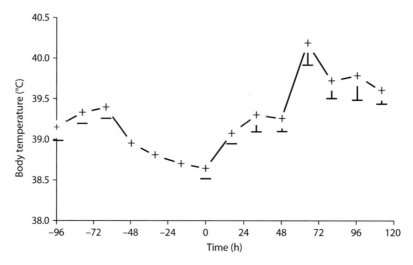

Fig. 3.6 Body temperature pattern in induced calving (parturition = 0 hour). (Adapted from Kornmatitsuk et al. [14] with permission of John Wiley and Sons)

Clinical signs, rectal temperature, hormonal profile or readings from electronic devices can serve as predictors for parturition (▶ Box. 3.4).

Box 3.4   Prediction of the Parturition Time (Hours Before Foetal Expulsion)

— Relaxation of the pelvic ligaments and vulva oedema. Within 12–22 h.

— Intense udder distension and teat filling. Within 12 h.

— Plasma P_4 level dropping to <1.2 ng/mL. Within 16 h.

— Intensification of calving behaviour (see ▶ Box. 3.3). Within 6 h.

— Body temperature. Measured as rectal or vaginal temperature. Decreasing of around 0.5–1 °C at 16 h. Needs be compared with the previous three days.

— Use of electronic health sensors, e.g. iVET® (▶ www.birth-monitoring.com); Cow Call® (▶ www.cowcall.com); C6 Birth Control® (▶ www.sisteck.com); RumiWatch® (▶ www.rumiwatch.ch); and IceRobotics® (▶ www.icerobotics. com) system.

Although a significant variation can be observed, some, or a combination of a few clinical signs, show a high positive and negative predictive values at 12–24 h before calving (true positive and negative values detected by a specific method, respectively): (a) combination of pelvic ligaments relaxation, teat filling and plasma $P_4 \leq 1.2$ ng/mL – negative predictive value within 12 h before calving = 96.8% [34]; (b) daily increase in pelvic ligaments relaxation (depth of sacrosciatic ligament measured between the sacrum and the tuber ischii) ≥5 mm, positive predictive value within 12 h before calving = 94.7% [30]; or (c) plasma P_4 < 1.2 g/mL, positive predictive value within 24 h before calving = 97.9% [19]. These prediction indicators are useful to mitigate the risk of dystocia, especially in primiparous cows, and to reduce perinatal mortality.

3.5   Stages of Parturition

Two main events are crucial for successful calving. Firstly, the birth canal needs to be adequately enlarged. The major barrier is the cervix which needs to be fully dilated.

On the other hand, sufficient expulsion force need be exerted to drive the foetus through the uterus and birth canal. In natural parturition, the expulsion of the foetus results from the combined forces of the myometrium (smooth muscle) and abdominal wall (skeletal muscles) contraction. In consequence, the disturbance of any of these two mechanisms may be a major cause of dystocia.

The parturition process is normally divided into three stages, which occur successively: (a) first stage (Stage I) or cervical dilatation phase; (b) second stage (Stage II) or foetal expulsion phase; and (c) third stage (Stage III) or foetal membranes expulsion phase. A functional definition, from a clinical point of view, is reported in ► Box 3.5, and the major events of each stage are described in ► Box 3.6. A 3D motion of Stage I and Stage II is integrated in ◘ Fig. 3.7.

> **Box 3.5 Clinical Identification (Definition) of the Stages of Labour for Obstetric Veterinarians**
> - *Stage I of labour* or *cervical dilatation phase*. Interval between more intense restlessness behaviour of the dam, or the onset of myometrial contractions, and the full cervical dilatation and/or appearance of the chorioallantoic (allantoic) sac at the vulva.
> - *Stage II of labour* or *foetal expulsion phase*. Interval between the appearance of the chorioallantoic sac, or amniotic sac if the previous one ruptures inside the birth canal, at the vulva and the complete expulsion of the foetus(es).
> - *Stage III of labour* or *placental expulsion phase*. Interval between the expulsion of the (last) foetus and placental membranes expulsion.

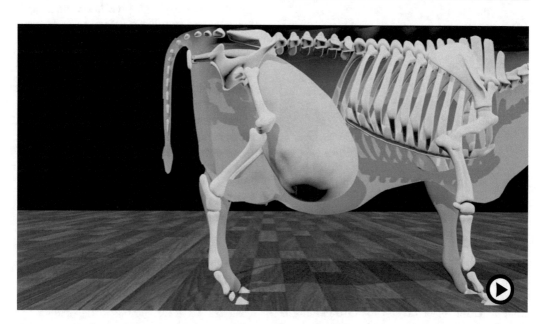

◘ **Fig. 3.7** Video box: 3D motion of the labour (Stage I and Stage II) in cows. Legend: During Stage I of labour, the myometrial contractions start impelling the foetus located in the pregnant uterine horn towards the pelvic inlet. The foetus acquires the calving posture extending both forelimbs, neck and head. Simultaneously, the cervical canal opens and the chorioallantoic membrane enters into the vagina without separation between maternal caruncles and foetal cotyledons. At the onset of the Stage II, the abdominal contractions start impelling the chorioallantoic trough the birth canal causing it to rupture (rupture of the first water sac). The foetus then adapts to the birth canal, and when entering, it contributes to the rupture of the amniotic membrane (rupture of the second water sac). Successive strong abdominal contractions complete the foetal expulsion rupturing the umbilical cord. Finally, the foetal membranes are expelled in the few hours after placental dehiscence, initiating the puerperal period that will last around 40 days. (Original from Jacqueline Zurowski) (► https://doi.org/10.1007/000-2qd)

3

Box 3.6   Main Events on Each Stage of Parturition

Stage I
- Onset of myometrial contractions.
- Restlessness behaviour.
- Change from foetal position to birth position.
- Cervix dilatation (including Fergusson reflex for mechanical enlarging).
- Chorioallantoic sac (i.e. first water bag) enters the vagina and usually ruptures.

Stage II
- Persistence of myometrial contractions.
- Onset of strong abdominal contractions (vagina distention).
- Rupture of chorioallantoic sac, if it did not occur previously.
- Amniotic sac (second water bag) enters in the vagina and ruptures.
- The foetus' limbs and head enter in the birth canal and progresses.
- Complete foetal expulsion occurs.

Stage III
- Persistence of mild myometrial and abdominal contractions.
- Loss of placental circulation due to rupture of the umbilical cord.
- Placenta dehiscence: separation of maternal caruncles from foetal cotyledons.
- Expulsion of the placenta.

3.5.1   Stage I (Dilatation Phase)

The average duration of the first stage is about 2.5 h but can go up to ≥6 h, especially in primiparous cows. For stillborn calves, the time interval from onset of labour until calving is completed is approximately twice as long as for live-born calves [4].

■ **Relaxation of the Cervix and Ligaments**
This phase is mainly characterized by cervical, vulva and pelvic ligaments/symphysis relaxation, i.e. relaxation and dilatation of the soft birth canal. As a first step, there is a decrease in the muscular tone of the smooth muscle fibres from the external *Os* progressing gradually towards the internal Os of the cervix. This is the follow-up of the cervix ripening and softening that occurs in the prodromal phase, under hormonal influence. Cervical full dilatation eventually results from uterine forces being exerted on the cervix viscoelastic matrix, i.e. hormonal and mechanical factors temporally contribute to this event. An imbalanced or incomplete hormonal milieu is an important cause for non-dilatation or insufficient cervical dilatation (cervical stenosis). In these cases, deficient vulvar dilatation is normally concomitant.

Although, under natural conditions, cervical ripening and softening is only complete at this stage, cervical matrix remodulation starts few weeks before. This cervical progressive remodulation, mainly caused by collagen alterations through the activity of collagenolytic enzymes, proteoglycan synthesis and pro-inflammatory cytokines, decreases cervical matrix resistance [35] and explains the appearance of a mucus plug hanging from the vulva one or two weeks before calving. In summary, cervical dilatation gradually starts during the prodromal period due to hormonal activity and progresses onto a fast dilatation phase near to calving mainly because of myometrial uterine contractions pulling the foetal membrane and foetus into the birth canal.

❯ **Important**
The diameter of a full dilated cervix varies between individuals but can reach more than 25 cm. An evaluation of cervical dilatation, as well as the comparison between cervical and pelvic diameter, in both cranial and caudal ends of the cervix can be made through vaginal palpation. However, the maximum cervical and vulvar dilatation normally occurs during the physical passage of the foetus, and it is the main cause of pain during parturition.

The physical presence of the foetus also contributes to the temporal pattern of cervical dilatation. Anterior presentation induces quicker and larger cervical dilatation at the onset of the second stage than does a posterior presentation, probably due to the higher foetal volume and conformation presented in the cervical canal. The importance of foetal presence (pressure) is evidenced in ewes presenting the false cervical non-dilation disturbance. In fact, in some cases with foetal malpresentation, the cervix fully dilates a few to several minutes after a circular massage and slight pressure on the internal Os by a hand introduced through the cervical canal. Sheep present a strong smooth muscle activity involved in the cervical dilatation process. Additionally, contraction of both circular and longitudinal smooth muscle fibres in the cervix also occurs, but their direct impact on cervical dilatation remains unclear and is probably not significant in cows.

■ **Uterine Contractions**

The rate and strength of myometrial contractions increase due to the influence of several hormones. In the early stage of the first phase of parturition, the circular and longitudinal myometrial uterine contractions drive the foetus towards the cervix internal *Os*. A total of five to seven contractions occur every 10–15 min., each one with an initial duration of 15–30 sec. Meanwhile, the foetus acquires its final disposition. The myometrial contractions increase the pressure of the chorioallantoic membrane and foetal structures (usually the hooves) onto the internal Os of the cervix, which starts to enlarge. The mechanical distension of the cervix, and subsequently of the vagina wall, also triggers neuro-sensorial stimuli to the spinal medulla, progressing to *nucleus tractus solitarii* and reaching the paraventricular nuclei in the hypothalamus. At this level, a pulsatile release of oxytocin by the neurohypophysis occurs, with this hormone reaching by bloodstream the oxytocin receptors, mainly located in the myometrium, thus promoting the contraction of the smooth muscles' fibres and an increase in the pressure exerted by the uterine contents onto the cervix and vagina – this is denominated the *Ferguson reflex*.

Inversely, some pharmacologically active substances, such as clenbuterol, acting at the

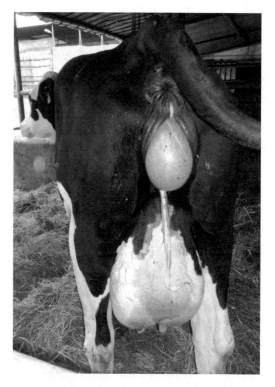

■ **Fig. 3.8** Intact amniotic sac outside of the vulva and ruptured chorioallantoic membrane pending on the udder. (Original from George Stilwell)

β2-adrenergic receptors of the myometrium, may delay the first two stages of labour, causing primary uterine inertia. This phenomenon involving the sympathoadrenal system is very well-known in mares, by which internal and external stressors may inhibit or delay parturition [21]. The same effect may occur in cows, but this is still not proven.

> **Important**

> The Stage I ends when the cervix is fully dilated and the chorioallantoic sac appears at the vulva, if meanwhile it was not ruptured (■ Fig. 3.8). When the chorioallantoic sac enters the vagina, it can rupture, even before reaching the vulva. The rupture of chorioallantoic sac is due to the pressure caused by myometrial and/or abdominal contractions and leads to the flow of the allantoid fluid, which also serve as lubricate for the birth canal. At this stage, the foetus is protected from pressure by the amniotic fluid, sometimes for several hours.

■ **Restlessness Behaviour**

Usually, the onset of calving is pronounced by the cow repeatedly lying down and standing, i.e. so-called restlessness behaviour. Other typical behaviours are isolation, smelling the floor, licking the limbs and flanks, vocalizing, defecating frequently, raising the tail and even presenting a nest-building-like action. The discomfort and the abdominal pain are mainly due to uterine contractions but also result from abdominal pressure being exerted on the cervix. The restlessness of the dam, particularity the postural changes, can start 12–24 h before parturition but normally increase in intensity in the last 2–4 h [31].

3.5.2 Stage II (Foetal Expulsion Phase)

The average duration of the second stage is approximately 30–70 min., again taking a little longer in the case of primiparous (◘ Fig. 3.9).

This stage is characterized by the entrance and progression of the foetus along the birth canal and finishing with its complete expulsion. The complete expulsion defines the "calving time". The duration can vary between several minutes after the amniotic sac emerges through the vulva, to up to 4–6 h, with heifers showing longer durations. Very old cows may also have a prolonged second stage mainly due to uterine inertia. When the amniotic sac is still intact, it is recommended to wait 90 min. after the onset of this stage before any human intervention is initiated.

At the onset of this stage, the foetus reaches and enters the cranial end of the birth canal. The dam is usually lying down. The slight protrusion of the amniotic sac through the vulva is considered a clinical indicator of the onset of this stage. While the amniotic sac remains intact, minimizing thermal and tactile stressors for the foetus, and the umbilical cord is free from obliteration or compression, maintaining the normal blood foetal partial pressures of O_2 (pO_2) and CO_2 (pCO_2), the foetus is under a favourable ambience even if

◘ **Fig. 3.9** Video box of the Stage II of labour. Legend: (timeline) Minute 0, the intact amniotic sac is pending outside the vulva; minute 69, the amniotic membrane has ruptured. Abdominal contractions occur. Both legs and nose of the foetus are observed outside of the vulva without evident progression at this time; minute 72, gentle traction of both legs is made. The heifer makes straining (successive abdominal contractions) to expel the foetus; minute 73, new gentle traction, simultaneously to abdominal contractions, is made to complete the foetal expulsion. (Original from Mário Martins (farmer) and João Simões) (► https://doi.org/10.1007/000-2qc)

other environmental stressors persist. In heifers, with longer parturition compared with multiparous dams, the intact chorioallantoic sac can be seen outside the vulva for some time, sometimes over 110 min. before parturition. The amniotic sac rupture and the first sighting of the calf hooves occur, in average, 66 ± 46 and 33 ± 14 min. before parturition is complete, respectively [14]. The presence of the amniotic sac at the vulva indicates calving eminence. Nevertheless, a high variation in the duration of this stage can happen, and sometimes the hooves can be observed inside the amniotic sac even though it remains intact.

Usually, the dam is recumbent throughout this stage, often changing from sternal to right or left lateral decubitus. The right lateral decubitus facilitates the foetal expulsion because the foetus acquires the same direction than the pelvic inclination. This position should be preferred during assisted calving of recumbent dams. However, it is usual that the dam will stand and lie down repeatedly due to discomfort and pain. It is not uncommon for the foetus to be expelled with the dam standing, especially in multiparous cows.

The pressure exerted by the foetus body as well as by the intact placental membranes, on the cervical and vaginal walls, stimulates, via neuro-sensitive paths, the onset of the abdominal contractions which start at this time or some time before in some cases. At this time, the number of myometrial contractions increases up to ten per 3–5 min. The amniotic sac appears in the vulva and ruptures due to the pressure: it is known as the rupture of the second water bag. The amniotic fluid has an ultimate utility: it will lubricate the birth canal, reducing friction between the vaginal walls and the foetus.

The duration and frequency of abdominal contractions are crucial to create enough pressure, up to approximately 70 kg, to push the foetus and fluid along the birth canal, as described in ◘ Fig. 3.10a for multiparous cows. Some slight differences in this pattern can be noted for nulliparous cows, mainly regarding the restlessness behaviour, usually causing a slight delay in the first stage duration (◘ Fig. 3.10b).

The progress of the calf in a natural calving usually follows the typical sequence: first front limbs (◘ Fig. 3.11) with nose-head lying on the carpal area. Two specific moments require extra force: (1) head exiting the vulva, concomitant to the shoulders passing the osseous birth canal, and (2) hindquarters passage, especially for wide foetal hips in some beef breeds (◘ Fig. 3.12).

This stage ends with complete deliver of a singleton or of the last foetus in multiple pregnancies.

3.5.3 Stage III (Placental Expulsion Phase)

This stage is characterized by placenta expulsion, which usually occurs in the first 6 h after foetus delivery. However, some authors will only consider retained placenta if expulsion does not happen before 12 or even 24 h after foetal expulsion. Some parts of the foetal membranes can adhere to the head of the newborn and should be immediately removed, especially if the nostrils are covered. In some cases, the placenta can adhere to the foetus being concomitantly expelled. In dizygotic twins, both placentas should be expelled after the end of Stage II.

In this phase, placental dehiscence is characterized by the separation of foetal cotyledon from maternal caruncle crypts microvilli in all 60–80 placentomes. The dehiscence occurs through a mixed action of hormonal, immune, biochemical and mechanical elements. The separation is mediated by relaxin hormone which induces collagenase activity (by matrix metalloproteinase activity requiring zinc and calcium ions) to degrade the collagen matrix in the connections. In cows, relaxin seems to be produced by the CL, with secretion being increased after $PGF_{2\alpha}$-induced luteolysis. The decrease in serotonin (stimulates placental cell proliferation during pregnancy) and P_4 (causes myometrial quiescence and suppresses collagenase activity) and the activity of major histocompatibility complex (initiates an inflammatory response at junction level) all contribute to induce this collagenase activity (◘ Fig. 3.13).

Placental separation before calving is finished is possible (◘ Fig. 3.14) and usually means the death of the foetus if not delivered quickly. So, if caruncles are seen exiting with the foetus, calving should be hasted.

The continued after-birth myometrial contractions, stimulated by oxytocin, as well as abdominal contractions, are the mechanical factors contributing to cotyledons separation and to the expulsion of the placental membranes. The physical presence of the membranes inside the cervical canal can slow down the cervical closure in the following days. Nevertheless, the local biochemical effects seem to be essential for a normal placental expulsion, meaning that very seldom placenta retention occurs due to uterine inertia. The third stage ends with placenta expulsion, being followed by uterine involution and lochia excretion, during the early postpartum.

3

a - Multiparous dairy cows

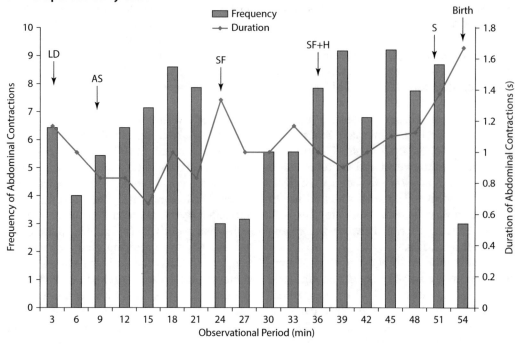

b - Primiparous dairy cows

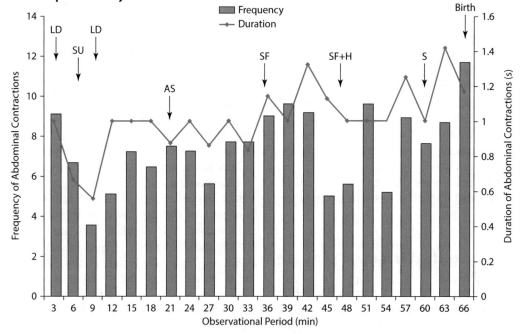

Fig. 3.10 Mean frequency and duration of abdominal contractions during the labour stage in multiparous **a**; $n = 5$ and primiparous **b**; $n = 5$ dairy cows during natural and eutocic calving. Legend: LD lying down; SU standing up; AS amniotic sac appearance; SF showing feet; SF + H showing feet and head; S shoulder. (Adapted from Schuenemann et al. [29] with permission of Elsevier)

3.6 Pain Caused by Parturition

Pain before, during and after labour is recognized as acute and is probably present in every mammal. Moreover, births associated with difficult parturitions or dystocia are cause of very severe pain for the mother and the newborn. This being said, it is surprising that there are very few studies looking at parturition pain and its short- and long-term effects on cattle. Recent literature review shows that pain management after calving is probably underused and that further research on the use of analgesics in the post-calving cow is required

◻ **Fig. 3.11** The foetal feet (forelimbs) progressing outside the vulva after the rupture of the second water bag which lubricate the birth canal. (Original from George Stilwell)

[17]. It would be expected that pain management after calving would be economically beneficial as a cow that has a better appetite after calving is likely to produce more milk and better nurse its' calf. Beneficial effect on welfare and milk yield of analgesia postpartum has been shown for primiparous animals [33].

During the Stage I, pain and discomfort are visceral in origin as they result from stretching and distension of the lower uterine segment and from cervical dilation. Control of this pain is very difficult and probably counterproductive. In the Stage II, somatic pain will predominate being caused by vulva, vagina and cervix broadening, perivaginal structures traction and pelvic and peritoneum distension. The degree of pain during Stage III has deserved very little attention. However, empiric observations show that cows with retained placenta change their behaviour and are seen more often with tail raised and straining.

The pain, after normal parturition, is probably negligible in multiparous cows but may have some negative impact on primiparous welfare and productivity [33]. Some degree of hypoalgesia (increase in nociceptive threshold), mediated via endogenous opioids and by changes in the ratio of sexual steroids, has been demonstrated in cows, probably as a defence mechanism [2].

As said before, all these events and the pain resulting from them are probably much

◻ **Fig. 3.12** The two largest diameters of the foetus: shoulders **a** and hindquarter **b** in some breed and individuals. Both the regions are commonly responsible for foetal retention in foetomaternal disproportion cases, in anterior or posterior presentation, respectively. (Original from Jacqueline Zurowski)

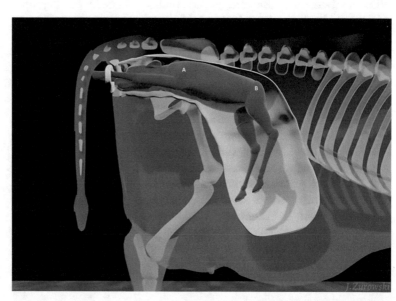

3

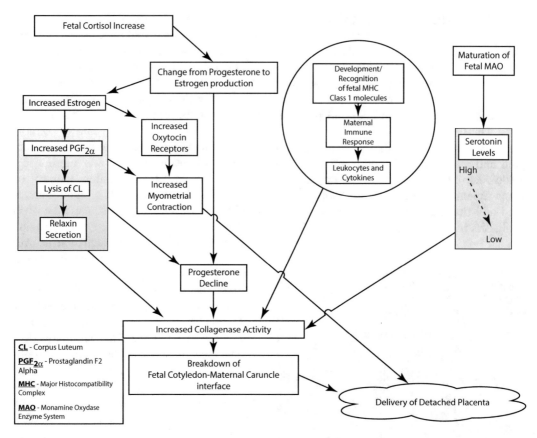

◙ **Fig. 3.13** Hormonal and biochemical leading to the dehiscence of placenta. (Reprinted from Beagley et al. [3] with permission of John Wiley and Sons)

more severe during and after prolonged or dystocic calving. Pain management in these situations, through epidural anaesthesia or other method, is considered good practice and should be mandatory. Further details of pain management around parturition will be presented in ▶ Chap. 5.

3.7 Foetal Physiologic Changes During Parturition

Foetal maturation involves several homeostatic systems at cardiovascular, respiratory, metabolic, haematological and thermogenic adaptation level. These events are supported by endocrine regulation. Although several studies have been published for cows, the sheep model is mostly used to acquire knowledge, being then adapted to other species [7]. Similar to maternal changes, cortisol also plays a major role in final foetal maturation and neonatal adaptation. It is responsible for the induction or improvement of other active substances or structures involved in neonatal adaption (▶ Box 3.7).

> **Box 3.7   Effect of Cortisol in Foetal Maturation and Survival**
> − Increase in β-adrenergic receptors.
> − Induction of the foetal lung clearance.
> − Maturation of the surfactant system.
> − Increase in thyroxine (T_4) to triiodothyronine (T_3) conversion.
> − Increase in catecholamines in several tissues.
> − Enzyme induction of the digestive tract.
> − Control energy substrate metabolism ($\uparrow$ gluconeogenesis; $\uparrow$ lipolysis; $\uparrow$ amino acids utilization; $\downarrow$ peripheral use of glucose).

◻ **Fig. 3.14** Premature separation of the placenta. Legend: **a** Foetal cotyledons hanging down from the vulva with a retained dead foetus into the birth canal. **b** The stillborn calf was removed with intact umbilical cord and detached placenta. (Original from George Stilwell)

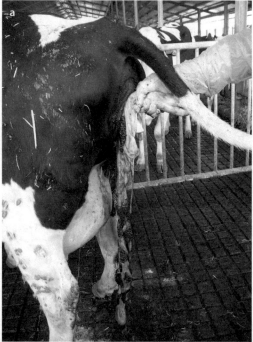

Catecholamines (norepinephrine, epinephrine and dopamine) are also primary mediators and play an essential role in the extrauterine adaptation of the calf. They are significantly released from the neonatal adrenal medulla and other sympathetic tissues in response to different stressors immediately after the birth. Catecholamines regulate blood pressure and contribute to initiate thermogenesis through brown fat metabolism and, probably together with corti-sol, energy metabolism adaption regarding serum glucose and free fatty acid levels.

At birth, the umbilical cord stretches and breaks during the last phase of the foetal expulsion, most times when the head is already outside the vulva. Several physical, endocrine and biochemical events, which initiate respiratory and cardiovascular changes, occur at this time. Vasocontraction of the umbilical vessels occurs in the following 2–5 min. (▶ Box 3.8).

3

Box 3.8   Induction of Umbilical Vessels Vasocontraction

- *Mechanicals factors.* Rupture of the umbilical cord.
- *Thermal factor.* The Wharton's jelly, which protected the umbilical vessels during intrauterine life, reacts to the low temperature (air).
- *Circular and longitudinal smooth muscles contraction.* Mediated by prostanoids (e.g. thromboxane A_2) and catecholamines (e.g. serotonin); ↑ pO_2: direct effect on contractions.

Box 3.9   Changes in the Lungs from Foetal to Neonatal Life

- *Neonatal circulation development* Decrease in pulmonary vascular resistance and pulmonary vasodilation → Increase in pulmonary blood flow.
- *Onset of breathing* Clearance of foetal lung fluid → Surfactant secretion is initiated → Ventilation of distal airspaces.

3.7.1   Respiratory Changes

After birth, some critical changes take place at the pulmonary level leading to breathing. The essential factors for a normal transition from intrauterine to extrauterine life of the newborn calf are summarized in ▶ Box 3.9. Effective coordination between the clearance of foetal lung fluid, surfactant secretion and the onset of respiratory movements (breathing) is required.

Under cortisol, thyroid hormones and catecholamines regulation, the secretion of foetal lung fluid, mediated by active chloride transport, is replaced by a reverse active absorption from type II pneumocyte under Na+, K+ and ATPase action. This activation removes the foetal lung fluid from the intraluminal to the interstitial space and posteriorly drains it into the lymphatic or blood vessels. This process is denominated *clearance of foetal lung fluid* and starts just before parturition. Type II cells also produce surfactant, constituted by lipids and proteins, which reduces the intraluminal surface tension in the lungs. The surfactant secretion, probably mediated by the effect of catecholamines on β-receptors, is stimulated during parturition to ensure enough at foetal deliver time.

During late pregnancy, the foetus is under a state of sleep with REM (rapid eye movement), non-REM (quiet sleep-associated to a low pO_2 ambience) or intermediate phases. In REM state, swallowing, licking and head or feet motion activities occur. Several hormones and endocrine mediators released during parturition, such as cortisol, catecholamines and endorphins, act to regulate neurophysiologic events at central and peripheral levels. The foetus acquires perception of nociceptive stimuli by the maturation of central and peripheral neuroanatomical structures. Still, most probably an effective state of consciousness (cortical awareness) only appears after the inspiration of fresh air. However, due to the lack of scientific evidence, this point remains poorly elucidated, and some degree of early cortical interpretation of pain may be possible.

From a clinical point of view, the onset of regular breathing is initiated with the rupture or obliteration of the umbilical cord. In the first moments, the absence of umbilical circulation causes a temporary blood pCO_2 increase and a pO_2 decrease, causing hypoxia and stress. These partial pressures, as well as subsequent low pH in blood and interstitial milieu (respiratory centre), blood vessels pressure and alveolar distension, stimulate peripheral chemoreceptors at the aortic arch and carotid body, and mechanical receptors, helping to activate and adjust the respiratory function of the newborn. Tactile and thermal stimulus, and eventually pain, also act at the central level to contribute to induce breathing and should be considered during the obstetrical approach. The onset of breathing stimulates foetal arousal and awareness states, causing lung inflation and increasing blood oxygen levels. With the first postnatal breath, the pulmonary vascular resistance decreases dramatically. This is caused by a combination of increased oxygen exposure as well as by ventilation itself. This step occurs typically within the first minute after birth but can happen when the foetus is still inside the birth canal, mainly due to a delay in the second stage of parturition. Usually, the foetus or newborn can survive if the cerebral anoxia persists for approximately 4 min., but neuronal sequalae (necrosis of neurons) are described when there is prolonged hypoxia or anoxia. If the onset of foetal breathing occurs when still inside the birth canal or if the amniotic fluid is not expelled from the upper respiratory tract after birth, a significant amount of this fluid can be inhaled to the lower respiratory tract (intra-luminal lobules and alveolar ducts) which can cause fatal neonatal aspiration pneumonia. Also, if breathing starts inside the birth canal, mechanically expansion of the lungs is restricted due to the thoracic compression, causing foetal suffering and eventually death.

3.7.2 Cardiovascular Changes

The intrauterine life of the foetus is characterized by a variable O_2-saturated environment (◘ Fig. 3.15) under low blood pressure, which is related with the amniotic fluid pressure, and high resistance in some organs, such as the lungs. A great part of the foetal tissues are supplied by a mix of arterial and venous blood flow, and, typically, the foetal caudal regions are poorly oxygenated. However, foetal haemoglobin, that has a very high oxygen affinity, together with shunts and a highly combined ventricular output, ensures adequate tissue oxygenation. Physical cardiovascular changes of the foetus occur when the umbilical cord is cut, causing the interruption of placental circulation. A new functional circulation – inversion of the intrauterine foetal circulation – of the newborn is quickly established.

Three foetal structures, producing shunts in some areas of the blood circulation network, are responsible for the particularities of foetal circulation: the ductus venosus and the right-to-left shunts, namely, the foramen oval and the ductus arteriosus. The shunts drain the great part of the blood, but a low blood flow goes into the vessels. The intra-uterine foetal circulation is characterized by the arterial blood supply from placentomes to the foetus by the (left) allantoic vein mainly reaching the ductus venosus which directly connects the right hepatic and portal veins (caudal vena cava). At this point, the blood enters in the left atrium by the foramen ovale, which directly communicates with the right atrium of the heart. Also, the ductus arteriosus directly connects the pulmonary arteria to the aorta.

When the umbilical cord ruptures, the systemic vascular resistance of the foetus increases due to the loss of the low placental vascular resistance. An increase in pressure is firstly observed in distal aortic, reacting to the loss of blood circulation in both placental arteries, as well as to the increase of blood flow from the lungs. In consequence, the pressure increases mainly in the left atrium, and the foramen oval closes against the atrial septum usually within minutes. Simultaneously, the dilatation of the pulmonary arteries occurs. Inversely, the constriction of the ductus venosus and ductus arteriosus occurs, and functional closures can be observed in the next few days. As consequence of all these immediate changes, the oxygenation of the tissue increases, in particular at cerebral and pulmonary level.

The morphological closure of these three structures takes place latter by the proliferation of endothelial or fibrous surrounding tissue. The persistence of the foramen oval or ductus arteriosus provokes functional disturb of blood flow and is considered congenital circulatory defects.

Congenital cardiac defects represent less than 3% of total congenital malformations. Ventricular septal defect is the most common congenital cardiac defect and is found concomitantly to others defects, such as dextroposition of the aorta, patent ductus arteriosus, persistent truncus arteriosus, persistent foramen ovale, transposition of the great vessels

3

□ **Fig. 3.15** Circulatory changes of the newborn calf. The blue arrows represent the loss of blood circulation due to the collapse of umbilical cord vessels and the shunts. The red arrows represent the new circulation pattern. (Original from João Simões)

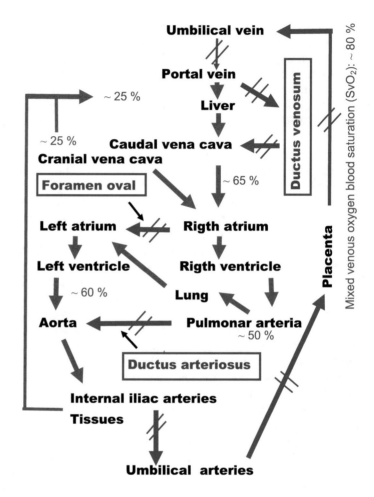

and anomalies of the tricuspid or pulmonary valve [6]. A complex case of congenital cardiac defect involving a complete dextroposition of the aorta in a limousine calf is described in □ Fig. 3.16. The significant hypertrophy of the right ventricle as well as the almost complete absence of left ventricle cavity was a secondary consequence of anatomical and functional alterations.

3.7.3 Neuroinhibition and Neurostimulation

As said earlier, the last months of pregnancy are characterized by states of foetal unconsciousness, which require less oxygen and pre-

vents excessive intra-uterine movement. Nine neuroinhibitory factors may contribute to this: adenosine, P_4, allopregnanolone, pregnanolone, prostaglandin D_2 and a placental neuroinhibitory peptide. Warmth, buoyancy and cushioned tactile stimulation join up to ensure this state of sleepiness and neuroinhibition [22].

Compression and hypoxia during labour play two crucial roles: firstly, it reinforces this sleep-like unconsciousness reducing movements during parturition that could endanger the foetus and the dam, and, secondly, it stimulates widespread noradrenergic activation within the brain by oestradiol-17β and noradrenalin [5, 23]. Compression of the thorax during labour may activate a yet undescribed

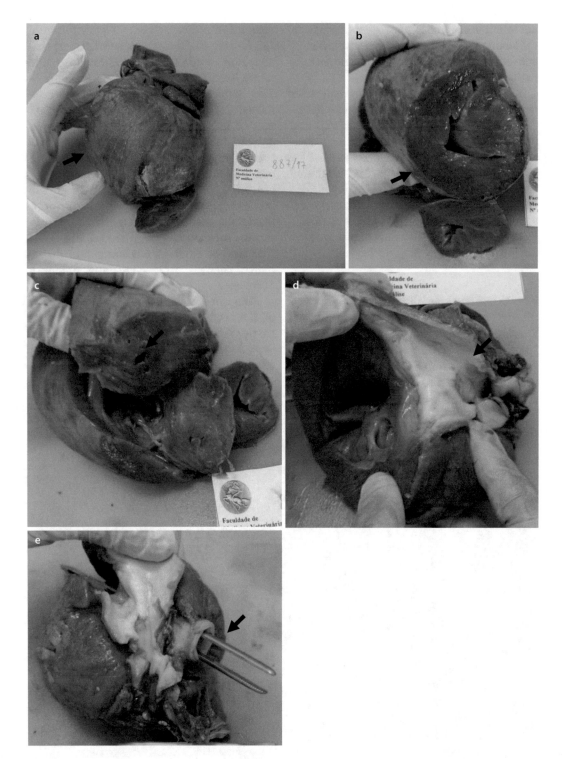

◘ Fig. 3.16 Complex malformation with aortic dextroposition and (high) ventricular communication of the calf's heart. Legend (arrows): **a** Globose heart with very hypertrophied right ventricle on the left side. **b** Apex cut with exceptionally hypertrophied right wall ventricle. The left ventricle cavity is not observed in this cut. **c** Cross section of the left ventricle, with a very small cavity. The gloved finger is inserted in the interventricular communication. **d** The aorta artery coming out of the right ventricle. **e** The aorta artery was sectioned and tweezers inserted into the pulmonary artery. (Original from Conceição Peleteiro (FMV-UL))

3

neuroinhibitory reflex that counterbalances the labour-related stimuli that promote neuroactivation via locus coeruleus-noradrenergic pathways [22].

Once compression is over, these and other neuroactivators, as well as environment stressors – lower temperature, hard surfaces and noises, induce postnatal consciousness associated with marked activity that is essential for precocious species survival [22, 32]. Simulating the compression followed by releasing it after 20 min. has shown to stimulate awareness and activity in newborn calves with maladjustment syndrome [32]. This technique is described in ▸ Chap. 5.

3.8 Maternal Behaviour

During the preparatory period, close to labour Stage I, the dam normally seeks isolation trying to ensure a secure and calm environment to calve and nurse the newborn calf. When possible, the cow will try to maintain visual and olfactory contact with the herd. In intensive production systems, special attention should be given to provide an adequate local for parturition (see ▸ Chap. 5). The expression of prepartum maternal behaviour is necessary to improve the welfare of the dam and also guarantee normal calving progression [26]. The presence of specific areas where pre-parturient animals can prepare for calving, i.e. maternities, is crucial, especially in large farms. In grazing beef herds, separation of cows into an appropriate pen improves calving and postpartum success for dam and offspring (◻ Fig. 3.17). The risk of external environmental disturbance, such as other cows, and predation decreases.

Maternal behaviour involved in the caring of the newborn is upregulated by hormonal changes during the peripartum period and is crucial for the development and even survival of the bovine calf in the wild or in extensive conditions. In normal condition, within few minutes after calving, the dam will start lick-

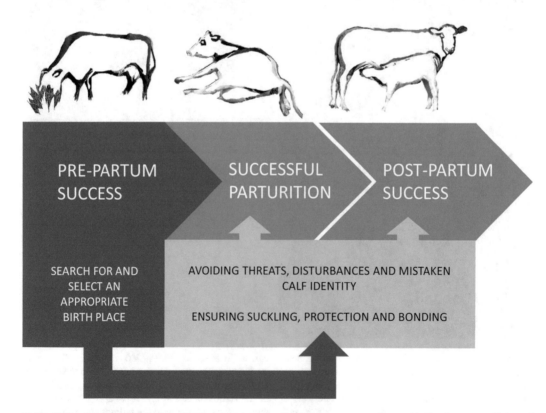

◻ **Fig. 3.17** Impact of birth site selection on calving and postpartum success in grazing beef systems. (Adapted from Rørvang et al. [26])

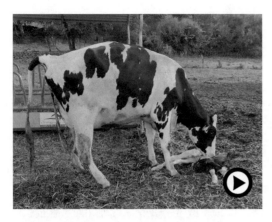

Fig. 3.18 Video box: Maternal (heifer) licking behaviour. (Original from João Simões) (► https://doi.org/10.1007/000-2qe)

Fig. 3.19 The dam eating her placenta. (Original from George Stilwell)

ing the newborn calf, establishing the so-called cow-calf bond. At this time, foetal fluid odours and calf movements are powerful stimuli for the maternal behaviour. The dam starts smelling and licking the calf (**Fig. 3.18**) within the first 1–7 min. after calving and persisting for up to 40 min. This stimulates calf activity, removes membranes eventually attached to the head/nostrils and dries its coat reducing evaporative heat loss. Licking probably also stimulates breathing, circulation, urination and defecation [36]. Under farm conditions, remaining with the dam for at least 24 h postpartum facilitates the calf to stand up, suck, defecate and urinate [24].

Licking may also have an evolutionary justification as it promotes the calf to stand and walk and eliminates odours that could eventually attract predators. This could be also the reason behind a frequent behaviour of periparturient cows – to lick and eventually swallow the placenta (**Fig. 3.19**).

In dairy farms, the newborn calf is separated from the dam within minutes to a few hours, due to handling and production reasons. In a few farms (e.g. organic milk production), the calf may stay for longer (days or even weeks). The ability of dairy cows to release either prolactin or oxytocin in response to machine milking is reduced in dairy cows maintained with their calves during the 1st week postpartum [1].

Removing the calf after bonding may induce acute stress in the cow and calf. If no contact is allowed between mother and young for a period of up to 5 h postpartum, this maternal bond is not formed in more than 50% of the animals [10]. In the case, the cow-calf bond is not established, and separation causes only mild responses in multiparous dairy [8, 9]. Krohn [16] concludes that separation after a short-term suckling period does not appear to be particularly traumatic for either the calf or the cow. In contrast, separation after a long-term suckling period influences the behaviour of both animals to a large extent. For example, the separation after 2 days caused disturbances in feeding and sleeping pattern in the cow for 2 days [27].

However, there is convincing evidence that maternal presence is important for social learning and decreases future fearfulness of others [15] and may facilitate earlier and larger intakes of colostrum which in turn leads to higher immunoglobulin concentrations in the calf.

Key Points
- Normal parturition is initiated by the activity of mature foetal hypothalamic-pituitary-adrenal axis.
- Foetal cortisol is the key hormone, inducing, directly or indirectly, the activity of several hormonal, biochemical and protein modulators with specific functions.

- A progressive placental and foetal maturation takes place towards the pregnancy term.
- A preparatory period with prodromal external signs can start a few days before parturition and usually intensifies its activity towards calving time.
- Parturition is characterized by three successive stages, with durations that can vary with individuals and is often longer in primiparous.
- A specific hormonal environment is required to prepare for the relaxation of the cervix and other structures of the birth channel. Myometrial and abdominal forces are both essential to push the foetus outside during parturition.
- During normal calving, the rupture of the umbilical cord initiates a cascade of biochemical and mechanical events responsible for effective respiratory and cardiovascular changes at the time of birth.
- Several neuroinhibitors factors keep foetus in an unconsciousness state during pregnancy and labour, while several neuroactivators ensure a rise in awareness and activity after birth.

? Questions

1. What initiates parturition in cows?
2. What is the normal length of parturition in cattle?
3. Can posterior presentation be considered a normal foetal orientation?
4. What are the roles of neuroinhibitors during pregnancy?

✓ Answers

1. The initiation of calving is a hormonal event which starts at foetal hypothalamic-pituitary-adrenal level leading to cortisol release, a key hormone, from the foetal adrenal glands. Cortisol is released in response to several stressors, in which fast foetal growth in the last 2 weeks of pregnancy and uterine distention seem to play a significant role. Foetal cortisol triggers, directly and indirectly, the release of several endogenous substances responsible for the opening and dilatation of the birth canal and for finalizing foetal maturation. One of these hormones, $PGF_{2\alpha}$, is produced by endometrium and causes a quick decrease in P_4, to <1 ng/mL, by functional luteolysis of CL, contributing to the loss of uterine quiescence and prompting myometrial contraction activity. This means that $PGF_{2\alpha}$ also assumes a crucial role in the initiation of parturition. This fact justifies the use of $PGF_{2\alpha}$ (as well as cortisol) for clinical induction of parturition at the final stages of pregnancy.

2. Parturition is divided into three successive stages, each one with a specific duration although with significant individual variation. The exact duration also depends on the definition of the onset and end of each phase. From a clinical point of view, Stage I varies typically between 2 and 6 h, starting with an increased restlessness behaviour, indicative of discomfort and more intense myometrial contractions, and ending when placental membrane appears at the vulva (onset of Stage II), indicative of a wide cervical dilatation. Usually, Stage II length varies from 20 to 80 min. ending with complete foetal expulsion and depends primarily on abdominal contractions, as well as the relative proportion between the maternal pelvis area and foetal size. Stage III is usually completed within the 6 h after foetal expulsion. So, a high percentage of dams conclude their parturition between 3 and 14 h, although in some cases the placental expulsion can take longer (up to 24 h after foetal expulsion). Calving in heifers usually takes 25–50% more time than in multiparous cows.

3. Yes. However, in nature, only a low proportion of singletons are expelled in this presentation. Also, the incidence of perinatal mortality, including stillbirths or weak newborns that die within

the first 48 h, increases. This is due to the fact that when the umbilical cord is compressed or even ruptured, the thorax and head are still inside the birth canal or even inside the uterus. This means that when respiration initiates, the lungs cannot expand sufficiently, and the nose may be immersed in fluids, thus leading to prolonged anoxia. Also, aspiration of amniotic fluid can occur, causing blockage and eventually fatal aspiration pneumonia (necrotizing pneumonia). This presentation is associated with intrauterine foetal orientation and is very common in the first foetus in case of twinning parturition, due to the complementary orientation between foetuses.

4. Several neuroinhibitors, such as adenosine, P_4, allopregnanolone and pregnanolone, cause a unconsciousness state that means less oxygen needed and less intrauterine movement. This state of sleepiness usually continues through labour, reducing movements that could cause trauma to the dam. Neuroactivators factors will then increase awareness and activity, leading the calf to rise and follow and nurse the dam.

References

1. Akers RM, Lefcourt AM. Effect of presence of calf on milking-induced release of prolactin and oxytocin during early lactation of dairy cows. J Dairy Sci. 1984;67(1):115–22. https://doi.org/10.3168/jds. S0022-0302(84)81274-6.

2. Aurich JE, Dobrinski I, Hoppen HO, Grunert E. Beta-endorphin and met-enkephalin in plasma of cattle during pregnancy, parturition and the neonatal period. J Reprod Fertil. 1990;89(2):605–12. https://doi.org/10.1530/jrf.0.0890605.

3. Beagley JC, Whitman KJ, Baptiste KE, Scherzer J. Physiology and treatment of retained foetal membranes in cattle. J Vet Intern Med. 2010;24(2):261–8. https://doi.org/10.1111/j.1939-1676.2010.0473.x.

4. Berglund B, Philipsson J, D0anell Ö. External signs of preparation for calving and course of parturition in Swedish dairy cattle breeds. Anim Reprod Sci.

1987;15(1–2):61–79. https://doi.org/10.1016/0378-4320(87)90006-6.

5. Berridge CW, Waterhouse BD. The locus coeruleus-noradrenergic system: modulation of behavioral state and state-dependent cognitive processes. Brain Res Brain Res Rev. 2003;42(1):33–84. https://doi.org/10.1016/s0165-0173(03)00143-7.

6. Buczinski S, Fecteau G, DiFruscia R. Ventricular septal defects in cattle: a retrospective study of 25 cases. Can Vet J. 2006;47(3):246–52.

7. Hillman NH, Kallapur SG, Jobe AH. Physiology of transition from intrauterine to extrauterine life. Clin Perinatol. 2012;39(4):769–83. https://doi.org/10.1016/j.clp.2012.09.009.

8. Hopstee H, O'Connell JM, Blokhuis HJ. Acute effects of cow-calf separation on heart rate, plasma cortisol and behaviour in multiparous dairy cows. Appl Anim Behav Sci. 1995;44(1):1–8. https://doi.org/10.1016/0168-1591(95)00581-C.

9. Houwing H, Humik JF, Lewis NJ. Behavior of periparturient dairy cows and their calves. Can J Anim Sci. 1990;70:355–62.

10. Hudson SJ, Mullord MM. Investigations of maternal bonding in dairy cattle. Appl Anim Ethol. 1977;3(3):271–6. https://doi.org/10.1016/0304-3762(77)90008-6.

11. Jackson PG, Cockcroft PD. Clinical examination of the female genital system. In: Jackson PG, Cockcroft PD, editors. Clinical examination of farm animals. Oxford, UK: Blackwel; 2002. p. 125–40. https://doi.org/10.1002/9780470752425.ch10.

12. Kindahl H, Kornmatitsuk B, Gustafsson H. The cow in endocrine focus before and after calving. Reprod Domest Anim. 2004;39(4):217–21. https://doi.org/10.1111/j.1439-0531.2004.00506.x.

13. Kolkman I, De Vliegher S, Hoflack G, Van Aert M, Laureyns J, Lips D, de Kruif A, Opsomer G. Protocol of the caesarean section as performed in daily bovine practice in Belgium. Reprod Domest Anim. 2007;42(6):583–9. https://doi.org/10.1111/j.1439-0531.2006.00825.x.

14. Kornmatitsuk B, Königsson K, Kindahl H, Gustafsson H, Forsberg M, Madej A. Clinical signs and hormonal changes in dairy heifers after induction of parturition with prostaglandin F2 alpha. J Vet Med A Physiol Pathol Clin Med. 2000;47(7):395–409.

15. Krohn CC, Foldager J, Mogensen L. Long-term effect of colostrum feeding methods on behaviour in female dairy calves. Acta Agric Scand. 1999;49(1):57–64. https://doi.org/10.1080/090647099421540.

16. Krohn CC. Effects of different suckling systems on milk production, udder health, reproduction, calf growth and some behavioural aspects in high producing dairy cows – a review. Appl Anim Behav Sci. 2001;72(3):271–80. https://doi.org/10.1016/S0168-1591(01)00117-4.

17. Laven R, Chambers P, Stafford K. Using non-steroidal anti-inflammatory drugs around calving: maximizing comfort, productivity and fertility.

Vet J. 2012;192(1):8–12. https://doi.org/10.1016/j.tvjl.2011.10.023.

18. Lombard JE, Garry FB, Tomlinson SM, Garber LP. Impacts of dystocia on health and survival of dairy calves. J Dairy Sci. 2007;90(4):1751–60. https://doi.org/10.3168/jds.2006-295.

19. Matsas DJ, Nebel RL, Pelzer KD. Evaluation of an on-farm blood progesterone test for predicting the day of parturition in cattle. Theriogenology. 1992;37(4):859–68. https://doi.org/10.1016/0093-691x(92)90047-u.

20. Mee JF. Managing the dairy cow at calving time. Vet Clin North Am Food Anim Pract. 2004;20(3):521–46. https://doi.org/10.1016/j.cvfa.2004.06.001.

21. Melchert M, Aurich C, Aurich J, Gautier C, Nagel C. External stress increases sympathoadrenal activity and prolongs the expulsive phase of foaling in pony mares. Theriogenology. 2019;128:110–5. https://doi.org/10.1016/j.theriogenology.2019.02.006.

22. Mellor D. Transitions in neuroinhibition and neuro-activation in neurologically mature young at birth, including the potential role of thoracic compression during labour, p. 7. In: Aleman M, Weich KM, Madigan JE. Survey of veterinarians using a novel physical compression squeeze procedure in the management of neonatal maladjustment syndrome in foals. Animals (Basel). 2017;7(9):69. https://doi.org/10.3390/ani7090069.

23. Mellor DJ, Gregory NG. Responsiveness, behavioural arousal and awareness in fetal and newborn lambs: experimental, practical and therapeutic implications. N Z Vet J. 2003;51(1):2–13. https://doi.org/10.1080/00480169.2003.36323.

24. Metz J, Metz JHM. Maternal influence on defecation and urination in the newborn calf. Appl Anim Behav Sci. 1986;16(4):325–33. https://doi.org/10.1016/0168-1591(86)90004-3.

25. Meyer CL, Berger PJ, Koehler KJ, Thompson JR, Sattler CG. Phenotypic trends in incidence of stillbirth for Holsteins in the United States. J Dairy Sci. 2001;84(2):515–23. https://doi.org/10.3168/jds.S0022-0302(01)74502-X.

26. Rørvang MV, Nielsen BL, Herskin MS, Jensen MB. Prepartum maternal behavior of domesticated cattle: a comparison with managed, feral, and wild ungulates. Front Vet Sci. 2018;5:45. https://doi.org/10.3389/fvets.2018.00045.

27. Ruckebusch Y. Feeding and sleep patterns of cows prior to and post parturition. Appl Anim Ethol. 1975;1(3):283–92. https://doi.org/10.1016/0304-3762(75)90021-8.

28. Saint-Dizier M, Chastant-Maillard S. Methods and on-farm devices to predict calving time in cattle. Vet J. 2015;205(3):349–56. https://doi.org/10.1016/j.tvjl.2015.05.006.

29. Schuenemann GM, Nieto I, Bas S, Galvão KN, Workman J. Assessment of calving progress and reference times for obstetric intervention during dystocia in Holstein dairy cows. J Dairy Sci. 2011;94(11):5494–501. https://doi.org/10.3168/jds.2011-4436.

30. Shah KD, Nakao T, Kubota H. Plasma estrone sulphate (E1S) and estradiol-17beta (E2beta) profiles during pregnancy and their relationship with the relaxation of sacrosciatic ligament, and prediction of calving time in Holstein-Friesian cattle. Anim Reprod Sci. 2006;95(1–2):38–53. https://doi.org/10.1016/j.anireprosci.2005.09.003.

31. Speroni M, Malacarne M, Righi F, Franceschi P, Summer A. Increasing of posture changes as indicator of imminent calving in dairy cows. Agriculture. 2018;8(11):182. https://doi.org/10.3390/agriculture8110182.

32. Stilwell G, Mellor DJ, Holdsworth SE. Potential benefit of a thoracic squeeze technique in two newborn calves delivered by caesarean section. N Z Vet J. 2020;68(1):65–8. https://doi.org/10.1080/00480169.2019.1670115.

33. Stilwell G, Schubert H, Broom DM. Short communication: effects of analgesic use postcalving on cow welfare and production. J Dairy Sci. 2014;97(2):888–91. https://doi.org/10.3168/jds.2013-7100.

34. Streyl D, Sauter-Louis C, Braunert A, Lange D, Weber F, Zerbe H. Establishment of a standard operating procedure for predicting the time of calving in cattle. J Vet Sci. 2011;12(2):177–85. https://doi.org/10.4142/jvs.2011.12.2.177.

35. Taverne MA, van der Weijden GC. Parturition in domestic animals: targets for future research. Reprod Domest Anim. 2008;43(Suppl 5):36–42. https://doi.org/10.1111/j.1439-0531.2008.01219.x.

36. von Keyserlingk MA, Weary DM. Maternal behavior in cattle. Horm Behav. 2007;52(1):106–13. https://doi.org/10.1016/j.yhbeh.2007.03.015.

37. Whittle WL, Patel FA, Alfaidy N, Holloway AC, Fraser M, Gyomorey S, Lye SJ, Gibb W, Challis JR. Glucocorticoid regulation of human and ovine parturition: the relationship between foetal hypothalamic-pituitary-adrenal axis activation and intrauterine prostaglandin production. Biol Reprod. 2001;64(4):1019–32. https://doi.org/10.1095/biolreprod64.4.1019.

38. Zhang WC, Nakao T, Moriyoshi M, Nakada K, Ribadu AY, Ohtaki T, Tanaka Y. Relationship of maternal plasma progesterone and estrone sulfate to dystocia in Holstein-Friesian heifers and cows. J Vet Med Sci. 1999;61(8):909–13. https://doi.org/10.1292/jvms.61.909.

Calving Guidelines

Schuenemann GM. Calving management practices for dairy herds. The Ohio State University; 2013. From: https://vet.osu.edu/sites/vet.osu.edu/files/legacy/userimages/u29/Calving%20Management%20Presentation.pdf. Assessed on 01 July 2020.

The National Animal Disease Information Service (NADIS). Calving part 1: The basics. UK; 2020. From: https://www.nadis.org.uk/disease-a-z/cattle/calving-module/calving-part-1-the-basics/. Assessed on 01 July 2020.

The National Animal Disease Information Service (NADIS). Calving cows. UK; 2020. From: https://www.dairynz.co.nz/animal/calves/calving-cows/. Assessed on 01 July 2020.

Suggested Reading

Johanson JM, Berger PJ. Birth weight as a predictor of calving ease and perinatal mortality in Holstein cattle. J Dairy Sci. 2003;86(11):3745–55. https://doi.org/10.3168/jds.S0022-0302(03)73981-2.

Miedema HM, Cockram MS, Dwyer CM, Macrae AI. Behavioural predictors of the start of normal and dystocic calving in dairy cows and heifers. Appl Anim Behav Sci. 2011;132(1):14–9. https://doi.org/10.1016/j.applanim.2011.03.003.

Nguyen PT, Conley AJ, Soboleva TK, Lee RS. Multilevel regulation of steroid synthesis and metabolism in the bovine placenta. Mol Reprod Dev. 2012;79(4):239–54. https://doi.org/10.1002/mrd.22021.

Shenavai S, Preissing S, Hoffmann B, Dilly M, Pfarrer C, Özalp GR, Caliskan C, Seyrek-Intas K, Schuler G. Investigations into the mechanisms controlling parturition in cattle. Reproduction. 2012;144(2):279–92. https://doi.org/10.1530/REP-11-0471.

Dystocia and Other Abnormal Occurrences During Calving

Contents

Electronic Supplementary Material The online version of this chapter (https://doi.org/10.1007/978-3-030-68168-5_4) contains supplementary material, which is available to authorized users. The videos can be accessed by scanning the related images with the SN More Media App.

Learning Objectives
- To define dystocia and the degrees of calving difficulty.
- To identify and classify dystocia from foetal or maternal origin, according to the immediate inability to complete Stage II of labour.
- To recognize foetal presentation, position and posture at calving and their relation with difficulty to complete calving.
- To establish relationships between proximal, intermediate and ultimate causes of dystocia.
- To evaluate the main risk factors for dystocia.

4.1 Introduction

The term *dystocia* derives from the Greek *dys* (difficult or abnormal) + *tokos* (parturition) + *ia*. In opposition to eutocia, it can be defined as a difficult birth normally associated with an abnormal duration of at least one of the labour stages. This definition should include any calving (1) causing maternal injuries; (2) reducing maternal reproductive potential, i.e. producing a negative impact on dam's fertility; or (3) reducing calf viability [49].

Dystocia can easily reach 10% of the total calvings in a herd, with a higher incidence in heifers and significant variation between breeds and even herds. In a study with beef cows, Waldner [58] found a prevalence of 8.5% (2558/29,970) for calving with assistance and 3.7% (1099/29,970) for severe dystocia. In Holstein dairy cows, López Helguera et al. [36] observed an overall dystocia prevalence of 17.8% (181/1019) with a significant difference between heifers (32.1%) and multiparous dams (8.5%; $P < 0.001$). However, in the USA, odds for dystocia show a decreased by 4.7% per year [31].

Mild cases of dystocia only require minor assistance with slight or no forced traction. Usually, correction of foetal malposition is assumed by only one person. These cases are usually solved by farmers and are considered as eutocia by some veterinarians under field conditions (see ► Chap. 3). These obstetrical manoeuvres do not cause any major adverse effect on dams or calves.

In contrast, moderate or severe dystocia needs a more intense and vigorous obstetrical intervention and probably veterinarian involvement. These dystocias are usually characterized (1) by a significant maternal-derived cause, which affects the anatomical uterine orientation or birth canal dimension, e.g. uterine torsion; (2) by a foetal malpresentation, malposition or malposture, requiring extensive corrections; or (3) by the need for intense traction, involving two to three persons (up to 150 kg of force) or a mechanical calf puller (up to 400 kg of force), that happens in oversized foetuses. Vaginal delivery after foetotomy and caesarean section (C-section) are obstetrical approaches most often related to severe dystocia.

Moderate to severe dystocia has a significant impact on individual's health and welfare and on farms' success. Severe dystocia is usually associated with pain, trauma and lesions to the dam or calf, reduced absorption of immunoglobulins by the calf, retained placenta and metritis, leading to other metabolic and infectious diseases of the dam and reproductive disturbs. Higher incidence of perinatal mortality (PM) is also observed. At reproductive level, dystocia delays uterine involution and ovarian activity resumption modifying progesterone (P_4) patterns. Calving-conception interval increase and reduced fertility are observed [56]. Daily milk production also decreases during early lactation and can cause a reduction in milk yield for the whole lactation.

This chapter mainly aims at identifying and classifying the different causes of dystocia and describing their risk factors, for primiparae and pluriparae animals.

4.2 Scoring the Degree of Dystocia

Over the years, many score points scales have been proposed to classify the degree of dystocia, so that an international uniformization is needed. The simplest table just uses two

4

□ **Table 4.1**   Different scales of calving difficulty scores used to assess parturient cows

Score points for assistance (1 = no assistance)				References
2	3	4	5	
One person	Two or more people	Mechanical extraction	Surgical procedures	Lombard et al. [35]
One person with minimal effort	One person with moderate effort	One person with considerable effort or two people	Mechanical extraction or C-section	Hiew et al. [26]
One person	One person + calf puller or >1 person	Veterinary assistance (caesarean section, C-section)		Mee [39, 42, 43]
Minor manual assistance	Mechanical extraction	C-section		Bellows and Lammoglia [5]
Easy pull	Mechanical assistance	C-section or embryotomy		Phocas and Laloë [48]
Easy	Difficult but without veterinary assistance	Difficult with veterinary assistance		Hansen et al. [23]
Mild dystocia	Severe dystocia	C-section		Tenhagen et al. [56]
Light traction	Heavy traction			Jacobsen et al. [30]
Assistance				Johanson and Berger [31]

scores: calving with or without assistance. This describes the absolute (in)ability of the dam to expel the foetus. However, calving difficulty is progressive, i.e. the relationship between the resistance to foetal passage through the birth canal and the forces needed to pull the foetus can vary. Furthermore, the cause of dystocia and foetal assessment or manipulation also influences the degree of calving difficulty. Therefore, in addition to traction force, some classifications also consider the use of different obstetrical interventions (e.g. vaginal delivery vs. C-section). Other factors such as need for professional assistance to calving (e.g. farmers vs. veterinarians) may be taken into consideration. Thus, distinct calving difficulty scores, from two to five points, have been proposed to classify the degree of dystocia (□ Table 4.1). The scoring definition and differences between the scales denote the influence of the risk factors and causes of dystocia, as well as the experience of the obstetrician. All these classi-

fication scores involve the unassisted (score 1) and assisted (score >1) calving. For scores above two points, the dystocia assistance is classified as minor, moderate or severe.

In current clinical practice, an unassisted parturition always scores 1 and is termed eutocic calving. Slightly assisted calving, originating a live or a dead calf, after a normal length Stage I or Stage II, can also be considered as eutocia (□ Fig. 4.1 – video box). These calvings are usually operated by the farmer and do not produce significant adverse effects on the dam or the calf. However, it should be said that it is relatively common for farmers to try to solve these dystocias by vaginal delivery. In consequence, recommendations, guidelines and good practices for calving management should be implemented in all farms (see ▶ Chap. 5).

The most straightforward dystocia classifications differentiate mild from moderate and severe dystocia. Additionally, some differen-

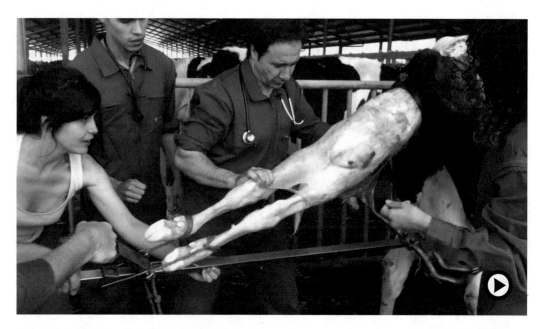

Fig. 4.1 Video box: Forced traction of a foetus in posterior presentation using a calving jack. (Original from George Stilwell) (▶ https://doi.org/10.1007/000-2qg)

tiation should be made within severe dystocia. Need for additional devices and methods, surgery or foetotomy should be considered for scoring dystocia when they directly result, or not, from an unsuccessful obstetrical intervention by one to three persons. In fact, severe dystocia classification can be influenced by external factors such as people's availability, foetal stress and viability and presence of experienced people or veterinarian. In this last case, the score may not be directly related to calving ease, originating a bias in the classification.

The degree of difficulty is primarily related to the forces needed to pull out the foetus. The difficulty to restore foetal disposition *in uterus* or unblock the birth canal, through obstetrical manipulations, should also be used to classify the severity of dystocia. Probably, the duration of assistance during foetal manipulation to change foetal maldisposition, to pull oversized foetuses or to rotate the uterus (e.g. <15 min., from 15 to 30 min. and >30 min.), needs to be taken into consideration to score each dystocia. Additionally, the duration of foetal forced traction can easily reach 20 min. before full dilation of the birth canal

occurs and to be able to synchronize traction with expulsive forces of the dam. For assistances, longer than 60 min. (including forced traction) with repeated handling of the birth canal and uterus will significantly increase the risk of soft tissue inflammation and rupture (lacerations) and will increase the probability of bacteria contaminating the uterus. In addition to the adverse effects to the dam, severe dystocia is usually associated with higher calf PM, including stillborn and calf death up to 48 h after birth (◻ Fig. 4.2).

A median of 2.4% stillborn (interquartile range: 1.3%, 3.9%) was observed in 29,970 full-term births from 203 beef cows' herds [58]. Stillborn incidence was significantly influenced by calving disturbs. Problematic calvings presented 8.6 times more likelihoods to deliver stillborn than normal calvings (95% IC odds ratio: 6.17, 11.0; $P < 0.001$). Also, the progressive degree of calving assistance, from no assistance (reference; $P < 0.001$), easy pull (odds ratio: 2.12, 95% IC: 1.59, 2.83), hard pull (odds ratio: 8.47; 95% IC: 6.45, 11.1) and C-section (odds ratio: 11.2; 95% IC: 7.14, 17.9), increased the probability for stillborn.

4

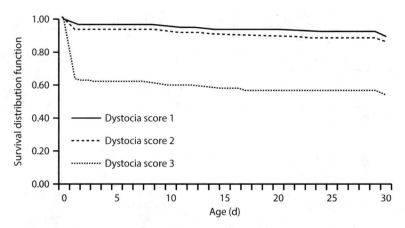

■ **Fig. 4.2** Survival plot function of perinatal and neonatal mortality according to the degree of birth difficulty. Legend: Score 1, unassisted births; Score 2, mild dystocia (assisted by one person without mechanical extraction); Score 3, severe dystocia (assisted by two or more persons, by mechanical extraction or with surgical procedures). The hazard ratio of neonatal mortality at 30-old-days was 5.4 (95% IC: 4.3 to 6.8; $P < 0.001$) for severe dystocia and 1.3 (95% IC: 1.0 to 1.7; $P < 0.05$) considering unassisted births as reference (hazard ratio = 1.0). (Adapted from Lombard et al. [35] with permission from Elsevier)

In this study, stillborn was defined as calves born dead or that died within 1 hour of birth. Mee [41] states that 75% of PM is expected in the first one hour after delivery, 15% during the next 48 h and 10% before calf delivers.

Elements such as assistance degree and the delivery of a dead or alive calf contribute to the distinction between mild, moderate, severe and extreme difficult calving. In this last case, a C-section or total foetotomy is usually the preferred solution since it comprises less complication to the dam.

Taking into account all these aspects, a four-score classification of dystocia based on calving difficulty and the degree of required assistance seems appropriate: (1) calving without assistance; (2) mild dystocia, calving assistance by one person without mechanical extraction (less than 15 min. of *in uterus* foetal manipulation); (3) moderate dystocia, calving assistance by two to three persons or with at least equivalent mechanical extraction forces (*in uterus* foetal manipulation between 15 and 30 min.); and (4) severe dystocia, *in uterus* foetal manipulation more than 30 min. (use of mechanical extraction forces; partial or total foetotomy or C-section is needed).

4.3 Classifying the Causes of Dystocia

From a clinical point of view, the causes of dystocia can be classified according to their origin as foetal, maternal or foetomaternal. This pragmatic classification can be useful in clinical settings and is correlated to the three main components of calving process: (1) foetal size (weight), shape and disposition; (2) birth canal adequacy; and (3) expulsive forces. This approach is further elucidated in ▶ Box 4.1 and ■ Fig. 4.3.

Box 4.1   Classification of Dystocia Causes, Regarding Obstetrical Approach During Stage II

Foetomaternal origin
— Foetopelvic disproportion.

Foetal origin
— Absolute oversize.
— Foetal maldisposition (malpresentation, malposition and or malposture), including twinning.
— Abnormal foetal shape due to congenital defects (infectious, toxic or genetic in origin).
— Foetal death (including cadaveric phenomena of putrefaction).

Maternal origin
(a) *Uterus origin*
    — Uterine torsion.
    — Uterine rupture.
    — Ectopic pregnancy (usually on oviduct).
    — Uterine hernia (abdominal floor rupture in late pregnancy).
(b) *Birth canal inadequacy*
    — Inadequate pelvis.
    — Soft birth canal inadequacy.
    — Insufficient or absent cervical dilatation.
    — Vaginal occlusion or obstruction.
    — Vulvar stenosis.
(c) *Expulsive forces disturbances*
    — Myometrial contractions disturbs, i.e. primary or secondary uterine inertia.
    — Abdominal contractions disturbances.

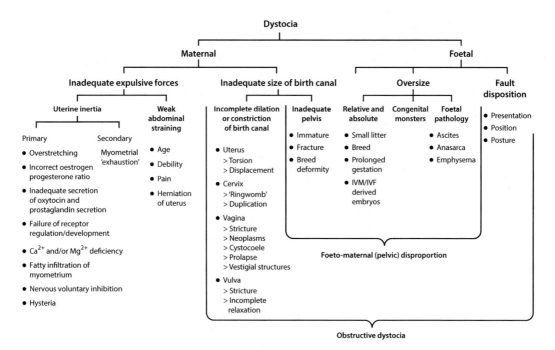

☐ **Fig. 4.3**   Maternal and foetal causes of dystocia. (Modified from Parkinson et al. [47] with permission from Elsevier). Legend: *IVM* in vitro maturation, *IVF* in vitro fertilization

4

Some of these causes may appear concomitantly. As an example, (1) insufficient cervical and vulvar dilatation can be caused by hormonal imbalance or uterine inertia, together with (2) a lack of abdominal contractions due to periparturient hypocalcaemia. Likewise, a dystocia can be a consequence of a previous one, i.e. ultimate causes of dystocia can successively originate intermediate and proximal (immediate) dystocia (▸ Box 4.2). In this last classification, some intermediate (e.g. pregnancy length) or ultimate (e.g. foetal gender, sire and dam breed) can also be considered as risk factors for specific dystocia causes, which can help the obstetrician to make a diagnosis during clinical examination. It is imperative that, during Stage II or even during Stage I of labour, all physical or/and physiological causes impeding normal birth are diagnosed and solved.

The incidence of different causes of dystocia varies between breeds (e.g. dairy or beef breeds) and parity, as described in ◻ Table 4.2. However, foetal causes (e.g. calf birthweight) prevail overall (◻ Fig. 4.4). Effects of inbreeding on incidence of dystocia are small [1].

In a recent study [12] involving 819 dystocias in dairy (Italian Frisian breed) and beef (Romagnola and Marchigiana breeds) cows, between 2005 and 2015, the prevalence of different causes was observed: 44.1% due to malposition; 35.9% due to macrosomia (foetal oversize) or foetopelvic disproportion; 2.3% caused by foetal malformations; 3.4% from pre-partum foetal death; 8.2% due to uterine torsion; 5.1% to uterine inertia; and 2.9% caused by cervical stenosis.

Box 4.2 Proximal, Intermediate and Ultimate Causes of Dystocia [43]

Proximal or immediate
- Foetopelvic disproportion.
- Abnormal foetal presentation, position and/or posture.
- Uterine inertia.
- Cervical and/or vulvar stenosis.
- Uterine torsion.

Intermediate
- Foetal oversize at birth.
- Undersize birth canal.
- Hypocalcaemia and hypomagnesaemia.
- Parturient stress.
- Gestation length.

Ultimate
- Foetal gender.
- Multiple foetuses.
- Foetal congenital abnormalities.
- Sire and dam breed.
- Parity.
- History of dystocia.
- Age, season, nutrition, exercise, disease, herd size, region and their interactions.

4.4 Foetopelvic Disproportion and Absolute Foetal Oversize

Foetopelvic disproportion is highly correlated to absolute foetal over size/weight. Odds of dystocia increase by 13%/kg increase in birth weight [31]. It remains as one of the most significant causes of dystocia in cattle, reaching up to 40% of difficult calvings.

Foetopelvic disproportion is really due to a relative disparity between foetal size and maternal pelvic area (size x conformation). However, it is not always easy to differentiate between both conditions, and some bias can occur in the classification as mild or moderate oversized foetuses.

In general, first-parity cows have a 4.7 times higher risk of dystocia than cows in later parities. [31]. Foetopelvic disproportion is more common in heifers in which the pelvis may not be fully developed at calving time. Heifers should reach 2/3 of their adult live weight at conception time (between 13- and 16-old-months in dairy breeds). This means that Holstein replacement heifers should weigh 580–635 kg at calving. Currently, due to nutritional management and genetic improvement, heifers of some dairy breeds or lineages such as Holstein, Frisian and Holstein-Friesian can calve earlier than

◻ Table 4.2 Associations between dystocia causes and parity, obstetrical treatment approach and mortality according to beef and dairy breeds

Variable		DP	BP	DM	BM
Foetal causes	Malposition (maldisposition)	1.02[a]	2.71	0.39[a]	1.27[a]
	Macrosomia	0.67[a]	2.39	0.96[a]	1.85
	Malformation	1.73	2.51	0.37[a]	1.69
	Prepartum death	0.74[a]	2.89	0.24[a]	1.55
Maternal causes	Uterine torsion	2.27	1.71	1.28[a]	1.38[a]
	Uterine atony	2.67	1.81	1.08[a]	1.48[a]
	Cervical stenosis	1.34[a]	2.06	1.21[a]	1.97
	Foetomaternal disproportion	0.69[a]	1.32[a]	2.53	2.03
Resolution method	Manual correction	0.66[a]	2.46	0.84[a]	1.74
	Caesarean section	2.83	1.69	0.7[a]	1.8
	Foetotomy	3.97	3.11	1.98	2.64
Mortality	Calf	0.52[a]	2.59	1.7	2.6
	Cow	1.81	2.3	0.81[a]	1.02[a]
	Calf and cow	1.6	2.34	1.97	0.6[a]

DP dairy primiparous, *BP* beef primiparous, *DM* dairy multiparous, *BM* beef multiparous

Data are expressed as the Euclidean distance between the group centroids and the coordinates of the outcomes

Adapted from De Amicis et al. [12] with permission from Elsevier

[a]Denotes within row high association (distances <1.5)

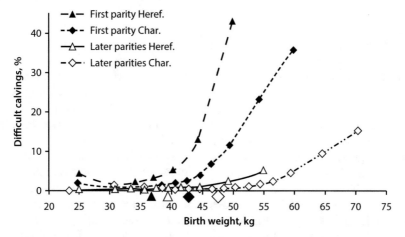

◻ Fig. 4.4 Relationships between calving difficulty and averaged birthweights in two beef breeds. (Adapted from Eriksson et al. [17] with permission from Oxford University Press). Legend: Mean birthweights, ▲ = first-parity Hereford (Heref.), △ = later-parity Hereford, ◆ = first-parity Charolais (Char.) and ◇ = later-parity Charolais

4

24-month-old. In fact, 24-months-old seems to be an adequate threshold to ensure a lesser risk of dystocia due to potential presence of underdeveloped pelvis. So, primipara dystocia is influenced by sire, weight at insemination and age, weight and body condition at calving. Overconditioned heifers have excessive deposits of fat in the pelvic canal which reduces its calibre and increases the difficulty of delivery, but calving difficulty is not diminished in underfed heifers [59].

Pelvimetry is a way to measure pelvic space for calving. Some authors [4, 19] state that pre-calving measurements have low correlations with pelvic area at calving and dystocia incidence, while other authors say that dystocia probability decreases with increase pelvic area [31].

> **Important**
>
> Foetal death can be the source or the consequence of dystocia. Foetuses are more likely to present maldisposition if death occurs before Stage II. Moderate, delayed or inappropriate calving assistance increases foetal stress and prolongs hypoxia and acidosis, so that perinatal or neonatal death prevalence increases.

4.5 Foetal Maldisposition

Foetal orientation should be classified based on three parameters: presentation, position and posture. Several nomenclatures have been proposed to characterize primary and secondary abnormal positions in different animal species. Primary faulty dispositions are the most significant alterations to identify and solve when intervening in a dystocia case. Unless there is a great disproportion between foetal size (very small foetuses or abortions) and the dam's birth canal diameter (large pelvis), none of these faulty foetal dispositions will allow a natural delivery of the foetus. Secondary faulty disposition is due to other relevant cause of dystocia, such as uterine torsion or the presence of twins.

(a) *Presentation*

Presentation is defined as the relationship between the cerebrospinal axis of the foetus and the longitudinal axis of the birth canal of the dam. Presentation can be denominated as longitudinal when both axes are (approximately) parallel. If the head or the tail enters first into the birth canal, the presentation is named *anterior longitudinal* or *posterior longitudinal*, respectively.

Presentations are considered abnormal when the cerebrospinal axis of the foetus is horizontally (transverse) or vertically (dog sitting) perpendicular to the axis of the maternal birth canal. The transverse presentation is denominated as *ventro-transverse* (*sterno-abdominal*) or *dorso-transverse* (*dorso-lumbar*) when the abdomen or the dorsum of the foetus is facing the birth canal, respectively. This relationship can also occur in the vertical presentation (*ventro-vertical* and *dorso-vertical presentations*). These presentations are difficult to maintain for a long time so an oblique presentation will usually result.

(b) *Position*

Position defines the relationship between the dorsum of the foetus and the birth canal/pelvis girdle or the uterus of the dam (◘ Fig. 4.5). Foetal position is designated as *dorsal* or *dorso-sacral*, when the dorsum of the foetus is alongside the dam's sacrum. This position can also be designated as *superior* when in relation to the uterus and *ventral* or *dorso-pubic* when the dorsum of the foetus is adjacent to the pubis (approximately 180 degrees from the dorsal position). This position can also be designated as *inferior* when referring to the uterus.

Dorso-iliac or lateral is used when the dorsum of the foetus is confronting the ilium (approximately 90 degrees from the dorsal position). Depending on the side the foetus dorsum is facing, this position can be denominated as *right lateral* or *right dorso-iliac* and *left lateral* or *left dorso-iliac*. Regarding the uterus walls, the same nomenclature (right and

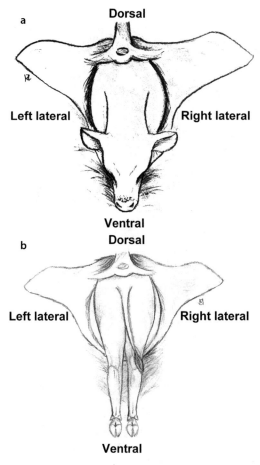

◘ Fig. 4.5 Faulty foetal positions (posterior views) at time labour. **a** Positions of anterior longitudinal presentation. **b** Positions of posterior longitudinal presentation. (Original From Soraia Marques)

left lateral position) can be used. The *right or left cephalo-iliac positions* occur in transverse presentations, as well as the *cephalo-sacral position* in vertical presentations.

(c) *Posture*

Foetal posture, also known as foetal attitude, is defined as the relationship between the forelimbs/hindlimbs or head/neck and the body of the foetus. The posture reports to relationships between the flexion (and its direction) or extension of the foetal limbs' joints and their relations to the birth canal (▶ Box 4.3). The deviation of the head (ventral or lateral) is caused by neck flexion in these directions. This situation usually occurs when the foetus

head is retained pre-pelvic at expulsion time, during unassisted or assisted calving.

> **Box 4.3   Unilateral (a) or Bilateral (b) Faulty Postures**
> **Forelimbs**
> – Carpal flexion (a or b).
> – Shoulder flexion (a or b).
> – Shoulder-elbow flexion or elbow lock posture (incomplete extension of limb; a or b).
>
> **Head (and neck)**
> – Left deviation (left lateral).
> – Right deviation (right lateral).
> – Downward deviation.
>
> **Hindlimbs**
> – Hock flexion (a or b).
> – Hip flexion (mostly b; sometimes a).

Foetal maldisposition can represent up to 30% of dystocia cases. Several abnormalities in foetal presentation, position and posture (◘ Fig. 4.6) may occur during the onset of Stage II or late Stage I. These faulty dispositions require obstetrical manoeuvres to change the foetus onto normal disposition (see ▶ Chap. 6). The normal foetal disposition is the anterior presentation, dorsal position and extended forelimbs and head. Although calving is usually easier in foetuses in anterior presentation, delivery can also occur in posterior presentation without needing obstetric assistance. In cattle, posterior presentation usually occurs in less than 5% of total calving's. However, the likelihood of delivering a stillborn increases.

Additionally to maldisposition presented in singletons, twinning may also cause dystocia due to simultaneous entrance, generally of three limbs, in the birth canal. Frequently, twin foetuses will show opposite longitudinal presentations due to their accommodation in the uterus, and not rarely one limb of the second foetus will enter and block the birth canal simultaneously with the two forelimbs or hindlimbs of the first foetus.

4

■ **Fig. 4.6** Illustrated faulty presentation **a**, position **b** and posture **c** in large animals (cows and horses) originating foetal dystocia. (Modified from Parkinson et al. [47] with from permission from Elsevier). Legend: Some faulty foetal dispositions are peculiar or more frequent in mares, such as bicornual ventro-vertical presentation

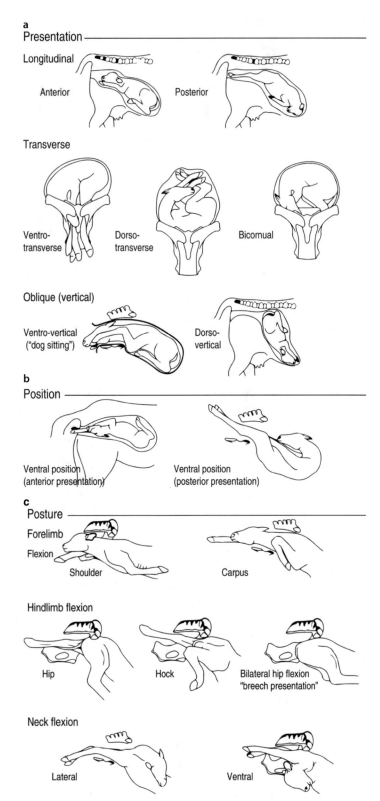

Foetal volume and hormonal changes, originated from the presence of multiple or not fully matured foetuses, can also adversely influence normal calving progress. Twining will usually cause shortening of pregnancy length in more than 4 days leading to premature parturition. In case of twins, individual calf weight and size are lower than in singletons.

4.6 Foetal Shape Alterations

Several disorders and congenital defects are responsible for abnormal foetal conformations, leading to physical obstruction that will hinder the foetus of entering the birth canal (► Box 4.4).

Box 4.4   Main Causes of Foetal Shape Alterations

Foetal death
- Autolysis followed by subcutaneous oedema and by emphysema (putrefaction and accumulation of gases after 24 h of death) due to cadaveric phenomena (◘ Fig. 4.7).

Dropsical conditions
- Anasarca (◘ Fig. 4.8a).
- Ascites.
- Hydrothorax.
- Hydrocephalus (◘ Fig. 4.8b).

Foetal monstrosities
- *Schistosoma reflexus* (skeleton anomalies; exposure of abdominal and thoracic content).
- *Perosomus elumbis* (vertebral agenesis and arthrogryposis).
- Double or conjoined monsters: craniopagus (cranius: dicephaly, polycephaly), thoracopagus (sternum), pygopagus (sacrum) and ischiopagus (pelvis).
- Dwarfism: Snorter (e.g. "Bulldog" syndrome), long head and compress forms (◘ Fig. 4.8c, d).
- Arthrogryposis.

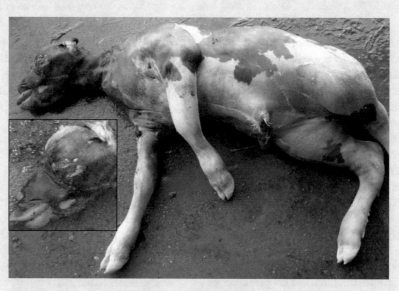

◘ **Fig. 4.7** Emphysematous foetus removed by caesarean section. The abdominal distension and subcutaneous emphysema due to anaerobic are evident. Also, note the corneal opacity which occurs between 24 and 48 hours after death. (Courtesy of António Carlos Ribeiro)

4

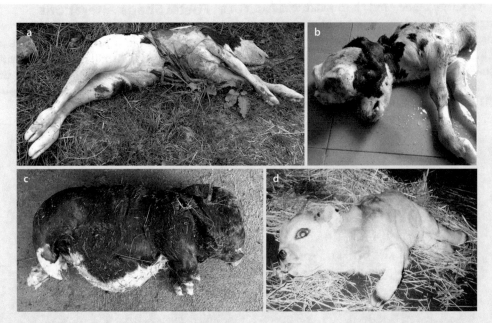

◪ **Fig. 4.8** Unusual congenital malformations presentations. **a** A Foetal generalized oedema (anasarca) of one of the twins (the second foetus was normal and alive; courtesy of Ana Paula Peixoto). **b** Hydrocephalus (courtesy of Ana Paula Peixoto). **c** Calf dwarfism: evident achondroplasia (courtesy of António Carlos Ribeiro). **d** Calf dwarfism: "Bulldog" calf. (Original from Humberto Tavares and João Simões)

One of the most frequent alterations in shape/size is caused by foetal death and autolysis when the foetus is not removed within a few hours after allantoid and amniotic sacs ruptures. Once the birth canal is open, external microbial contamination occurs and causes foetal tissue putrefaction and subcutaneous and abdominal accumulation of fluids and gases (generalized anasarca and emphysema). These alterations are fully detectable from approximately 24 h after death, by palpation or by detecting a foul-smelling vulvar discharge. The cow will show evident signs of discomfort (e.g. tenesmus, tail continuously raised) and most times also systemic signs of bacteraemia and toxaemia. The presence of the emphysematous foetus also originates a progressive friable uterine wall.

Dropsical conditions (foetal hydrops) also provoke enlargement of internal cavities and tissues. They can be due to inflammatory process related to a systemic infectious disease, but in most cases, it is the result of alterations in body fluids reab-sorption by the foetal lymphatic system or of a placental defect. Hydramnios will usually result from a defective calf (usually there is a defect in swallowing and recycling amniotic fluid). Hydroallantois is associated with structural or functional changes in the placenta (chorioallantoic membranes) with excessive production or imbalanced production/removal of allantoid fluid. As much as 25–30 L will accumulate in uterus with cows showing a typical pear-shaped abdomen, discomfort and an empty udder (see ► Chap. 2). The foetus will not show any evident defects (except, sometimes, hydronephrosis).

Musculoskeletal defects due to monstrosities are rare but may have an important impact on cow's health and welfare. Most times foetus will die at calving or short time after. *Schistosoma reflexus* syndrome (◪ Fig. 4.9 – video box), *perosomus elumbis* (◪ Fig. 4.10) and double or conjoined monsters (◪ Fig. 4.11) are the most commonly referred occurrences. Usually, in the case of twins from

Fig. 4.9 Video box: Schistosoma reflexus alive after caesarean section. Legend: Exposure of abdominal and thoracic content. The video shows the *musculoskeletal response* to external tactile stimulus and breathing and cardiac movements of two schistosoma reflexus. (Original from Carlos Cabral, Carlos Martins and João Simões) (► https://doi.org/10.1007/000-2qf)

Fig. 4.11 Thoracopagus conjoined twins. Both heads were removed by foetotomy before to caesarean section. The main finding at necropsy was the detection of a single heart shared by both foetuses. (Courtesy of Ana Paula Peixoto)

Fig. 4.10 Perosomus elumbis. Congenital defects (chondrodysplasia) can be the origin of very difficult calving, especially if the foetus is alive, precluding partial or total foetotomy. Not rarely perosomus elumbis calves are born alive. Euthanasia by barbiturate IV injection is the only acceptable course of action. (Original from George Stilwell)

Fig. 4.12 Severe arthrogryposis in a newborn calf. Calving was difficult, but foetotomy was not needed because it was a small calf. (Original from George Stilwell)

a single zygote, the axis of each embryo is not completed during embryonic development.

Some of these defects are also associated with arthrogryposis. The affected joints (limbs and vertebral column) become permanently fixed in a bent (flexed) or straightened (extended) position. Sometimes arthrogryposis is the single apparent malformation causing dystocia (▢ Fig. 4.12). Infectious (e.g. Schmallenberg, bovine viral disease or Akabane virus), toxic (e.g. anagyrine in lupines) or genetic causes explain most cases.

4.7 Inadequate Size of the Birth Canal

Anatomical and functional abnormalities of the pelvis or the soft birth canal can cause mild to severe dystocia. A birth canal diameter up to 20 cm (mean width and height intra-

4

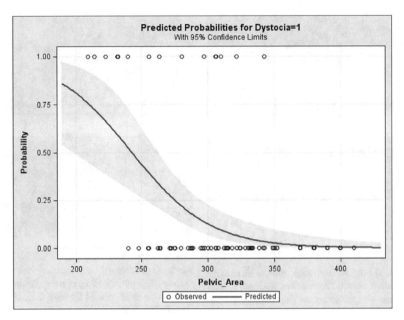

◘ Fig. 4.13 Relationship between the maternal intra-pelvic area (cm²) and probability to occur dystocia in Holstein-Friesian cows. The thick solid line represents the logistic regression line. Blue shaded area: 95% confidence interval. Open circles: the intrapelvic area for cows with (P ¼ 1.00) or without (P ¼ 0.00) dystocia. A total of 103 primiparous and multiparous late-gestation cows were used. Maternal intrapelvic area (cm²) = Intra-pelvic height X Intrapelvic width X Pi /4. (Adapted from Hiew et al. [26] with permission from Elsevier)

pelvic diameters), sometimes reaching 25 cm or more for large breeds, is adequate to deliver a foetus weighing from 30 kg up to 60 kg. The decrease in pelvic area is inversely related to dystocia (◘ Fig. 4.13), even if the pattern is not linear. The intrapelvic area seems be the best predictor of dystocia [26].

The small size of the pelvis is a significant cause of foetopelvic disproportion. However, abnormal pelvic conformation, pelvic fracture and exostoses or sacral displacement and luxation can also cause foetal retention. In addition to insufficient pelvis development in heifers, other anatomical abnormalities may cause obstruction or occlusion of the birth canal. The obstruction or occlusion of three independent structures, i.e. cervix, vagina and vulva, may also be responsible for inadequate size of the soft birth canal. Birth canal obstruction is usually related to the presence of blocked foetal parts, while birth canal occlusion is usually caused by the presence of an external mass (e.g. pelvic fat, abscesses) or the closure of its lumen (e.g. uterine torsion, cervix stenosis). Trauma or displacements

occurring during pregnancy or at previous calvings, such as a pelvic fracture or vaginal prolapse, may also create functional obstruction.

Together with pelvis inadequate size in heifers, incomplete or insufficient cervical dilatation and uterine torsion are the two major causes of dystocia of maternal origin. Incomplete cervical dilatation or even non-dilatation is usually due to hormonal imbalances during the preparatory period or Stage II of labour, limiting or delaying complete softening and ripening of the cervix. When this occurs, the vulva often also fails to adequately dilate. Other causes for inadequate cervical dilatation are the presence of fibrous connective tissue due to lacerations in the previous calving or even as a sequel of cervicitis, which can be enough to cause fibrosis. Less frequent causes are congenital stenosis of the cervix as well as hypoplasia of the vagina and vulva. Pre-cervical (*torsio uteri*), cervical (intra-cervical) and post-cervical (*torsio vaginae*) uterine torsion always causes birth canal's occlusion (see ▶ Chap. 2).

4.8 Inadequate Forces

Two distinct forces are involved in the mechanical segment of calving. The first one is the myometrial contraction, which drives the foetus towards and into the birth canal. Once the foetus reaches the pelvic inlet, periodical cycles of strong abdominal contractions push the foetus along the pelvis and through the vulva to complete delivery.

Weak or lack of myometrial contractions are designated *uterine inertia* and explain up to 10% of dystocia cases. Uterine inertia is classified as primary or secondary. The most common primary cause is hypocalcaemia, mainly found in high-producing dairy cows. However, there are several other causes such as hormonal disturbances, inherited weakness or degeneration of myometrial smooth muscle, toxic infections and senility. Because of the adrenergic system involvement in myometrial contractions, anxiety of the dam can also contribute to block smooth muscle contractibility. Adrenaline, released in stressful conditions, is also a potent inhibitor of oxytocin release. The secondary uterine inertia is mainly caused by the smooth muscle contractibility exhaustion, usually due to foetal progression block-ing. Repeated and inadequate use of oxytocin is an iatrogenic cause of uterine inertia.

4.9 Risk Factors of Dystocia

Several other factors are indirectly associated with dystocia (▶ Box 4.5). In general, they predispose to dystocia and so are considered risk factors (◘ Table 4.3). Most of them are more or less manageable, mitigating their adverse effects on the dam and the foetus. Examples of modifiable risk factors are dam's age at first calving, sire characteristics, calf breed, breeding method (e.g. embryo or nuclear transfer and natural vs. artificial insemination), gestational nutrition, foetomaternal health status, gestation length and calving management [40]. Even sex can be controlled through the use of sexed semen.

Dams presenting difficult birth have more chance (odds ratio = 2.4; 95% CI: 2.15–2.78) of showing dystocia in the next calving when compared with dams without dystocia history [39]. This fact is very relevant, once a greater surveillance of these periparturient cows is required at calving time. This procedure should be included in calving management plan of the farm.

Box 4.5   Main Factors Affecting Dystocia Prevalence (Adapted from Zaborski et al. [60])

Phenotypic factors
- *Related to calf*
  - Calf birth weight.
  - Multiple foetuses (twins, triplets, etc.).
  - Antenatal and intranatal foetal death.
  - Acquired physical defects or deformities.
- *vcRelated to cow*
  - Pelvic area.
  - Weight at calving.
  - Body condition during the dry period and at calving.
- *Related to pregnancy*
  - Gestation length.
  - Calving assistance.

Non-genetic factors
- Dam age at calving.
- Parity.
- Calf sex.

- Season.
- Local of calving.
- Herd management.
- Disease and disorders of the dam (e.g. hypocalcaemia, fatty liver syndrome, vaginal prolapse).
- Nutrition in late pregnancy.
- Level of hormones in the periparturient period.
- In vitro production of embryos and embryo cloning by nucleus transfer (e.g. large offspring syndrome).

Genetic factors
- Dam, sire and calf breed.
- Inbreeding.
- Muscular hypertrophy.
- Genetic congenital defects.
- Trait selection and quantitative trait loci.

4

☐ **Table 4.3** Risk factors quantification (odds ratio) of dystocia in dairy and beef cows

Factor	Comparison	Odds ratio	Interpretation	References
Calf's birth weight	Linear trend	1.13	13% increase in odds for dystocia per Kg increase in BW (95%CI: 1.11,1.15; $P < 0.001$).	Johanson and Berger [31] (dairy cows)
Sex	Female vs. male	1.25	25% higher odds for dystocia in males than females (95%CI: 1.06,1.48; $P < 0.01$).	Johanson and Berger [31]
		1.67	67% higher odds for dystocia in males than females (95%CI: 1.07,2.60; $P < 0.05$).	López Helguera et al. [36] (dairy cows)
		2.30	2.30 times higher odds for dystocia in males than females (95%CI: 2.10,2.53; $P < 0.05$).	Waldner [58] (beef cows)
	Singleton female vs. singleton male	0.81	19% lower odds for dystocia in singleton females than singleton males (95%CI: 0.74,0.88; $P < 0.05$).	Atashi et al. [3] (dairy cows)
Twinning	Twins vs. singleton	5.47	5.5 times higher odds for dystocia in twins than singletons (95%CI: 0.98,30.52; $P = 0.05$).	López Helguera et al. [36] (dairy cows)
		3.57	3.6 times higher odds for dystocia in twins than singletons (95%CI: 3.06, 4.17; $P < 0.001$).	Waldner [58]
	Twins: male-male, female-female or male-female vs. singleton male	33 38 43	33 (95%CI: 11.3,99), 38 (95%CI: 12.2117) or 43 (95%CI: 14.2129) times higher odds for dystocia for twins male-male, male-female or female-female, respectively, than singleton male ($P < 0.05$).	Atashi et al. [3]
Perinatal mortality	Dead vs. alive	2.46	Death foetuses within the first 48 h (including stillbirths) present 2.5 times higher odds for dystocia than alive foetuses (95%CI: 1.84,3.28; $P < 0.001$).	Johanson and Berger [31]
Parity	Primiparous (heifers) vs. multiparous cows	4.72	4.7 times higher odds for dystocia in first than later parities (95%CI: 3.77,5.91; $P < 0.001$).	Johanson and Berger [31]
		9.67	9.7 times higher odds for dystocia in first than later parities (95%CI: 4.03,23.20; $P < 0.001$).	Hohnholz et al. [27] (beef cows)
Body condition	BCS gain vs. no BCS	0.83	17% decrease in odds for dystocia in dams with BCS gain than no BCS change from pregnancy testing to pre-calving (95%CI: 0.72,0.96; $P < 0.01$).	Waldner [58]
		1.30	30% increase in odds for dystocia per kg increase in BW (95%CI: 1.14,1.44; $P < 0.05$).	Hohnholz et al. [27]
	In singletons, linear trend	1.10	10% increase in odds for dystocia per kg increase in BW (95%CI: 1.08,1.11; $P < 0.05$).	Atashi et al. [3]
	In twin births, linear trend	1.02	2% increase in odds for dystocia per kg increase in BW (95%CI: 0.99,1.05; $P < 0.05$).	Atashi et al. [3]

⬛ **Table 4.3** (continued)

Factor	Comparison	Odds ratio	Interpretation	References
Pelvic area	Linear trend	0.89	11% decrease in odds for dystocia per dm^2 increase in pelvic area (95%CI: 0.86,0.92; $P < 0.001$).	Johanson and Berger [31]
Season	Winter vs. summer	1.15	15% higher odds for dystocia in winter than summer (95%CI: 0.98,1.36; $P = 0.08$).	Johanson and Berger [31]
	Spring vs. summer	0.89	11% lower odds for dystocia in spring than summer (95%CI: 0.79–0.99; $P < 0.05$).	Atashi et al. [3]
	Spring vs. fall	0.78	22% lower odds for dystocia in spring than summer (95%CI: 0.69–0.87; $P < 0.05$).	Atashi et al. [3]

BW body weight, *BCS* body condition score

4.9.1 Calf Birth Weight

Several dystocia risk factors can distinctly affect primiparous (heifers) and multiparous cows. Also, some factors can predispose to others. Calf birth weight is the most significant factor influencing dystocia, mainly due to the concurrent increment in dimension of certain body elements (e.g. hindquarters). However, foetal size can cause dystocia independently of birth weight, so that both factors have to be taken into account in dystocia occurrence, mainly in heifers. In dairy breeds, calf birth weight gradually increases from first parity for both male and female foetuses, reaching the highest values at third parity [32]. Similar calf birth weight pattern can be observed in beef breeds.

Calf birth weight usually ranges from 5% to 10% of the dam's weight, and foetal genetics is responsible for 60% of the total weight. In ⬛ Tables 4.4 and 4.5, the influence of some phenotypic and other non-genetic factors on birthweight is presented. These factors can work as predictors to estimate the calf birth weight, before or at peripartum period. They can contribute to making decisions to improve both dam and calf health and welfare.

In multiparous dairy cows, birth weight may be lower in calves from high yielding cows or from dams with short dry period. These occurrences can be related to partitioning of energy and other resources between dam and foetus.

Some modern beef breeds show an extremely high level of muscle development (double-muscling) that predisposes to dystocia. In the case of the Belgian Blue breed, this is due to the selection of natural mutation in the myostatin gene reducing the production of this protein that normally inhibits muscle development. Because of the very high probability of dystocia, most Belgian Blue cows are submitted to a C-section as soon as they go into Stage I of labour.

Also, new alterations deriving from new technology tools, such as embryo or nucleus transfer, also expose embryos to an altered environment and can result in muscle overdevelopment or developmental abnormalities (e.g. large offspring syndrome).

4.9.2 Foetal Sex

Males have a higher risk of dystocia than females. On average, males weigh more 1–3 kg than females, and pregnancy length is usually 1.5 days longer. Both occurrences are probably due to hormonal difference such as highest genetic insulin resistance of females causing lower trophic effects on cells when compared

4

□ **Table 4.4** Predictors of birth weight of calves born to heifers

Predictor	Comparison	No.	Estimate (kg)	P-value
Intercept		540	41.3	<0.001
Calf sex	Male	264	2.57	<0.001
	Female	276	Referent	
Season of calving	Summer and fall	325	−2.23	<0.001
	Winter and spring	215	Referent	
Gestation length (d)	Short (265 to 275)	136	−5.01	<0.001
	Medium (276 to 285)	379	−2.18	0.011
	Long (286 to 295)	25	Referent	
Heart girth (cm)	Linear	540	0.52	0.016
Wither height (cm)	Linear	540	0.54	0.007
Diagonal length (cm)	Linear	540	0.73	<0.001
Age at calving (mo)	Very young (20.3 to <22)	98	2.75	<0.001
	Young (22 to <23.5)	145	3.29	<0.001
	Standard (23.5 to <25.5)	198	2.35	<0.001
	Old (25.5 to 37.3)	99	Referent	

Adapted from Kamal et al. [32] with permission from Elsevier

to males. Physical differences between males and females (e.g. greater muscle and bone development) can also predispose to dystocia.

4.9.3 Parity

The prevalence of dystocia usually decreases along with parity number, even if some causes of dystocia are related to high milk yield. Despite shorter pregnancy length and lighter foetuses, high dystocia prevalence in heifers is partially explained by low pelvic measures and slower and sometimes incomplete soft tissue dilatation during Stage II.

In contrast, energy and other nutrients partitioning are different in developing heifers. This is probably due to hormonal homeostasis of the somatotropic axis mainly regarding growth hormone and insulin. Mature cows seem to have more resistance to these hormones as well as to placental lactogen, explaining why they have larger and heavier foetuses than heifers.

4.9.4 Pregnancy Length

Pregnancy length is indirectly related to calving ease. In the last week of pregnancy, the estimated daily increase in foetus weight is approximately 0.5 kg. A positive correlation ($P < 0.001$) between foetus body weight and pregnancy length was observed in heifers ($R = 0.35$) and multiparous cows ($R = 0.34$) [36]. Usually pregnancy length is shorter in very young or in too old heifers. Immaturity in young or overcondition in older heifers seems to predispose to dystocia.

4.9.5 Nutrition

Dams should present an adequate and stable body condition throughout the last trimester of pregnancy, when an increment of two-thirds of foetal weight occurs. Low body condition

◻ **Table 4.5** Predictors of birth weight of calves born to adult cows

Predictor	Comparison	No.	Estimate (kg)	*P*-value
Intercept		1054	44.1	<0.001
Calf sex	Male	520	3.51	<0.001
	Female	534	Referent	
Season of calving	Summer and fall	570	−1.12	<0.001
	Winter and spring	484	Referent	
Gestation length (d)	Short (265 to 275)	173	−4.96	<0.001
	Medium (276 to 285)	765	−2.52	<0.001
	Long (286 to 295)	116	Referent	
Parity of the cows	2 and 3	762	1.02	0.005
	4 to 9	292	Referent	
Heart girth (cm)	Linear	1054	0.19	0.63
MGEST[a] (kg)	Low (1400 to <5400)	270	−1.16	0.021
	Intermediate (5400 to <6500)	334	−0.74	0.083
	High (6500 to <7200)	222	−0.43	0.34
	Very high (7200 to <11,600)	228	Referent	
Heart girth × MGEST (cm·kg)	Low (1400 to <5400)	270	0.97	0.039
	Intermediate (5400 to <6500)	334	0.10	0.84
	High (6500 to <7200)	222	1.11	0.034
	Very high (7200 to 11,600)	228	Referent	
Dry period (d)	Long (55 to 275)	336	1.14	0.021
	Standard (45 to 54)	469	1.60	<0.001
	Short (3 to 44)	249	Referent	

Adapted from Kamal et al. [32] with permission from Elsevier
[a]MGEST = cumulative milk production during gestation from conception to drying off

scores, lower than 2.5 points (5-point scale), or significant loss of the dam's body condition, as well as severe restrictions in protein, can lead to foetus and placenta undergrowth, increasing the risk of dystocia (e.g. uterine inertia and inadequate relaxation of the pelvic ligaments) and PM. These nutritional deficits can also significantly affect proper skeletal development of pregnant heifers. Good transition nutrition in dry dairy cows reduces the incidence of dystocia [52, 55].

In extensive systems (pasture-based production), care should be taken to supplement pregnant heifers when grass is scarce. Hunger or nutritional imbalances will usually affect the development of the future dam more

4

than the foetus, eventually leading to foetal-maternal disproportion dystocia.

In contrast, a significant increase in body condition due to overfeeding leads to an overweighed and oversized foetus associated with an excessive accumulation of fat in pelvic depots of the dam, causing partial occlusion of the birth canal.

4.9.6 Climate/Season

High temperatures and humidity variations (e.g. heat stress) have a negative effect on foetal growth, especially during the second half of pregnancy, and can be related to a decrease in nutrients and energy (glucose) provided to the foetus. Season and climate conditions can affect pregnancy and calving by:

1. Decreasing dry matter intake and reducing uterine blood flow and placental function, in case of heat stress.
2. Longer days can stimulate plasma prolactin release, increasing nutrients and energy shift towards milk yield in dairy cows.
3. Shortening pregnancy length in 1–2 days.

4.9.7 Other Relevant Risk Factors

- **Cow Body Weight and Size**

Cow body weight and size are inversely related to dystocia incidence in both beef and dairy cows. Heifers and adult cows presenting sub-optimal skeletal size at calving time are more likely to develop dystocia. Heifers' age at first calving is associated with more difficult calving mainly due to insufficient skeletal and pelvic development (pelvic area, such as reported previously), even if smaller cows produce smaller calves mitigating dystocia occurrence.

- **Concomitant Diseases**

Concomitant diseases also may play an essential role at calving time. Some puerperal metabolic disorders are common in this period. *Subclinical or clinical hypocalcaemia* causes a decrease in contractility of the smooth and skeletal muscle involved in parturition, mainly in high-producing dairy cattle. The lack of

myometrial contractions induces primary uterine and cervical inertia. Abdominal contractions stop due to the incapability of the skeletal muscles to contract. In consequence, lack of muscular tone disrupts or prevents calf delivery, unless intravenous calcium is given. Also, *fatty liver syndrome* mainly related to twining in beef cows (pregnancy toxaemia) can induce constraints at this time due to its general debilitating effects. In the recumbent cow with fatty liver syndrome, a C-section should be immediately performed to save at least the calf's life. *Vaginal prolapse* is a relatively frequent disorder, which can occur/worsen some days before calving. At this stage, the increase of oestrogens induces a relaxation of the smooth birth canal, and the vaginal wall is exteriorized through the vulva. Vaginal prolapse causes a functional occlusion or even an obstruction of the birth channel when inflammation and oedema of the vaginal wall are extensive. Predisposing factors are age (older cows have very lax ligaments and very large vulva), previous lesions of the vulva and vagina or sloping floor in stalls leading to abdominal viscera pressing the vagina when the cow lies down.

Lameness, severe mastitis and other causes of pain or toxaemia can also delay or disturb parturition leading to dystocia.

- **Regular Exercise**

Regular *exercise* consisting of walking during the last pregnancy period can have a positive effect on calving, probably due to an increase in muscle tone resulting in stronger contractions during labour. Cows that exercised during the dry period had fewer calving-related problems than those that did not exercise [22]. Hereford heifers in confinement during parturition had more dystocia (mainly due to vulva constriction) and stillbirths, compared with animals left to calve in either the paddock or a large yard [15].

- **Imbalanced Hormone Profile**

Since hormonal changes are essential to prepare for calving, temporary imbalanced hormone profile induces morphological and physiological disturbs. The moment when progesterone decreases, as well as the change in the progesterone/oestrogen ratio, seems to be a key factor unblocking myometrial activity.

■ **Genetic Factors**

As we described previously, dam and calf's shape and size are primarily determined by their genotype, and so genetic factors are the basis and may show significant heritability with dystocia. In consequence, different dystocia incidence can be expected according to individuals' breed, inbreeding or crossbreeding.

■ **Mental Stress**

Social isolation, human presence or presence of other animals (e.g. dogs) may adversely affect the progress of calving resulting in dystocia. Equally, the movement of animals, particularly nervous cows and heifers, may suspend calving behaviour possibly for hours [41].

4.10 Consequences of Dystocia

4.10.1 Dam

Dystocia consequences to the dam ranges from discomfort because of small lacerations, haematomas or other trauma to the vulva to paralysis of the obturator/sciatic nerve and downer cow syndrome [16]. Calving paralysis occurs more frequently in heifers although it occasionally is observed in multiparous cows following seemingly normal parturition.

Dystocia affects adversely milk, protein and fat yield, reproduction indexes, disease incidence, culling and mortality [56]. Placenta retention is also more frequent in animals with difficult and assisted calving.

Dystocia requiring C-section or total foetotomy following unsuccessful extraction causes enhanced postpartum adrenocortical function [46]. Compared with cows with normal calving, cows with dystocia showed higher plasma glucocorticoids after parturition [25, 33]. Depression or exhaustion of the adrenal cortex, if it occurs following adrenocortical hyperactivity, may cause many postpartum diseases [46]. In a study, vasopressin levels were higher in heifers that needed assis-

tance than in those that did not, indicating that this hormone is released in order to deal with pain and stress associated with difficult labour [28].

4.10.2 Calf Trauma and Perinatal Mortality

Even mild dystocia has been shown to impact calf's health and survival [21]. Calving ease is a trait considered to be correlated with calf mortality within the first 48 hours.[1] In a US study, female calves born alive after severe dystocia had greater odds of treatment for respiratory disease (odds ratio = 1.7), digestive disease (odds ratio = 1.3) and overall mortality (odds ratio = 6.7) [35].

More parturitions of primiparous Holstein cows result in calf death, compared with multiparous cows [35, 44]. More than half of calves born after dystocia in heifers, compared with less than one third of calves born to multiparous dams, need resuscitation measures and assistance. Also, a larger percentage of male calves (40.0%) will require assistance compared with female calves.

Calf welfare issues include hypoxia/anoxia, acidosis and severe trauma such as broken vertebrae, ribs or limbs and death [16, 53]. Also, neuropathies due to hyperextension of femoral nerve during incorrect forced traction when stifle lock or hip lock in anterior presentation occur (❑ Fig. 4.14) [29].

Prolonged labour, partial umbilical compression and trauma during obstetric manoeuvres are the major causes of low calf vitality and PM. A progressive degree of hypoxia or anoxia takes place and can cause immediate foetal death. Additionally, oxygen deprivation induces foetal acidosis affecting vital organs functionality and originating a weak or a dead calf. Meconium staining of the foetal fluids is an indicator of intrauterine hypoxia and foetus stress.

Up to 40% of veterinary-assisted deliveries may result in rib fractures and up to 10% in

---

1 Most authors will consider PM calves born alive but that die before 48 h of age.

4

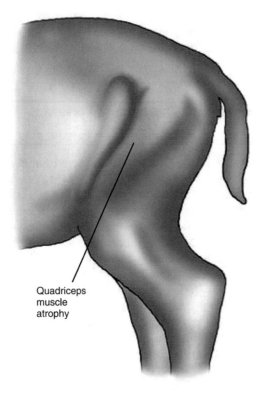

Quadriceps
muscle
atrophy

◻ Fig. 4.14 Illustration of severe muscle atrophy of the quadriceps muscles consequent to femoral nerve paralysis due to nerve hyperextension. The calf can be recumbent or even presenting a frog-like posture. (Adapted from Jackson and Cockcroft [29] with permission from John Wiley and Sons)

vertebral fractures [57]. It has been shown that rib fractures leading to trachea collapse and stenosis may occur during difficult delivery of calves [18, 50]. The prevalence of rib fractures was 23% in 235 dead calves after severe dystocia [53]. Rib fractures may cause lung disease (e.g. lung contusion or pneumonia).

Limb fractures are also common when excessive force is exerted, especially in disproportion origin dystocia. The most common fractures include the metacarpus and metatarsus (approximately 50%), tibia (approximately 12%), radius and ulna (approximately 7%) and humerus (<5%). Use of mechanical extractors is very often associated with such fractures. If not adequately assisted immediately, most of these fractures will complicate leading to the need to amputate (◻ Fig. 4.15) or euthanize the newborn.

Independently of the primary cause, calves delivered in dystocia settings are reported to increase events of respiratory diseases and mortality up to 30 days of age [35]. Calves born from difficult calving correlate to increased age at first calving [24]. In a study in Portugal, records from 4537 calves born over 1 year in 6 dairy farms showed a global PM prevalence of 9.9%. The risk factors associated with a higher PM were difficult calving in winter,

◻ Fig. 4.15 Two female calves whose front limbs had to be amputated because of complicated fractures after the incompetent use of a foetal extractor. (Original from George Stilwell)

male or twin calves and being born from first calving [51]. In summary, calves born by difficult calving have an increase incidence of PM, respiratory disease and other diseases, reduced fertility and long-term mortality.

Significant basic risk factors associated with PM following all calvings include genetic variables such as sire, sire breed, dam breed, inbreeding and gestation length [31, 41]. Important non-genetic variables include year, season, beginning and end of the seasonal calving period, calving environment and pre-calving nutrition [41]. Although it may reduce the probability of dystocia, shorter length of gestation increases the probability of stillborn [37, 38, 44]. Induction of calving is also associated with a substantially lower calf survival.

PM is also associated with calf compromised adaptation to cold stress [41]. Dystocia-born and body condition of four or above are risk factors for stillborn [10].

The primary determinant of whether a herd has high or low PM is not management factors before calving but rather calving management [14]. Calves born in maternity pens have a significantly lower mortality rate [41]. Ninety percent of calves that die in the perinatal period are alive at the start of calving [42], and two-thirds of calf mortality within the first 48 hours occur at calving [9]. Difficult births tend to result in PM 2.7 times more often than unassisted births [31]. Heavy calves have a 9.6% higher probability of PM than lighter ones [31].

Case 6.1 Practical Guidelines to Control Modifiable Risk Factors of Dystocia and Perinatal Mortality

A Dairy Farmers Association requested a protocol with general guidelines to mitigate risk factors of dystocia and PM. Very few dairy producers incorporate breeding strategies to decrease dystocia or have delivery management protocols.

The size of farms varied from 100 to 600 lactating cows, with Holstein-Frisian breed predominating in semi-intensive or intensive system production. The following basic guidelines were suggested:

Age at first calving 23–27 months, average of 24 months; to ensure adequate heifer weight and pelvis development. Heifers should never be inseminated at first oestrous, which normally occurs at 11–12-old-months in dairy heifers. Having too old heifers (≥ 28-old-months) calving should be avoided as the probability of dystocia and stillborn increases.

Selective breeding Pure Holsteins cows are more likely to present dystocia or PM than other breeds. Selection of sires for heifers significantly reduces dystocia [2]. Sire should be selected according to the breed and individual calving ease indexes, especially for heifers. Crossbreed younger and smaller animals with other dairy breeds (e.g.

Jersey, Scandinavian Red or Norwegian Red) will reduce the risk for foetal-maternal disproportion origin dystocia. Because sire predicted transmitting ability (PTA) for perinatal survival and calving ease is demonstrated [45], first-parity records should be used for genetic evaluation of bulls for calving performance [54].

Artificial insemination should prevail, and unexpected PM prevalence associated to individual sires should be analysed. Only multiparous cows should be inseminated with male sexed beef semen.

Infectious disease control A veterinary health programme should be devised and implemented. All diseases with implication on foetal development, e.g. bovine viral diarrhoea virus, neosporosis (*Neospora caninum*) and brucellosis (*Brucella abortus*), should be monitored, as well as leptospirosis (*Leptospira interrogans* serovar Hardjo) in grazing systems. Vaccination should be considered when applicable.

Prolonged pregnancy length Calving induction (dexamethasone + $PGF_{2\alpha}$) for multiparous cows at 290 days and for primiparous at 280 days should be considered. Retained placenta preva-

4

lence should be monitored and prevention measures applied if necessary.

Calving management   Move dam to calving facilities 1–2 days prior to the expected date of pregnancy term (277 days for multiparous and 274 days for heifers) or whenever signs of imminent calving are evident. Do not move dams during late Stage I due to increased risk of dystocia (2.5 times) due to a prolonged Stage II of labour. Supervision of Stage I and Stage II every 60 and 20 min., respectively, should be implemented. In case of prolonged Stage I (more than approximately 5 hours) or Stage II (more than 80 min.), obstetrical examination of the cow should be performed and assistance to complete calving provided if deemed necessary.

4.11 Breeding Cows for Ease Calving

Selection of sires with low birth weight is much more effective than selection of replacement heifers based on yearling pelvic area, in reducing both the incidence and severity of dystocia in first-calf heifers [11]. First-parity records should be used for genetic evaluation of bulls for calving performance [54]. Sire PTA for perinatal survival and calving ease is an useful tool to reach this goal [45].

To prevent dystocia, two easy genetic calving indexes were developed, mainly for heifers (▶ Box 4.6), to use in breeding programs. Both Direct (foetal) Calving Index and Maternal Calving Index show low heritability (up to $h^2 = 0.20$). They are genetically antagonistic (genetic correlations up to −0.63), i.e. small newborn females will be dams presenting more difficult in calving [13]. Nevertheless, these indexes are a useful tool to move towards a reduction in calving difficulty in heifers [7]. Additionally, calf shape, at shoulder and hip levels, can be influenced by individual characteristics and breed sire selection [3]. Estimated breeding values (EBVs) are worldwide used to improve genetic merit and phenotype of bull progeny according to selected traits. Half of the genetic merit is transmitted to bull's offspring. The traits' selection also involves Calving Index and Maternal Calving Index. These are calculated using Best Linear Unbiased Prediction (BLUP; [8]) animal models for females calving at 24 months age. Scale between −5 and 5 (breed average = 0; positive values correspond to % of improved easy calving from a specify bull) are usually reported. In near future, genomic-enhanced breeding values method [6] will accelerate and increase the accuracy of breeding genetic predictions.

Box 4.6   Easy Calving Indexes Prediction

Direct Calving Index (dCE %)
- Refers to how easily a calf will be born (e.g. calf size and weight, according to statistical models).
- Reports the calving ease of a bull's progeny.
- More appropriate to improve immediate calving.

Maternal Calving Index (mCE %)
- Refers to the ability of the dam to easily deliver a calf (e.g. pelvic dimensions and shape, and response to parturition signalling).
- Reports the calving ease of a bull's daughters.
- More appropriate to obtain a next generation of heifers with more adequate pelvis.

Expected progeny differences (EPDs) has been used to genetically improve beef breeds to meet current and future beef production demands, specifically in the USA as EBVs are used in the most of all other countries. Similar to EBVs, EPDs (EPDs = ½ EBVs values) are predictions of the genetic transmitting ability of a sire to its descendants and are used as selection tools. Usually, birth weight, growth, maternal and carcass traits are determined. The across-breed EPDs are used to compare individuals from different breeds. The adjustment factors are useful to commercial producers who may wish to compare the genetic merit of bulls from more than one breed in crossbreeding programs. In ▢ Table 4.6, the weight means for 2017-born animals of 18 beef breeds are presented. In some breeds (e.g. Gelbvieh and Simmental breeds), the EPDs of calving ease have also been calculated [20]. This allows for some prediction related to calf weight at birth and can be used to manage (increase) calving ease.

Several intrinsic (e.g. genetic and sex) and extrinsic (e.g. nutrition and pregnancy length) factors, as well as uterine environment, influence the absolute size of the foetus. Both relative and absolute foetal oversizes are major causes of stillborn. In Holstein breed, the optimal foetal weight varies between approximately 42 and 45 kg at birth time. Values higher than 45 kg increase the risk of stillborn (▢ Fig. 4.16).

The crossbreeding of Holstein cows may increase calving ease. For example, crosses with other dairy (e.g. Brown Swiss or Scandinavian Red breeds) or with beef (e.g. Limousine breed) sires may lead to lighter calves.

▢ **Table 4.6** Calf birth weight adjustment factors for expected progeny differences in 18 beef cattle to calculate across breed expected progeny differences (EPDs) in 2017-born animals

Calf birth weight

Breed	Breed of sire means [a]		Expected progeny differences [b]	
	lb	kg	lb	kg
Angus	85.3	38.7	0.0	0.0
Hereford	88.0	39.9	1.0	0.5
Red Angus	85.2	38.7	2.5	1.1
Shorthorn	90.0	40.8	4.2	1.9
South Devon	88.2	40.0	2.3	1.0
Beefmaster	88.5	40.1	4.0	1.8
Brahman	95.5	43.1	9.7	4.4
Brangus	87.6	39.7	2.7	1.2
Santa Gertrudis	89.2	40.5	4.9	2.2
Braunvieh	89.0	40.4	1.9	0.9
Charolais	90.6	41.1	6.2	2.8
Chiangus	88.3	40.1	2.5	1.1
Gelbvieh	87.3	39.6	3.3	1.5
Limousin	87.3	39.6	2.2	1.0
Maine-Anjou	87.6	39.7	1.6	0.7
Salers	85.8	38.9	0.6	0.3
Simmental	88.4	40.1	2.5	1.1
Tarentaise	87.0	39.5	2.5	1.1

Adapted from Kuehn and Thallman [34]

Data retrieved from Beef Improvement Federation (▶ https://beefimprovement.org/), assessed on 20-03-2020

[a]Breed of sire means (one half of full breed effect) for 2017-born animals: predict differences when bulls from two different breeds are mated to cows of a third, unrelated breed

[b]Adjustment factors to add to EPDs updated in December 2019. The Angus is the reference. The adjustment factors were calculated at the US Meat Animal Research Center (▶ https://www.ars.usda.gov/plains-area/clay-center-ne/marc/)

4

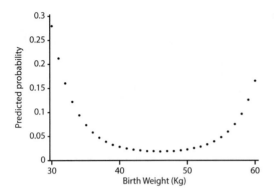

Fig. 4.16 Quadratic relationship between a calf's birth weight and its predicted probability of being stillborn in Holstein cows. (Adapted from Villettaz Robichaud et al. (2017) with permission of Elsevier)

Key Points

- Dystocia, as defined as difficult or abnormal calving, can reach 10% of total parturitions with a higher incidence in heifers.
- Moderate and severe dystocia can cause maternal and offspring trauma, disease and death.
- There are several maternal and foetal causes of dystocia, or combination between both.
- Oversize/overweight, alteration in shape or disposition of the foetus, birth canal inadequacy and disturbances in expulsive forces are the main factors responsible for abnormal calvings.
- Unsuccessful prolonged Stage II usually leads to secondary uterine inertia due to the smooth muscle contractibility exhaustion.
- Risk factors of dystocia can be quantified and modified by reproductive management and implementation of herd health programmes.

❓ Questions

1. How do you define the degree of dystocia?
2. Describe the potential causes of dystocia due to twinning.
3. What are the relationships between PM and dystocia?

✅ Answers

1. The degree of dystocia is usually evaluated using a 3–5 score scale. These scores are based on the degree of forced traction and obstetrical treatment option (vaginal delivery, foetotomy or C-section). Score 1 usually represents unassisted calving and score 2 a mild dystocia assisted by one person using light or no traction. Some veterinarians consider score 2 as eutocia. Moderate and severe dystocias are classified into two or three additional scores according to the different type of obstetric assistance: (a) moderate traction made by one or two persons; (b) strong traction made by two or three people or, as an alternative, mechanical traction (calf puller); and finally, (c) the use of foetotomy or C-section after irreducible dystocia. Additionally, obstetrical assistance duration and potential adverse effects on dam and foetus caused by foetal manipulation should also be taken into consideration in order to classify the degree of dystocia.

2. There are different causes related to calving difficulty in dams carrying twins: (1) foetal maldisposition can occur similarly to singletons. In twinning, the relatively small intrauterine space can contribute for limbs and head's retention or flexion. (2) Both foetuses can try to simultaneously enter into the dam's pelvis and block the birth canal; (3) excessive foetal load causes additional effort resulting in dam fatigue and uterine inertia; and (4) the pregnancy length is shortened approximately 4 days and can result in foetal immaturity and an increase in PM.

3. PM is defined as the occurrence of stillborn or calf death during the first 48 h after delivery and can cause or be caused by dystocia. It has been shown that dystocia more than duplicates the chance of producing a dead calf. On the

other hand, occurrence of foetal death during Stage I or Stage II can provoke a difficult calving. This is initially due to passive progression, eventually associated with a relatively dry birth canal. Posteriorly, foetus size increases, and shape changes due to cadaveric phenomena (putrefaction and emphysema). Severe dystocia significantly increases PM (up to two days after calving) and neonatal mortality (up to 30 days after calving). The occurrence of stillbirths and weak calves is primarily, but not exclusively, related to hypoxia and neonatal asphyxia consequent to prolonged labour.

References

1. Adamec V, Cassell BG, Smith EP, Pearson RE. Effects of inbreeding in the dam on dystocia and stillbirths in US Holsteins. J Dairy Sci. 2006;89(1):307–14. https://doi.org/10.3168/jds.S0022-0302(06)72095-1.
2. Anderson KJ, Brinks JS, LeFever DG, Odde KG. A strategy for minimizing calving difficulty. Vet Med. 1993;88:778–81.
3. Atashi H, Abdolmohammadi A, Dadpasand M, Asaadi A. Prevalence, risk factors and consequent effect of dystocia in Holstein dairy cows in Iran. Asian-Australas J Anim Sci. 2012;25(4):447–51. https://doi.org/10.5713/ajas.2011.11303.
4. Basarab JA, Rutter LM, Day PA. The efficacy of predicting dystocia in yearling beef heifers: II. Using discriminant analysis. J Anim Sci. 1993;71(6):1372–80. https://doi.org/10.2527/1993.7161372x.
5. Bellows RA, Lammoglia MA. Effects of severity of dystocia on cold tolerance and serum concentrations of glucose and cortisol in neonatal beef calves. Theriogenology. 2000;53(3):803–13. https://doi.org/10.1016/S0093-691X(99)00275-7.
6. Bengtsson C, Stålhammar H, Strandberg E, Eriksson S, Fikse WF. Association of genomically enhanced and parent average breeding values with cow performance in Nordic dairy cattle. J Dairy Sci. 2020;103(7):6383–91. https://doi.org/10.3168/jds.2019-17963.
7. Bennett GL, Thallman RM, Snelling WM, Kuehn LA. Experimental selection for calving ease and postnatal growth in seven cattle populations. II. Phenotypic differences. J Anim Sci. 2008;86(9):2103–14. https://doi.org/10.2527/jas.2007-0768.
8. Bennett GL. Experimental selection for calving ease and postnatal growth in seven cattle populations. I. Changes in estimated breeding values. J Anim Sci. 2008;86(9):2093–102. https://doi.org/10.2527/jas.2007-0767.
9. Berglund B, Steinbock L, Elvander M. Causes of stillbirth and time of death in Swedish Holstein calves examined post mortem. Acta Vet Scand. 2003;44(3–4):111–20. https://doi.org/10.1186/1751-0147-44-111.
10. Chassagne M, Barnouin J, Chacornac JP. Risk factors for stillbirth in Holstein heifers under field conditions in France: a prospective survey. Theriogenology. 1999;51(8):1477–88. https://doi.org/10.1016/s0093-691x(99)00091-6.
11. Cook BR, Tess MW, Kress DD. Effects of selection strategies using heifer pelvic area and sire birth weight expected progeny difference on dystocia in first-calf heifers. J Anim Sci. 1993;71(3):602–7. https://doi.org/10.2527/1993.713602x.
12. De Amicis I, Veronesi MC, Robbe D, Gloria A, Carluccio A. Prevalence, causes, resolution and consequences of bovine dystocia in Italy. Theriogenology. 2018;107:104–8. https://doi.org/10.1016/j.theriogenology.2017.11.001.
13. Dekkers JC. Optimal breeding strategies for calving ease. J Dairy Sci. 1994;77(11):3441–53. https://doi.org/10.3168/jds.S0022-0302(94)77287-8.
14. Drew B. Factors affecting calving rates and dystokia in Friesian dairy heifers, the results of a large scale field trial. Ir Grassl Anim Prod J. 1986;20:98–104.
15. Dufty JH. The influence of various degrees of confinement and supervision on the incidence of dystokia and stillbirths in Hereford heifers. N Z Vet J. 1981;29(4):44–8. https://doi.org/10.1080/00480169.1981.34796.
16. Egan J, Leonard N, Griffin J, Hanlon A, Poole D. A survey of some factors relevant to animal welfare on 249 dairy farms in the Republic of Ireland Part 1: Data on housing, calving and calf husbandry. Ir Vet J. 2001;54:388–92.
17. Eriksson S, Näisholm A, Johansson K, Philipsson J. Genetic parameters for calving difficulty, stillbirth, and birth weight for Hereford and Charolais at first and later parities. J Anim Sci. 2004;82(2):375–83. https://doi.org/10.2527/2004.822375x.
18. Fingland RB, Rings DM, Vestweber JG. The etiology and surgical management of tracheal collapse in calves. Vet Surg. 1990;19(5):371–9. https://doi.org/10.1111/j.1532-950x.1990.tb01211.x.
19. Gaines JD, Peschel D, Kauffman RG, et al. Pelvic growth, calf birth weight and dystocia in Holstein x Hereford heifers. Theriogenology. 1993;40(1):33–41. https://doi.org/10.1016/0093-691x(93)90339-7.
20. Garrick DJ. The nature, scope and impact of genomic prediction in beef cattle in the United States. Genet Sel Evol. 2011;43(1):17. https://doi.org/10.1186/1297-9686-43-17.

4

21. Garry FB. Animal well-being in the US dairy industry. In: Benson GJ, Rollin BE, editors. The well-being of farm animals – challenges and solutions. Iowa, USA: Blackwell Publishing Professional; 2004. p. 207–40. https://doi.org/10.1002/9780470344859.ch11.

22. Gustafson GM. Effects of daily exercise on the health of tied dairy cows. Prev Vet Med. 1993;17(3–4):209–23. https://doi.org/10.1016/0167-5877(93)90030-W.

23. Hansen M, Misztal I, Lund MS, Pedersen J, Christensen LG. Undesired phenotypic and genetic trend for stillbirth in Danish Holsteins. J Dairy Sci. 2004;87(5):1477–86. https://doi.org/10.3168/jds.S0022-0302(04)73299-3.

24. Heinrichs AJ, Heinrichs BS, Harel O, Rogers GW, Place NT. A prospective study of calf factors affecting age, body size, and body condition score at first calving of Holstein dairy heifers. J Dairy Sci. 2005;88(8):2828–35. https://doi.org/10.3168/jds.S0022-0302(05)72963-5.

25. Heuwieser W, Hartig U, Offeney F, Grunert E. Significance of glucocorticoids as a parameter of stress in cattle in the periparturient period. J Vet Med Ser A. 1987;34(3):178–87.

26. Hiew MW, Megahed AA, Townsend JR, Singleton WL, Constable PD. Clinical utility of calf front hoof circumference and maternal intrapelvic area in predicting dystocia in 103 late gestation Holstein-Friesian heifers and cows. Theriogenology. 2016;85(3):384–95. https://doi.org/10.1016/j.theriogenology.2015.08.017.

27. Hohnholz T, Volkmann N, Gillandt K, Waßmuth R, Kemper N. Risk factors for dystocia and perinatal mortality in extensively kept Angus suckler cows in Germany. Agriculture. 2019;9(4):85. https://doi.org/10.3390/agriculture9040085.

28. Hydbring E, Madej A, MacDonald E, Drugge-Boholm G, Berglund B, Olsson K. Hormonal changes during parturition in heifers and goats are related to the phases and severity of labour. J Endocrinol. 1999;160(1):75–85. https://doi.org/10.1677/joe.0.1600075.

29. Jackson PGG, Cockcroft PD. Clinical examination of the nervous system. In: PGG J, Cockcroft PD, editors. Clinical examination of farm animals. Oxford, UK: Blackwell; 2002. p. 198–216. https://doi.org/10.1002/9780470752425.ch14.

30. Jacobsen H, Schmidt M, Hom P, Sangild PT, Greve T, Callesen H. Ease of calving, blood chemistry, insulin and bovine growth hormone of newborn calves derived from embryos produced in vitro in culture systems with serum and co-culture or with PVA. Theriogenology. 2000;54(1):147–58. https://doi.org/10.1016/s0093-691x(00)00333-2.

31. Johanson JM, Berger P. Birth weight as a predictor of calving ease and perinatal mortality in Holstein cattle. J Dairy Sci. 2003;86(11):3745–55. https://doi.org/10.3168/jds.S0022-0302(03)73981-2.

32. Kamal MM, Van Eetvelde M, Depreester E, Hostens M, Vandaele L, Opsomer G. Age at calving in heifers and level of milk production during gestation in cows are associated with the birth size of Holstein calves. J Dairy Sci. 2014;97(9):5448–58. https://doi.org/10.3168/jds.2014-7898.

33. Kornmatitsuk B, Veronesi MC, Madej A, Dahl E, Ropstad E, Beckers JF, Forsberg M, Gustafsson H, Kindahl H. Hormonal measurements in late pregnancy and parturition in dairy cows–possible tools to monitor foetal well being. Anim Reprod Sci. 2002;72(3–4):153–64. https://doi.org/10.1016/s0378-4320(02)00092-1.

34. Kuehn L, Thallman M. 2019 across-breed EPD table and improvements. 2020. Retrieved from https://beefimprovement.org/wp-content/uploads/2019/12/19_ABEPDpressreleaseandfactsheet.pdf. Accessed on 20 Mar 2020.

35. Lombard JE, Garry FB, Tomlinson SM, Garber LP. Impacts of dystocia on health and survival of dairy calves. J Dairy Sci. 2007;90(4):1751–60. https://doi.org/10.3168/jds.2006-295.

36. López Helguera I, Behrouzi A, Kastelic JP, Colazo MG. Risk factors associated with dystocia in a tie stall dairy herd. Can J Anim Sci. 2016;96(2):135–42. https://doi.org/10.1139/cjas-2015-0104.

37. Mansell PD, Cameron AR, Taylor DP, Malmo J. Induction of parturition in dairy cattle and its effects on health and subsequent lactation and reproductive performance. Aust Vet J. 2006;84(9):312–6. https://doi.org/10.1111/j.1751-0813.2006.00031.x.

38. Martinez ML, Freeman AE, Berger PJ. Factors affecting calf livability for Holsteins. J Dairy Sci. 1983;66(11):2400–7. https://doi.org/10.3168/jds.S0022-0302(83)82098-0.

39. Mee JF, Berry DP, Cromie AR. Risk factors for calving assistance and dystocia in pasture-based Holstein-Friesian heifers and cows in Ireland. Vet J. 2011;187(2):189–94. https://doi.org/10.1016/j.tvjl.2009.11.018.

40. Mee JF, Sánchez-Miguel C, Doherty M. Influence of modifiable risk factors on the incidence of stillbirth/perinatal mortality in dairy cattle. Vet J. 2014;199(1):19–23. https://doi.org/10.1016/j.tvjl.2013.08.004.

41. Mee JF. Managing the dairy cow at calving time. Vet Clin North Am Food Anim Pract. 2004;20(3):521–46. https://doi.org/10.1016/j.cvfa.2004.06.001.

42. Mee JF. Newborn dairy calf management. Vet Clin North Am Food Anim Pract. 2008a;24(1):1–17. https://doi.org/10.1016/j.cvfa.2007.10.002.

43. Mee JF. Prevalence and risk factors for dystocia in dairy cattle: a review. Vet J. 2008b;176(1):93–101. https://doi.org/10.1016/j.tvjl.2007.12.032.

44. Meyer CL, Berger PJ, Koehler KJ. Interactions among factors affecting stillbirths in holstein cattle in the United States. J Dairy Sci. 2000;83(11):2657–63. https://doi.org/10.3168/jds.S0022-0302(00)75159-9.

45. Meyer CL, Berger PJ, Thompson JR, Sattler CG. Genetic evaluation of Holstein sires and maternal grandsires in the United States for perinatal survival. J Dairy Sci. 2001;84(5):1246–54. https://doi.org/10.3168/jds.S0022-0302(01)74586-9.

46. Nakao T, Grunert E. Effects of dystocia on postpartum adrenocortical function in dairy cows. J Dairy Sci. 1990;73(10):2801–6. https://doi.org/10.3168/jds.S0022-0302(90)78967-9.

47. Parkinson TJ, Vermunt JJ, Noakes DE. Approach to an obstetrical case. In: Noakes DE, Parkinson TJ, GCW E, editors. Veterinary reproduction and obstetrics. 10th ed. Amsterdam, NL: Elsevier; 2019. p. 203–13. https://doi.org/10.1016/B978-0-7020-7233-8.00011-2.

48. Phocas F, Laloë D. Genetic parameters for birth and weaning traits in French specialized beef cattle breeds. Livest Prod Sci. 2004;89:121–8. https://doi.org/10.1016/j.livprodsci.2004.02.007.

49. Rice LE. Dystocia-related risk factors. Vet Clin North Am Food Anim Pract. 1994;10(1):53–68. https://doi.org/10.1016/s0749-0720(15)30589-2.

50. Rings DM. Tracheal collapse. Vet Clin North Am Food Anim Pract. 1995;11(1):171–5. https://doi.org/10.1016/s0749-0720(15)30515-6.

51. Rodrigues TC, Nunes BMR, Carolino N, Carreira MC, Stilwell G. Perinatal mortality in Portuguese dairy herds. Rev Port Ciênc Vet. 2014;109(589–590):26–32.

52. Rogers PA. Perinatal calf deaths. Vet Rec. 1996;138(12):287.

53. Schuijt G. Iatrogenic fractures of ribs and vertebrae during delivery in perinatally dying calves: 235 cases (1978–1988). J Am Vet Med Assoc. 1990;197(9):1196–202.

54. Steinbock L, Näsholm A, Berglund B, Johansson K, Philipsson J. Genetic effects on stillbirth and calving difficulty in Swedish Holsteins at first and second calving. J Dairy Sci. 2003;86(6):2228–35. https://doi.org/10.3168/jds.S0022-0302(03)73813-2.

55. Studer E. A veterinary perspective of on-farm evaluation of nutrition and reproduction. J Dairy Sci. 1998;81(3):872–6. https://doi.org/10.3168/jds.S0022-0302(98)75645-0.

56. Tenhagen BA, Helmbold A, Heuwieser W. Effect of various degrees of dystocia in dairy cattle on calf viability, milk production, fertility and culling. J Vet Med A Physiol Pathol Clin Med. 2007;54(2):98–102. https://doi.org/10.1111/j.1439-0442.2007.00850.x.

57. Tyler HD. Calf development and birth. In: Raising dairy replacements. midwest plan service; north central regional extension publication NCR-205, 1–9, 2003, Iowa state university, Iowa, US.

58. Waldner C. Cow attributes, herd management and environmental factors associated with the risk of calf death at or within one hour of birth and the risk of dystocia in cow-calf herds in Western Canada. Livest Sci. 2014;63:126–39. https://doi.org/10.1016/j.livsci.2014.01.032.

59. Youngquist RS. Parturition and dystocia. In: Youngquist RS, editor. Current therapy in large animal theriogenology. 1st ed. Philadelphia: Saunder Company; 1998. p. 309–23.

60. Zaborski D, Grzesiak W, Szatkowska I, Dybus A, Muszynska M, Jedrzejczak M. Factors affecting dystocia in cattle. Reprod Domest Anim. 2009;44:540–51. https://doi.org/10.1111/j.1439-0531.2008.01123.x.

Dystocia Guidelines

Sprott LR. Recognizing and handling calving problems. USA: Texas A&M AgrilLife System; 2020. From: https://agrilifeextension.tamu.edu/library/ranching/recognizing-and-handling-calving-problems/. Accessed on 18 Sept 2020.

Understanding EBVs and Selection Indexes. Angus Australia, Armidale, AUSTRALIA; 2020. From: https://www.angusaustralia.com.au/education/breeding-and-genetics/understanding-ebvs-and-indexes/. Accessed on 17 Sept 2020.

Suggested Reading

Barrier AC, Haskell MJ, Birch S, Bagnall A, Bell DJ, Dickinson J, Macrae AI, Dwyer CM. The impact of dystocia on dairy calf health, welfare, performance and survival. Vet J. 2013;195(1):86–90. https://doi.org/10.1016/j.tvjl.2012.07.031.

Barrier AC, Mason C, Dwyer CM, Haskell MJ, Macrae AI. Stillbirth in dairy calves is influenced independently by dystocia and body shape. Vet J. 2013;197(2):220–3. https://doi.org/10.1016/j.tvjl.2012.12.019.

Berry DP, Amer PR, Evans RD, Byrne T, Cromie AR, Hely F. A breeding index to rank beef bulls for use on dairy females to maximize profit. J Dairy Sci. 2019;102(11):10056–72. https://doi.org/10.3168/jds.2019-16912.

Assisted Vaginal Delivery and Newborn Calf Care

Contents

The original version of this chapter was revised. The correction to this chapter can be found at https://doi.org/10.1007/978-3-030-68168-5_12

Electronic Supplementary Material The online version of this chapter (https://doi.org/10.1007/978-3-030-68168-5_5) contains supplementary material, which is available to authorized users. The videos can be accessed by scanning the related images with the SN More Media App.

© Springer Nature Switzerland AG 2021, corrected publication 2022
J. Simões, G. Stilwell, *Calving Management and Newborn Calf Care*,
https://doi.org/10.1007/978-3-030-68168-5_5

5

🔁 Learning Objectives

- To elaborate calving protocols and to recognize reference landmarks and guidelines for an obstetrical intervention in farms.
- To identify all procedures for a full obstetrical examination.
- To manage foetus extraction by vaginal delivery.
- To assess viability and vitality of the calf and know resuscitation techniques.
- How to recognize and guarantee post-calving health and welfare of the dam.
- To describe effective measures for the resuscitation and care of the newborn.

5.1 Introduction

Obstetrical assistance can be defined as any intervention during labour to help the dam with or without dystocia, to deliver a calf. In the previous two chapters, normal calving process and abnormal occurrences resulting in dystocia were described. Usually, assistance is given when Stage II does not initiate or is prolonged, or when the dam shows an abnormal calving behaviour or signs of extreme discomfort. As a rule, humans should not interfere in the course of a normal calving. However, it is appropriate for farms to establish intervention protocols that include surveillance and monitorization procedures from Stage I of labour.

Ideally, obstetrical assistance should mimic natural calving to increase the possibility of obtaining healthy calf and dam. Even if periodical observation of the dam during the preparatory period and/or Stage I is done, a previous obstetrical examination to exclude dystocia risks and to determine if obstetrical intervention is needed is recommended. Some decades ago, the so-called Utrecht method was developed as a good practice for calf delivery. The guidelines of this method are as follows: (1) assess proper dilatation of the soft birth canal; (2) assure correct foetal disposition; (3) lie down the cow in right lateral recumbency; (4) and synchronize obstetrical traction with the dam abdominal contractions. The goal is to minimize trauma and unnecessary stress to the dam and foetus.

This chapter describes the main diagnosis procedures in the obstetrical approach as well as good practices used for a successful vaginal delivery. Specific obstetrical manoeuvres to solve dystocia of foetal origin are described in the next chapter.

5.2 Calving Settings

Usually, the periparturient cow looks for a secluded place where it goes through the different stages of calving without being disturbed by other animals. Here it will express the normal restlessness and discomfort behaviour, caused by the pain involved in the dilatation and expulsion phases. Whenever possible, an isolated, clean and comfortable site (i.e. maternity) should be provided as soon as the animal shows the behaviour of trying to get away from the herd, that usually occurs about 36 h before calving (a few hour more in heifers). Close and knowledgeable monitoring should be ensured so as to avoid moving a cow during labour. In fact, Proudfoot et al. [20] observed that cows which were moved to individual maternity in late Stage I would prolong Stage II of labour (total duration of 91 min.), compared to cows which were moved before labour started or in early Stage I.

A sufficient number of individual maternity pens or enough space in common maternities need to be provided according to the size and seasonal calving distribution in the herd [24] (▶ Box 5.1).

Box 5.1   Maternity Pens
- Area: at least 16 m² per cow.
- Floor: non-slippery floor; if necessary, cover the floor with straw (15–25 cm deep) and keep it dry and clean.
- Environment: well-ventilated; adequate lighting (◘ Fig. 5.1).
- Chute or headgate for animal restraint.
- A clean water source to wash the perineal region of the dam.
- High roof to be able to rise by the hips a downer cow.

◘ Fig. 5.1 A spacious, comfortable, clean, well-ventilated and well-illuminated, group maternity. (Original from George Stilwell)

5.3 The Appropriate Time for Obstetrical Intervention

Since the duration of Stage I varies significantly between individuals, up to 12 h, supervision every 1–3 h is appropriate to detect when abdominal contractions start. The transition between the preparatory period to calving and the onset of Stage I is not always evident. Prediction is mainly achieved by the frequency of restlessness behaviours and discomfort intensity shown by the dam, e.g. tail constantly raised, rising and lying down, costal decubitus and overall hyperactivity. In some cases, the farmer or stockperson should perform a vaginal palpation to use cervical dilatation degree as an additional predictor to access the onset of Stage II. Dams already presenting evident cervical dilatation should be monitored every hour [14].

Obstetrical assistance should be provided whenever the calf delivery is not completed after two hours of the emergence of the first (chorioallantoic membrane) or second (amniotic membrane) water bag outside the vulva or if the signs of calving progress are not evident in a 30-min. interval during

Stage II. Rupture of the chorioallantoic membrane occurs inside the vagina in approximately half of the parturitions, and only the second water bag will be observed intact outside of the vulva. Usually, the amniotic membrane appears outside of vulva about one hour after the rupture of the chorioallantoic membrane and indicates that foetal delivery is eminent.

> **Tip**
>
> Dystocia from foetal origin can cause a delay between the rupture of the allanto-chorion and the amnion or foetus hooves appearance at the vulva.

Stockpersons should be trained to manage calvings according to specific and predetermined protocols. As reported by Schuenemann et al. [23], all staff in a farm should be able to identify the imminent signs of birth, to recognize signs of normal calving progression and to determine the landmarks and reference times to initiate calving assistance. These stockpersons also need to acquire skills and competences to assist calvings and solve easy to moderate dystocia and to know when to seek help (i.e. veterinarian). Thus, they should be able to assess normal and abnormal foetal dispositions, as well as the degree of birth canal dilation. Finally, they should be at ease with the procedures eventually needed to ensure the correct vaginal delivery of the calf. The main topics regarding stockperson training are presented in ▶ Box 5.2.

Box 5.2   Topics and Reference Landmarks for Dairy Personnel Training to Manage Calvings

Description of the birth canal, foetopelvic disproportion and strategies to correct abnormal presentations, postures or positions.

Behaviour of the cow and of the first-calf heifer before and during labour. Imminent signs of birth, assessment of calving progress and time spent in labour (i.e. normal versus dystocic births).

Reference landmarks for normal births: (1) mean time from the appearance of the amniotic sac (AS; 70 min.) or feet (65 min.) outside the vulva to birth; (2) mean time (≤2 h) that a cow or first-calf heifer (3–4 h) spend in labour; (3) signs of calving progress (evident every 15–20 min.); and (4) frequency of observation (every 1–2 h).

Reference signs to determine the appropriate time for obstetric intervention during difficult births (i.e. when and how it is appropriate to assist the cow or first-calf heifer). Guidelines provided: (1) for first-calf heifers when the feet-nose of the calf are visible outside the vulva, help the final expulsion efforts; (2) for backward presentations, start traction as soon as hooves are accessible; (3) when malposture is evident (e.g. only one foot of the calf is visible outside the vulva) after the appearance of AS or for uterine torsion (where nothing is visible outside the vulva), immediate assistance is required; and (4) immediately after delivery, examine the cow or heifer to determine the presence of a second calf in case of multiple births (twins or triplets).

Hygiene practices for assisted births (i.e. sanitation of perineum region, disinfection of the obstetric chains, use of lubricant and disposable gloves).

Best communication practices within the farm team (i.e. when to call for help, between work shifts) and with the herd veterinarian.

Newborn care practices such as navel disinfection, assessment of calf vigour and feeding of colostrum.

Record-keeping (i.e. score of calving difficulty, sex of calf, identification of the dam and calf, birth date, start and end time of labour and born alive or dead).

Review of calving protocols (e.g. what to look for or monitor before and during calving and why is it important?).

Strategies to correct malpostures (e.g. appearance of only one foot of the calf outside the vulva) and uterine torsion were discussed.
Placement of obstetric chains and considerations for human or mechanical extraction (force).

Adapted from Schuenemann et al. [23] with permission from Elsevier

---

5.4 Obstetrical Examination

The cow should be moved into a clean and well-lighted area with a non-slippery floor, for observation. This can be the maternity, if it exists and has suitable handling facilities. Because docile animals, such as dairy cows, are usually familiarized with human contact, simple head restraint may be sufficient to allow for an obstetrical examination while ensuring animal and human safety. Additional physical and chemical methods can be necessary to restrain more aggressive cows. The use of a halter with a long lead rope facilitates dam's handling during obstetrical manipulations. This simple physical restraint can be posteriorly complemented using the Burley method or other techniques to lie down the dam for foetal extraction, if deemed necessary.

Physical restraint should be first tried so to as to avoid the use of tranquillizers. This way the cow is more likely to remain standing during the obstetrical examination, mitigating the risk of unintended recumbency. Also, several tranquillizers have adverse effects on the foetus. Xylazine can reduce the flow and availability of oxygenated blood in the uterus potentially causing foetal physiologic distress [9]. The combined use of different pharmacological, such as alpha-2-adrenergic agonists (xylazine and detomidine), or other sedatives (acepromazine), analgesic (butorphanol) and anaesthetic (ketamine) substances seems to be the better choice to limit these adverse effects. One well-established protocol is 0.02 mg/kg of xylazine + 0.04 mg/kg of ketamine + 0.01 mg/kg of butorphanol. With these doses, the dam usually remains standing. The intensity of sedative effects depends on the type, quantity and associations between pharmacological substances (tranquillizers, anaesthetics and analgesics). All routes of administration can be used. However, subcutaneous route prolongs the length of the sedative effect and is the preferable route, unless a quick onset of sedation (approximately 1 min) is necessary in which case intravenous administration should be used. The α2-adrenergic agonists are route- and dose-dependent, causing several degrees of sedation, muscle relaxation and analgesia. Ketamine is a dissociative anaesthetic, also route- and dose-dependent, but it induces potent analgesic effects at low (subanaesthetic) doses, such as previously reported. The normal doses to promote recumbency and a short-term (up to 15 min.) general anaesthesia are 0.025–0.05 mg/kg of xylazine + 0.3–0.5 mg/kg of ketamine + 0.05–0.1 mg/kg of butorphanol, given intravenously [1].

Beforehand, adequate equipment (▶ Box 5.3; also see ▶ Chap. 7) should be prepared to ensure a clean and safe vaginal and uterine examination, as well as a transrectal examination when necessary.

Box 5.3 Basic Equipment and Material for Obstetrical Examination and Intervention
- One or two clean and disinfected buckets.
- Hot water.
- Neutral soap.
- Disinfectant (e.g. iodine or chlorhexidine-based solution).
- Obstetrical lubricant (e.g. carboxymethylcellulose gel, liquid vaseline or mineral oil as the alternative).
- Obstetrical long sleeves.
- Transrectal sleeves.
- Two obstetrical chains or ropes.
- Obstetrical hooks for death foetus (Krey-Schöttler double hook and two Ostertag's blunt eye hook).
- Calf puller.
- Local anaesthetic for epidural.

Tip

If there is danger of the cow slipping during parturition, the hindlimbs can be hoppled above the fetlock, to prevent trauma when falling. This procedure is particularly important in cows with large oedematous udder or in heifers.

After a general external examination, the tail is tied (◘ Fig. 5.2), and the perineum and vulvar area should be cleaned, washed and disinfected. Furthermore, the obstetrician should also wash both hands and arms before putting on obstetrical sleeves. These procedures are indispensable to reduce contamination of the birth canal and uterus. Obstetrical long sleeves can be waived, but this malpractice can lead to human infection by zoonotic bacteria or fungi (e.g. brucellosis or Q fever) transmitted by an infected placenta, foetal fluids or foetus. If gloves are to be discarded, the hands and arms of the obstetrician should be carefully washed and disinfected before and after the obstetrical intervention.

5

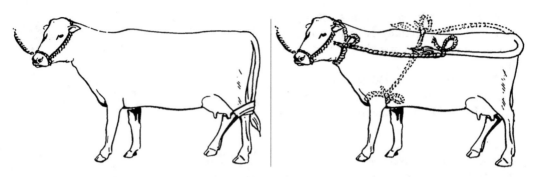

◘ Fig. 5.2 Two different techniques to restraint the tail preventing vulvar bacterial contamination during vaginal manipulation. (Modified from Ames [2] with permission from John Wiley and Sons)

The abundant use of obstetrical lubricant is always necessary to facilitate the introduction of hands and arms into the birth channel, preventing (1) potential inflammation and lacerations of the vaginal wall and (2) the introduction of environmental bacteria into the vagina and uterus.

> **Tip**
>
> Especially in heifers, pouring hot water on the vulva, perineum, base of the tail and even over the rump (◘ Fig. 5.3) will cause vasodilatation and increase significantly soft tissues stretching capability, preventing tears and injuries from occurring to the vulva and vagina.

> ❯ **Important**
>
> Ensure a clean environment: tie or push away the tail of the cow from the vulvar area; use warm water and soap to clean the vulva, anus and perineum area; wash hands and arms and put on the obstetrical gloves; lubricate the gloves; and slowly insert wedge-shaped fingers into the vagina (◘ Fig. 5.4).

Firstly, the vulva should be inspected for its integrity and the presence or absence of placental annexes and foetal parts (▸ Box 5.4). This step can be done mainly by inspection and palpation. Further obstetrical intervention depends on whether the foetus is present or not in the birth canal (▸ Box 5.5). The presence of placental annexes or foetal parts

◘ Fig. 5.3 Pouring hot water over the perineum area during the Stage I of parturition will increase tissue elasticity and prevent tears during the expulsion stage. (Original from George Stilwell)

is the immediate evidence that the cow is under Stage II of labour. In that case, methodical vaginal palpation should take place to evaluate the integrity of the water sacs as well as the degree of foetal vitality or suspicion/ evidence of foetal death (▸ Box 5.6). The colour, odour and contents of the amniotic fluid, which can be quickly evaluated when the amnion is ruptured, are also indicative of anomalies. The normal amniotic fluid has a

Fig. 5.4 Washing, disinfecting and the use of gloves and lubricant are ways of ensuring the protection of both cow and obstetrician. (Original from George Stilwell)

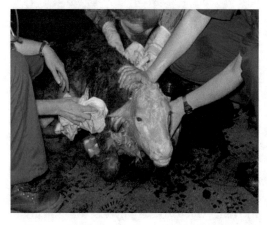

Fig. 5.5 Meconium staining in a crossbreed calf (Holstein × Montbéliarde) after a difficult parturition. (Original from George Stilwell)

milky white colour and a slightly sweet smell. Its colour can change to meconium staining (mustard or brown; ■ Fig. 5.5) in the presence of free foetal meconium or red in case of placental haemorrhage or late foetal death. A fetid smell can be evident even in the early stage of foetal death.

> **Box 5.4   Vulvar Inspection**
> ▬ Is the vulva fully dilated?
> ▬ Are there any evidence of vulvar injuries?
> ▬ Is the odour sweet (normal) or foul (infection)?
> ▬ Presence of first or second water sac? Is it the chorioallantois or amnion membrane?
>
> ▬ Is one or both already ruptured?
> ▬ Are the membranes still fresh or decayed?
> ▬ Are the foetus hooves observed through the vulva?
> ▬ Do these correspond to the front or the hind limbs?
> ▬ Is the nose/head present in the vagina?
> ▬ Is there evidence of foetal movements? Does the foetus react to interdigital pinching?

> **Box 5.5   Obstetrical Evaluation of the Birth Canal**
> Presence of the foetus (check for):
> ▬ Vulvar and vaginal integrity and relaxation.
> ▬ Foetal membrane integrity.
> ▬ Foetal viability and vitality (foetal life and stress).
> ▬ Foetal presentation, position and posture.
> ▬ The relative proportion between birth canal diameter and foetal size (foetopelvic disproportion).
>
> Absence of the foetus (check for):
> ▬ Vulvar and vaginal integrity and relaxation.
> ▬ Foetal membranes integrity, if present.
> ▬ Cervix identification difficult or impossible (e.g. uterine torsion; vaginal rupture).
> ▬ Degree of cervical dilatation (e.g. cervix stenosis).

5

Box 5.6   Checking for Viability and Vitality
- Withdrawal reflex – Pinching interdigital area of the hoof causes a withdrawal movement of the limb.
- Suckle reflex – touching the mouth (mouth closes or tongue moves).
- Palpebral reflex – touching the palpebra (eye closes or head moves).
- Rectal sphincter reflex – introducing a finger in the rectum (contracture of the anal sphincter muscle).
- Heartbeat or umbilical pulse: slight compression on the chest or touching the umbilical cord (detection of movements/pulsation).

    Note – interdigital pinching of the hind limbs will sometimes result in apparently no reaction or withdrawal reflex, even in a live foetus. This is due to the compression of the foetus hindquarters in the pelvic canal that does not allow for much movement.

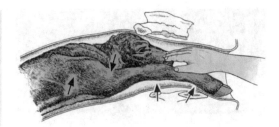
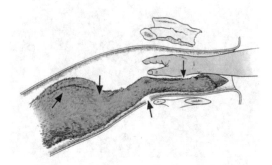

☐ **Fig. 5.6** Identification of forelimbs (top) or hindlimbs (button) by systematic palpation of joints. (Modified from Jackson [10] with permission from Elsevier)

Tip

Crossing limbs in anterior foetal presentation is a sign of foetopelvic disproportion because the maternal pelvis puts pressure on the shoulders deviating the limbs towards the midline.

After vulvar inspection, the foetal presentation, position and posture need to be evaluated for any significant alteration (see ▶ Chaps. 3 and 4). Identification of forelimbs or hindlimbs in the vagina is achieved by systematic palpation and assessment of flexion direction of the different joints (☐ Fig. 5.6). This action identifies anterior or posterior foetal presentation and can be complemented by the identification of the presence or absence of head or tail, respectively. Also, the relative position of the hooves is informative – when the foetus is in anterior presentation and dorsal position, the hooves' soles are directed ventrally. On the other hand, when the foetus is in posterior presentation and dorsal position, the hooves' soles are facing upwards. However, in case of ventral position, the hooves' sole direction is inverted, so this information should always be complemented with a thorough foetal position evaluation.

When, at the end of Stage I, the placental membranes or any part of the foetus body are not visible at the vulva, special attention must be given to vulvar and vaginal integrity and dilatation, the possible presence of acquired stenosis (e.g. fibrosis) or of a mass or uterine torsion, which cause physical or functional occlusion or obstruction.

Further examination should seek to identify the cervix and to assess its dilatation degree. An intraluminal diameter up to 3–4 cm at the external Os can be observed in a non-dilated cervix. In these situations, one or two fingers may be inserted in the posterior cervical lumen, but usually it is difficult or not possible to pass the cervical canal reaching the intrauterine milieu. Cervical non-dilation can also be easily diagnosed by transrectal palpation. In less severe cases, a small cervix dilatation from 3 to 10 cm in diameter will be present but is still considered an insufficient cervical dilatation – the hand may reach the intrauterine cavity and touch foetal parts, or placental membranes, but it is evident that the foetus will not pass through.

The management of an inadequate or a non-dilated cervix depends on the stage and duration of parturition, on foetal vitality and on integrity of placental membranes. In most cases of insufficient dilatations, placental membranes will remain intact and the foetus will still be alive. So, if Stages I and II are not delayed more than 6 and 2 h, respectively, cervical dilatation progress should be re-evaluated every 30 min., awaiting for dilatation by hormonal influence as well as mechanical effect directly promoted by the foetus on the birth canal. Some drugs, e.g. synthetic prostaglandin E_1/E_2 (misoprostol/dinoprostone), were tested in dams to improve cervical relaxation in those cases, and they seem to have a positive effect [8, 12]. The use of oxytocin induces muscular myometrial contractions, pushing the foetus into the cervical internal Os, and as a result, it can also promote cervical dilatation. Although it is used to induce the parturition in mares and rabbits, in cattle, oxytocin should be used with care – it causes strong myometrial contractions and increases the risk of uterine rupture and foetus death. Because of that, the use of oxytocin should be avoided until an appropriate cervical dilatation is obtained. When the amniotic membrane is ruptured, mechanical stimuli along the cervix can be tried for up to 15 min. using the hands. The drug denaverine (400 mg per animal) may be injected or even rubbed along the cervical ring, increasing dilatation in some cases. However, if the foetus is alive and showing signs of stress, a C-section will be more appropriate.

The size of the foetus needs to be compared with the size of the birth canal to ensure its passage [6]. The relationship between the coronary band of the foetal hoof and maternal pelvic measures can give an idea as to the ease in calving (▶ Box 5.7). Some recent studies tried to increase the prediction of dystocia. This prediction was based on the significant but low ($r < 0.5$) correlations between the calf birth weight and metacarpus/metatarsus bone thickness [33] or hoof circumference [7].

> **Box 5.7 Prediction of Calving Difficulty Score According to Ko and Ruble [11]**
> ▬ Predicted calving difficulty score = [(Hoof circumference – Pelvic Height + 3.5) + (Hoof circumference – Pelvic Width + 3.5)]/2.
>   Where the score 0.00–4.00 means eutocia (unassisted calving); 4.01–5.50 predicts manual assistance required; 5.51–6.50 suggests mechanical assistance (call puller) required; and ≥6.51 = caesarean section required.

To accurately predict dystocia, Hiew et al. [7] measured the hoof circumference (cm) at the coronary band level of the forelimbs and the intrapelvic area (cm^2). The optimal cut-off point to predict calving difficulty score was 0.07 cm/cm^2 with 50% of sensitivity and 93% of specificity; i.e. half of the dystocia cannot be truly predicted (false negatives) if we wish to keep a high specificity. One of the best practical predictors of foetopelvic disproportion, at the time of intervention, is the evaluation the free space between the pelvic inlet and the foetus at the shoulder or hip level, for anterior and posterior presentation, respectively (◘ Fig. 5.7). Usually, foetus can be delivered if the obstetrician's hand can easily pass through the birth canal, between foetus and the cows' pelvic bones, into the uterus.

5.5 Assisted Foetal Traction

Once the longitudinal alignment of the foetus into the anterior part of the birth canal is ensured, and foetopelvic disproportion is not present or suspected, the obstetrician can prepare the traction procedures. In anterior pre-

5

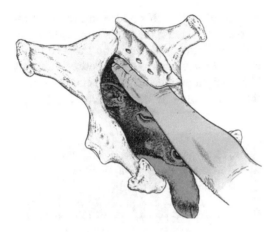

□ Fig. 5.7 Comparison between the foetal and pelvic size at pelvic inlet. (Modified from Jackson [10] with permission from Elsevier)

sentation and dorsal position, the extension of both forelimbs and head should be confirmed before traction commences.

In most cases, traction may be done with the cow standing. However, if the cow is lying or there is a risk of falling down, traction should be done with the cow in right lateral decubitus.

To force a cow to lie down, the Burley method or Reuff's method is the recommended technique (□ Fig. 5.8). The trachea and mammary veins should not be under prolonged pressure during the procedure, and both rope ends can be easily used to tie the forelimbs of the dam in a flexed position. The Reuff's double half-hitch method (rope squeeze technique) is also an adequate alternative to get the dam to lie down. In both methods, one person fixes and turns the head of the dam to the opposite side of the desired lateral decubitus, while the rope ends are pulled.

Tips

By principle and in a non-complicated delivery, assistance to vaginal delivery should only start when both pasterns are extended approximately 10 cm beyond the vulva.

5.5.1 Stretching the Vulva

Adequate degree of soft tissue dilatation makes foetal passage easier, preventing vulvar injuries. The relaxation of vulvovaginal sphincter acquires greater importance in heifers than in multiparous cows. Vulvar dilatation may be insufficient because of hormonal unbalance or because of acquired morphologic abnormalities such as fibrosis, caused by vulvar lacerations during previous calving, for example.

If necessary, the vulva should be mechanically extended by the obstetrician, usually for just a few minutes, but up to 20 min. can be required in heifers. The vulva should be stretched in an oblique direction towards ventral and dorsal lip commissures, using the fingers. Another viable option is to use the elbows when the obstetrician inserts both forearms into the vagina.

5.5.2 Applying Obstetrical Chains or Ropes

When the limbs are outside the vulva, an obstetrical chain or rope must be placed in each member, if not previously applied to correct mispositions. A double half-hitch method should be used. The chain should be first placed above the fetlock joint, on the narrowest part of the metacarpus/metatarsus, followed by half-hitch on the pastern joint, between the dewclaws and the hoof (□ Fig. 5.9). This procedure distributes the traction forces applied in each segment of the member and fixes the chains when they are not tensed (the chain should stay straight between the two loops), avoiding foot injuries or leg fractures. Furthermore, the hoop link should be placed on the dorsal side of each member. The limbs should be kept extended in order to prevent soft tissue lacerations of the birth channel or compression of pelvic nerves, such as the obturator or sciatic plexus.

Burley method

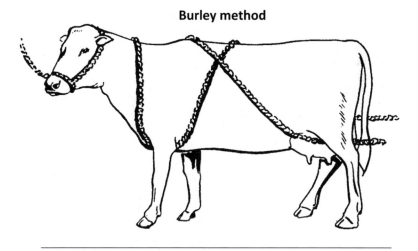

Reuff's method

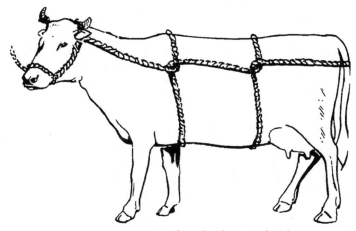

Rope anchored to horns or head

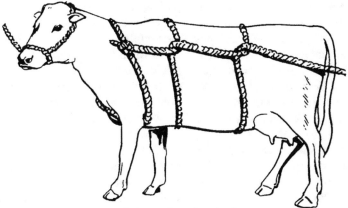

Rope anchored under forearm

⬛ **Fig. 5.8** Burley method (flying W technique) for casting the dam. (Modified from Ames [2] with permission from John Wiley and Sons)

◘ Fig. 5.9 Double half-hitch method. Adapted from Norman and Youngquist [17] with permission from Elsevier. (Original art by Mr. Don Conner)

❯ **Important**

If the obstetrical chains or ropes are placed only on the pasterns, they can become loose when not being pulled. If they are placed only above fetlock join, they can provoke fractures of metacarpal and/or metatarsal bones when too much force is applied.

5.5.3 **Pulling the Foetus**

■ **Anterior Presentation**

Initially, both limbs are simultaneously pulled, trying to accommodate the head into the pelvis. This procedure allows the obstetrician to assess if the head is not retained inside the uterus or if a foetopelvic disproportion is not present.

By vaginal palpation, the obstetrician evaluates whether the head is not retained at the bony pelvic inlet level, i.e. the nose follows the progressive movements into the birth canal, and the nostrils can be easily identified at the carpal level of both limbs. If the head stays retained in the uterus, a halter must be inserted between lip commissures and the backhead (occipital area). For dead foetuses, double orbital hooks attached to a rope can be quickly fixed into the eye sockets. At this point, a slight but progressive traction force should be applied to the head.

❯ **Important**

In anterior foetal presentation, the head should be resting on the carpus of both hindlimbs when entering the birth canal.

For dams remaining in standing position, a supplementary force needs to be applied in order to pull the foetus (against gravity), whereas in lying cows, approximately 30% less force is needed. So, recumbency of the cow on the right side permits the longitudinal axe of the foetus to stay aligned with the longitudinal axis of the birth canal, which is very helpful. This alignment of the pelvic angle facilitates the foetus' entrance and progress into and along the birth canal. Another significant advantage of the recumbent position is the possibility to bring the dam's hindlimbs forward, facilitating the obtention of the adequate alignment between the pelvic angle and the foetal axis. In this case, the dam must be placed on sternal decubitus. The force exerted on the foetus should be horizontally directed or slightly upwards regarding the dam's dorsum in order to ensure this alignment until the thorax arises outside the vulva.

Until the entire head enters the birth canal, application of traction forces should be progressive, alternating between forelimbs and synchronizing it with abdominal contractions. The bony pelvic inlet has a rigid conformation, and only minimal dilatation, mainly from the sacroiliac ligaments. So, the bony pelvic inlet represents the main area of the hard birth canal, where the foetus needs to enter, adapt and progress. On the other hand, there are two critical foetus areas which present higher perimeter: the shoulders and the hips.

The shoulders region represents the largest circumference. Special attention must be taken when the foetus pasterns fully arise outside the vulva. Usually, the shoulders will have passed through the bony pelvic inlet when both pasterns reach approximately 15 cm or more outside the vulva.

The greatest forces must be applied when the head and elbows, as well as the thorax, enter into the pelvis of the dam, progress and arise outside the vulva [3, 28] (◘ Fig. 5.10). A slight rotation of the foetus up to 45° is required to adjust the conformation of the foetal shoulders/thorax to the birth canal (◘ Fig. 5.11a). Alternate traction of forelimbs is the appropriate procedure to reduce foetal diameter at this level (◘ Fig. 5.12),

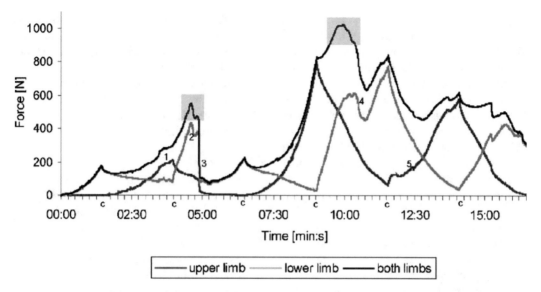

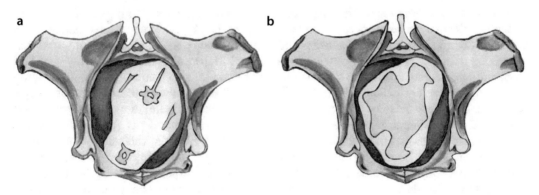

Fig. 5.10 Force-by-time diagram of the extraction of foetus applying alternate-forelimb traction. Legend: The grey areas represent the first and second peak forces applied to forelimbs. C = change of traction between limbs; entry of the first (1) and second (2) olecranon into the pelvis; the exit of the head from the pelvis (3); the exit of the left (4) and right (5) olecranon from the pelvis; a 46.7-kg female foetus, down in right side, was used in an in vitro study. (Adapted from Becker et al. [3] with permission from Elsevier)

Fig. 5.11 An approximately 45° rotation of the foetus should be made to adapt its thorax **a** and hips **b** into the oblique diameter of the pelvis. Adapted from Norman and Youngquist [17] with permission from Elsevier. (Original art by Ms. Carmen Reed)

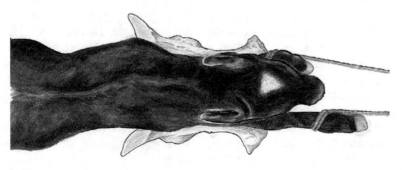

Fig. 5.12 Alternate traction in each forelimb to reduce the foetal diameter at elbows level. Modified from Norman and Youngquist [17] with permission from Elsevier. (Original art by Ms. Carmen Reed)

5

requiring less amount of force to pull the calf. For cows lying on the right side, the left limb should be the first member to pull, since it seems to be easier and less stressful for the foetus to initiate progression.

Once both shoulders enter the pelvis, and the head emerges outside the vulva, both limbs should to be pulled simultaneously. When the chest is also outside of the vulva, the extraction forces should be exerted downwards in the direction of the ventral side of the dam.

> ❯ **Important**
> The delivery should not be rushed. One must pull the foetus and pause synchronously with the dam's efforts and allow time for the natural rotation of the foetal pelvis. If nostrils and tongue are seen blueish (cyanosis), when still inside the vagina, the head should be pushed in again and traction reinitiated after a few seconds.

The foetal hips are the second limiting area, which can assume greater importance in beef than dairy cows. It is known that genetics and hereditary influence foetus conformation. The horizontal axis is larger than the vertical axis at the femurs' greater trochanter level. In consequence, an approximately 45-degree rotation of the foetal hips is necessary for its adaptation to the largest oblique diameter of the bony pelvic inlet (◻ Fig. 5.11b). This procedure prevents hip lock, in which the foetus hip is jammed inside the dam's pelvic bony canal causing important injury to the dam's nerves.

■ **Posterior Presentation**

In this foetal presentation, the traction of the hind limbs should also alternate so that it promotes the decrease in foetal pelvis' diameter as it is passing the anterior brim of the dam's pelvis. Also, the foetus should rotate slightly, allowing a better adaptation between both maternal and foetal pelvis. After the foetal pelvis enters the birth canal, both hind limbs should be pulled simultaneously. At this point, smooth but steady progress of the foetus through the birth canal is necessary to ensure its quick delivery, up to 2 min., preventing foe-

tal suffocation. In posterior presentation, the umbilical cord may become pinched or even ruptured while the head of the foetus is still inside the birth canal. A delay in foetal extraction originates neonatal asphyxia and increases the probability of calf death by asphyxia or by amniotic fluid aspiration and consequent aspiration pneumonia.

> ❯ **Important**
> In posterior presentation, and after the obstetrician confirms dorsal position and limbs extension, the foetus should be pulled. Plenty lubricant should be used. Obstetric manoeuvres to place the foetus in anterior presentation are strenuous and may result in uterine trauma with lacerations and neonatal asphyxia.

During calving, the sum of uterine and abdominal forces applied by the dam on the foetus can reach approximately 70–75 kg. One or two people are enough to pull the foetus, and usually, the contribution of a third person (up to 150 kg of force, approximately 50 kg each person) is not advantageous and can even be harmful to the calf. Murray et al. [16] observed a significant negative interaction ($P = 0.002$) between the number of people pulling the foetus and duration of calving on the predicted blood pH (acidaemia) of the calf – an increase in calf acidaemia occurred as the length of calving extends. This correlation was more pronounced as the number of people exerting force on foetus also increased, from one to two and from two to three people.

As an alternative, a calf puller (calf jack) can be used (◻ Fig. 5.13). The calf puller exerts up to 400 kg of force on the foetus and should be manipulated with caution by experienced people. This obstetrical instrument is helpful, mainly when no additional people are present to help the obstetrician. However, when inadequately used, without respecting the traction principles abovementioned, the excessive force can provoke: in the dam, nerve compression, lacerations or even pelvic fractures, and in the calf, limb fractures and nerve injuries.

Due to the conformation of the forehead, which usually is broad and moderately dished,

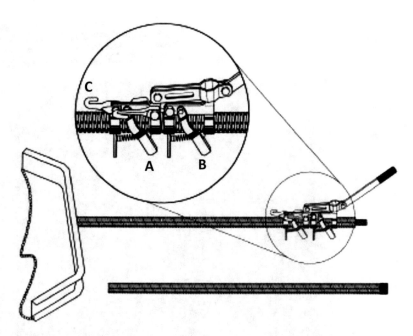

Fig. 5.13 Calvin jack with two hooks (A and B) with alternate movement. A third hook (C) serves to fix a halter, if necessary. (Modified from Jorgensen Laboratories, Inc. (USA) with permission)

special attention should be given when the nose arises outside. The forehead protrudes upwards and can get stuck in the vulva dorsal lip commissure. In posterior presentation, the tail base can also cause a similar effect. Usually, some manual stretching has to be applied to the dorsal lip commissure for forehead expulsion. Nevertheless, this condition can be mitigated if the obstetrician places the hands on the forehead or on the tail head, protecting the vulvar lips. Also, during the passage of the foetal hips, some care is needed to mitigate potential vulvar lacerations.

Episiotomy should be performed in the presence of irreducible vulvar stenosis or immature vulvar development in heifers. In fact, whenever there is a risk of laceration or severe trauma, episiotomy should be performed. After epidural anaesthesia, one or even two dorso-lateral (10 and 12 o'clock) incisions, between 1 and 2 cm deep, should be made. In addition to widening the vulvar diameter, these incisions allow the orientation of the potential tear caused by the passage of the foetus. At the end of the operation, these incisions should be sutured in order to prevent vulvar coaptation' alterations, which originate pneumovagina.

5.6 Immediate Post-calving Assistance of the Dam

Once the foetus is delivered and breathing is confirmed, vaginal, cervical and uterine walls need to be inspected by vaginal palpation to evaluate their integrity. At this moment, the obstetrician must confirm whether any additional foetus is present. Also, the integrity of the vulvar region should be evaluated.

Especially in heifers and after sturdy traction, rupture of perivaginal arteries can occur (Fig. 5.14). This is diagnosed by looking for a squirt sensation on the dorsal or lateral walls. Forceps have to be applied immediately or death is probable.

Usually, the placenta stays attached to maternal caruncles, with natural expulsion occurring within 6 h after the end of Stage II. In some abnormal and rare cases, the placenta may be expelled immediately after or even alongside with the calf.

The administration of 3–4 ml (30–40 IU, i.m.) of oxytocin promotes endometrium contractions which also compresses the uterine small blood vessels. These contractions allow the expulsion of intrauterine air, fluids

◻ **Fig. 5.14** Significant haemorrhage in a heifer after pudendal artery rupture during mechanical traction. A forceps was applied to the bleeding vessel and kept for 3 days. Recovery was uneventful. (Original from George Stilwell)

5

and blood, as well as the attenuation of potential uterine bleeding originated by imperceptible lacerations. Additionally, exogenous oxytocin can have a positive effect on placenta expulsion and in uterine prolapse prevention.

The use of intrauterine antibiotic therapy is done by some practitioners after prolonged obstetrical assistance or when cleanliness is compromised. Usually, a broad-spectrum anti-biotic, such as oxytetracycline (up to 4 g.), is placed between the endometrium and placental membranes. The systemic administration of a broad-spectrum antibiotic and an analgesic (e.g. non-steroidal anti-inflammatory drugs) is indicated for dams presenting lacerations or other trauma such as bruises [19]. Moderate to severe lacerations of the uterus, vagina and vulva always need to be sutured, but mild lacerations may stay unsutured.

Fig. 5.15 Positioning the calf with the hind limbs to each side of the body ensures more thoracic space and easy breathing. (Original from George Stilwell)

5.7 Newborn Calf Care

5.7.1 Immediate Assistance and Calf Resuscitation

Immediately after extraction, vital signs of the newborn should be evaluated. Ventilatory assessment is the most preeminent aspect to take in consideration. The priority is to ensure that the lungs insufflate. Breathing movements are naturally incited by hypercapnia and usually start within the first 30 sec. after delivery. Calves should be positioned in sternal decubitus with the hind legs stretch to each side of the body, ensuring less pressure on the diaphragm so that the lungs fill up easily (■ Fig. 5.15). Note that, due to hypoxia occurrence during delivery, the calf may first show a panting breathing, reaching 40–60 breaths per minute.

After prolonged or difficult calving, some calves may not show the inbreathing reflex, and so this should be stimulated (▶ Box 5.8). In simple cases, just pushing a straw inside the nostrils initiates the gasp reflex or causes sneezing or coughing reflexes. Empirical evidence shows that pouring cold water over the calf may also stimulate breathing. Wiping the calf with straw or a blanket is recommended to only to dry the calf but also to stimulate neonatal respiratory function.

Uystepruyst et al. [29] demonstrated the effectiveness of three stimulating methods on calves' respiratory and metabolic adaptation during the first 24 h: (1) immediate pharyngeal and nasal suction using hand-powered vacuum pump; (2) cold water (around 5 °C) poured immediately after delivery over the newborn's head; and (3) the use of infrared radiant heater for 24 h after birth. These methods can be used together or alone.

In more severe cases, artificial insufflation of the lung may be necessary. This can be done through the use of resuscitator device that is attached to the nose and blows air (■ Fig. 5.16) or by intubation. In the first case, air will more easily flow to the stomach, so care should be taken to gently compress the oesophagus behind the trachea. Intubation is done in the same way as for the use of volatile general anaesthesia in surgery – visualize the glottis and insert an adequate size tube. Finally, chemical stimulation may be achieved by injecting doxa-

Fig. 5.16 Resuscitator device that is used to blow air into the lungs or to suck fluids for the upper respiratory tract. Note that gentle squeezing of the oesophagus with two fingers is done to prevent air being blown into the stomach. (Original from George Stilwell)

Fig. 5.17 Video box: After ensuring breathing movements (or before in very severe obstruction), the airways may have to be cleared. The calf may have to be hanged upside down for no more than 60–90 sec. Original picture from George Stilwell. (Original video from António Carlos Ribeiro) (► https://doi.org/10.1007/000-2qh)

pram hydrochloride (acts on peripheral carotid chemoreceptors), intravenously or intralingual.

Once the lungs are filled and breathing movements are evident, clearing the upper airways may be needed. This can be simply done with the hands immediately after delivery, by using a syringe with a tube attached to suck small amounts from the nostrils and upper airways or by carefully lifting the hindlimbs and keeping the calf upside down for approximately 1 min. This position allows the fluid present in the upper and lower respiratory airways to drain by gravity. The calf should not be swung but gently placed upside down over a gate (Fig. 5.17 – video box) or even over the mother's back if she is lying. The calf should be closely monitored as the pressure of viscera on the diaphragm may cause respiratory distress.

Once nasal secretions and/or fluids are removed, the respiratory movements are again evaluated by inspection and auscultation. Auscultation of the heart (100–150 beats per minute is normal) and a thorough physical examination should also be performed at this time, to identify trauma (e.g. rib or limb fractures), haemorrhages, signs of acidosis, etc.

Finally, the calf should be dried with a towel or dry straw, especially if it is going to be separated from the dam and if environmental temperature is low.

Box 5.8   Neonate Breathing Stimulation
Nostril physical stimulation: using semi-rigid material (e.g. straw or a gloved fingertip) and repeatedly touching the nasal mucosa to stimulate coughing or sneezing.

Thermal stimulation: pouring a bucket of cold water over the calf's head to stimulate head movements, sudden inbreathing or coughing.

Chemical stimulation: 20 mg of doxapram hydrochloride by sublingual or intralingual route.

Use of a calf resuscitator: Inflating the lungs or induce coughing or sneezing.

Box 5.9 Treatment for Hypovolemic or Septic Shock [21]

Ensuring adequate tissue perfusion and electrolyte and energetic provision

- 7.2% NaCl, 4–5 mL/kg over 4–5 min. (60 mL/kg/h) – rapid resuscitation of severely hypovolemic calves. No effect on blood pH. For full and durable effect, it has to be followed by isotonic fluids i.v. or oral fluids, with electrolytes and energetic and alkalinizing products.
- Crystalloid solutions (0.9% NaCl and acetated; lactated Ringer's solution for metabolic acidosis (pH <7.2): 25–75 mL/kg/h), especially during the first 30–60 min.
- 5% glucose: 2.2–4.4 mL/kg/h.
- 8.4% sodium bicarbonate – rapid alkalization especially in presence of severe acidaemia (pH <7.2). Rate of infusion should be less than 1 mL/kg/h.

Septicaemic colostrum-deprived calves
- Plasma transfusion: 10–15 mL/kg.
- Whole-blood transfusion: 10–30 mL/kg.

Hypothermic calves (<34.5 °C): exogenous heath sources
- Radiant heaters and hot-water bottle.
- Immersion in a warm-water bath at 38 °C.
- Air temperature varies from 20 to 25 °C; initially up to 40 °C (infrared lamps).
- Warm intravenous or oral solutions (approximately 39 °C).

Hypotension: use of inotropic and vasopressor drugs (intravenous infusion with cardiac monitorization)
- Dopamine: 1–5 µg/kg/min. Can be added to NaCl, 5% glucose and Ringer's solution but not to alkaline solutions (chemical instability); stimulation of the cardiac contractibility and heart rate (cardiac beta1-receptors); release of norepinephrine (dopamine is its precursor) from terminal nerves.
- Dobutamine: 1–3 µg/kg/min.
- Norepinephrine: 0.01–1 µg/kg/min. (if the previous ones fail).

For septicaemia or toxaemia: nonsteroidal anti-inflammatory drugs (analgesic, anti-inflammatory and antiendotoxic proprieties). Be careful in very dehydrated animals as the anti-prostaglandin effect may be deleterious to kidney function.
- Flunixin meglumine: initial dose of 2.2 mg/kg followed by 1.1 mg/kg three times a day (TID).
- Ketoprofen at 3 mg/kg once a day (QD).

For severe shock (weak pulse, increased capillary refill time, lateral recumbency and coma): single intravenous dose of short-acting soluble corticosteroids (cells membrane-stabilizers reducing enzymes release)
- Dexamethasone: 2.2 mg/kg.
- Prednisolone sodium succinate: 1.1 mg/kg.

Calf vitality evaluation, by assessing depression degree and level of activity, is an important part of the general health status exam. Although the Apgar score (◻ Table 5.1) can easily be adapted from human newborns to evaluate calf viability and vitality, a proper and more complete vigour score using a 4-point scale, and presenting more parameters, was developed by Murray-Kerr [15] (◻ Table 5.2). The Calf VIGOR Scoring System, developed at the University of Guelph, is designed to be applied to calves shortly after birth to assess their visual appearance, initiation of movement, general responsiveness, oxygenation and heart and respiration rates. The input categories include meconium staining, appearance of the tongue and time to initiation of selective calf movements such as sitting, standing attempts, suckling, head shaking, tongue withdrawal after a

5

◘ Table 5.1 Apgar score for newborn calves

Variables	Score		
	0	**1**	**2**
Heart rate	Absent	Bradycardia/irregular <120 bpm	Normal/regular 120–220 bpm
Respiratory rate and effort	Absent	Irregular <35 mpm	Regular 35–90 mpm
Muscle tone	Flaccidity	Some flexion	Flexion
Irritability reflex	Absent	Some movement	Hyperactivity
Mucous colour	Cyanotic	Pale	Normal

Legend: *bpm* beats per min., *mpm* movements per min.
Adapted from Vannucchi et al. [30] with permission from Elsevier

◘ Table 5.2 Vigour score measurements for calf neonatal viability and vitality assessment

	Points towards total score/category			
Item	**0**	**1**	**2**	**3**
Meconium staining	No stain	Anal area only	Extended over the body	Completely covered
Responsiveness				
Shake	Vigorous	Moderate	Twitches	No response
Eye reflex[a]	Active blink	Slow blink	No response	–
Tongue pinch	Active withdraw	Attempt withdraw	Twitches	No response
Tongue swelling	No swelling	Protruding and not swollen	Protruding and swollen	Head, tongue swollen
Heart rate[a]	80–100 beats/min.	<80 beats/min.	>100 beats/min.	–
Respiration rate[a]	24–36 breaths/min.	<24 breaths/min.	>36 breaths/min.	–
Mucous membrane colour	Bright pink	Light pink	Red	White/grey

[a]Scored on a 3-point scale only. Lower score indicates greater vigour (modified from [15]). From Villettaz Robichaud et al. [32] with permission from Elsevier

pinch and eye blinking. Mucous membrane colour and heart and respiration rates are additional clinical signs that are assessed and assigned scores ranging from 0 to 3. The higher the score, the more vigorous the calf. Very recently, the Calf VIGOR Score app was developed at the University of Wisconsin School of Veterinary Medicine (please refer to the SVM Dairy Apps at the end of this chapter: Murray-Kerr et al., 2020).

The calf should be evaluated immediately after birth, usually within the first minute of life. A second evaluation after 5 min. is suitable to confirm the initial assessment. Assessing the presence and strength of suckle reflex is a practical procedure to evaluate the level of stress and neonatal hypoxia. Special care with supportive therapy and even environmental temperature control is required for calves presenting a weak or no suckle reflex. Usually, the calf

▣ Table 5.3 Changes in neonatal rectal temperature, venous blood pH and acid base variables in 12 healthy (pH >7.2) newborn calves within 24 h

Parameters	≤30 min.	6 h	24 h
BPM (1/min.)	44.9 ± 11.3	41.3 ± 7.2^a	$53.1 \pm 1 0.7^b$
TT	1.42 ± 0.33	1.48 ± 0.24^a	1.19 ± 0.29^b
Cdyn (L kPa^{-1})	0.80 ± 0.29^c	$1 0.58 \pm 0.64^d$	1.43 ± 0.51^d
Pmax	$0.66 \pm 0.13'$	0.37 ± 0.19^a	0.38 ± 0.08^b
Pmin	-0.28 ± 0.16	-0.24 ± 0.30	-0.09 ± 0.12
pH	7.26 ± 0.03^a	7.55 ± 0.04^b	7.39 ± 0.0^b
pO$_2$ (mm Hg)	45.2 ± 18^a	68.4 ± 15^b	78.8 ± 15.7^b
pCO$_2$ (mm Hg)	59.8 ± 3.9^a	50.4 ± 4.0^b	46.39 ± 3.9^b
Base excess (mmol/L)	-2.3 ± 2.8^e	$1 0.0 \pm 2. 1^f$	2.0 ± 2.3^f
HCO$_3^-$ (mmol/L)	24.9 ± 2.8	26.7 ± 1.6	26.7 ± 1.9
O$_2$ Sat (%)	60.0 ± 20.9^a	88.2 ± 8.3^b	92.7 ± 3.5^b

Legend: Values with different superscripts in the same line are significantly different: [a,b]$P < 0.05$, [c,d]$P < 0.01$, [e,f]$P \leq 0.001$

TT total respiratory time, *Cdyn* dynamic lung compliance, *Pmax* pressure maximum, *Pmin* pressure minimum
Adapted from Varga et al. [31] with permission from John Wiley and Sons

acquires sternal recumbency immediately or a few minutes after birth. Depressed, stressed or injured calves take longer to get to this position.

During foetal delivery process, circulatory and respiratory changes occur inducing hypoxia and variable degrees of metabolic and respiratory acidosis. Both hypoxia and acidosis play a key role in early foetal circulation and pulmonary ventilation of the normal newborn (▣ Table 5.3). Metabolic acidosis is usually controlled within the first 2–6 h, but respiratory acidosis may last from 6 to 48 h according to the intensity of hypoxia or anoxia developed before, during or immediately after delivery.

The treatment of a calf in shock during the perinatal or neonatal periods, independently of the cause (e.g. hypovolaemia, septicaemia, toxaemia), should be considered an emergency. Immediate and competent veterinary assistance is crucial (▶ Box 5.9; ▣ Fig. 5.18).

❯ Important

Fluid deficit is calculated in percentage according to the dehydration degree (5–12%) × BW (kg) of the calf. Approximately 1/3 of the volume can be done in first 2–3 h (corresponding to the fluid of the extracellular compartment). The daily fluid maintenance is 0.05 L × BW (kg). No more than 80 mL/kg/h of isotonic fluids should be intravenously administered, to reduce the risk of pulmonary oedema. Normal rate of intravenous fluid administration is 20 mL/kg/h and can be complemented by oral fluid administration. Alternatively, a hyperosmotic saline solution (7.2% NaCl) can be given for quick resuscitation. This will draw fluids into the intravascular compartment from the interstitial and gastrointestinal compartment. Beneficial effect will only last a few minutes unless isotonic or oral fluids are administered soon after.

◘ Fig. 5.18 Keeping warm a calf in shock. Note that the infusion tube is going through a bucket with warm water where an additional fluid bottle is also being kept warm. (Original from George Stilwell)

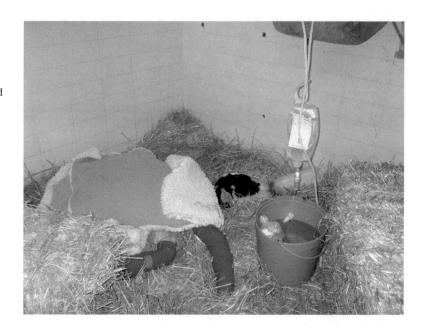

5

The bicarbonate requirements (mmol) are calculated according to BW (kg) × base deficit (mmol/L) × 0.6 (L/kg). Usually, in field conditions, the approximate base deficit (mmol/l) is obtained according to the percentage of dehydration (for diarrheic calves): dehydration of 5, 9, 11 and >12% corresponds to 5, 10, 15, 20 and >20 mmol/L, respectively. Isotonic (1.9%) or hypertonic (2.1% or 4.2%) bicarbonate solution should also be administered intravenously at 1–3 L/h rate, although acidaemia is usually more severe in older calves (over 7 days of age). A recommended version of the decision tree to treat metabolic acidosis is presented in ◘ Fig. 5.19.

5.7.2 Disinfection of the Umbilical Cord

The contamination of the umbilical cord by bacteria from the environment (e.g. *Escherichia coli*, *Trueperella pyogenes*, *Proteus* spp., *Staphylococcus* spp. and *Streptococcus* spp.) can cause internal inflammation and infection: omphalitis comprising all umbilical structures (umbilical arteries/omphaloarteritis, umbilical veins/omphalophlebitis, urachus

and connective tissues). Omphalitis can affect more than 20% of calves in some farms and is a predisposing factor for patent urachus, septic polyarthritis, pneumonia, meningitis and diarrhoea, mainly during the first two weeks of life. Animals, immune-compromised due to failure of passive immune transfer (see below in Colostrum), are particularly susceptible to these sequelae.

Navel disinfection, consisting in a single immersion of the umbilicus immediately after calf birth, will prevent contamination and will accelerate the healing of the umbilical cord. It is classically disinfected with 7% iodine tincture that would kill most pathogens in 15 min. to several hours, but this is currently not recommended because alcohol delays drying and umbilical cord separation [22]. Some other less concentrated disinfecting solutions that do not cause tissue inflammation (2–4% chlorhexidine, 10% trisodium citrate and 0.1% of chlorine-based solution) are considered preferable [22, 34]. During the second week after birth, the normal appearance is a dried umbilical cord.

A good alternative (or in complement with disinfection) is closing the umbilical cord by tightly tying it with a string (◘ Fig. 5.20). In our experience, it is a very efficient way of preventing navel infection (George Stilwell, personal communication). However, this method

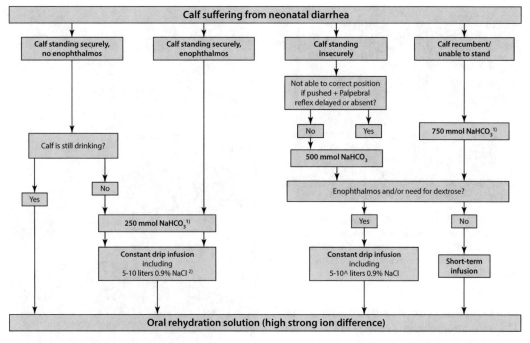

| Calf suffering from neonatal diarrhea |

| Calf standing securely, no enophthalmos | Calf standing securely, enophthalmos | Calf standing insecurely | Calf recumbent/ unable to stand |

Not able to correct position if pushed + Palpebral reflex delayed or absent?

No — Yes

750 mmol NaHCO₃[1]

Calf is still drinking?

500 mmol NaHCO₃

No

Yes

Enophthalmos and/or need for dextrose?

250 mmol NaHCO₃[1]

Yes — No

Constant drip infusion including 5-10 liters 0.9% NaCl [2]

Constant drip infusion including 5-10^ liters 0.9% NaCl

Short-term infusion

Oral rehydration solution (high strong ion difference)

Fig. 5.19 Decision tree to treat metabolic acidosis in calves in field practice. Legend: (1) Represents the intended amount of sodium bicarbonate. (2) An infusion volume of 10 L is recommended for calves with estimated enophthalmos ≥7 mm. Examination of the posture/ability to stand includes lifting of the animal if it is not able or willing to stand up. The term enophthalmos is defined as a visible gap between the eyeball and caruncula lacrimalis, which corresponds to a measured eyeball recession of at least 3–4 mm. (Adapted from Trefz et al. [27])

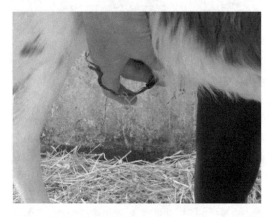

Fig. 5.20 A dried umbilical cord (2nd week of calf's live) to which a string was tied just after delivery. (Original from George Stilwell)

should not be used if contamination of the umbilicus is already possible (e.g. unassisted calving in a dirty environment).

Suture of the skin and eventually ligation of blood vessels may be necessary when the umbilical cord ruptures very short.

Antimicrobials should be given to these animals as peritonitis is expected (Fig. 5.21).

5.7.3 Colostrum

In late pregnancy, the colostrogenesis starts with the development and differentiation of mammary acinar gland forming colostrum. This colostrum is rich in nutrients, active compound and immunoglobulins which arise from systemic and local sources. It gradually changes to milk within the 4 days after parturition. Placental barrier prevents the passage of immunoglobulins during the intrauterine life of the foetus. In consequence, the calf is hypo- or agammaglobulinemic at birth. It needs to ingest colostrum during the first hours of life, because enterocyte pinocytosis, through which immunoglobulins reach the bloodstream, decrease progressively from 6 h

◘ Fig. 5.21 Umbilical cord ruptured internally – haemorrhage was copious, but it was not possible to identify any blood vessel for ligation. The abdomen was sutured and blood transfusion was given. (Original from George Stilwell)

after birth being completely null after 24–48 h. The immunoglobulin G1 (IgG1) represents around 90% of total immunoglobulins present in colostrum, while the remaining 10% is constituted by IgM, IgA and IgG2. The mean first-milking colostrum yield can vary from 2.7 L (range from 0.6 to 5.6 L; mean IgG/IgG1 = 99 mg/mL ranged from 31 to 200 mg/mL) and IgG/IgG1 for beef-suckler to 6.7 L (range from 3.7 to 9.5 L; mean IgG/IgG1 = 66 mg/mL ranged from 27 to 117 mg/mL) for dairy cows, considering a density of colostrum around 1.05 g/mL [13].

The route of ingestion may also play a role. Suckling from the dam (usual in beef but not in dairy calves) or from a bottle (◘ Fig. 5.22) is probably the best way to achieve good IgG absorption. Intubation may be stressful and should only be performed by trained staff but is sometimes the best way to get 10% of body weight (BW) into a recently calved animal (◘ Fig. 5.23). Many dairy farms will nowadays use the bottle with a teat but will change to intubation if the calf does not ingest close to 10% of its weight. Having a calf drink colostrum from a bucket is not acceptable.

Colostrum is also a very nutritious feed crucial for the first hours/days of life. Colostrum is much richer than milk in total solids (>23%) and especially in protein (14%) and fat (6.7%).

◘ Fig. 5.22 In dairy calves, colostrum is mostly given by bottle with a teat. Patience and proficiency are essential for this vital procedure. (Original from George Stilwell)

◘ Fig. 5.23 Intubation is sometimes needed for very large calves or those born from difficult-assisted calving. The delivery of 3–4 litres of colostrum within 4–6 h after birth is essential for these stressed animals. (Original from George Stilwell)

❯ **Important**
Calves should ingest 10–12% of their BW of a clean and warm colostrum (39–40 °C) within the first 6 h of life.

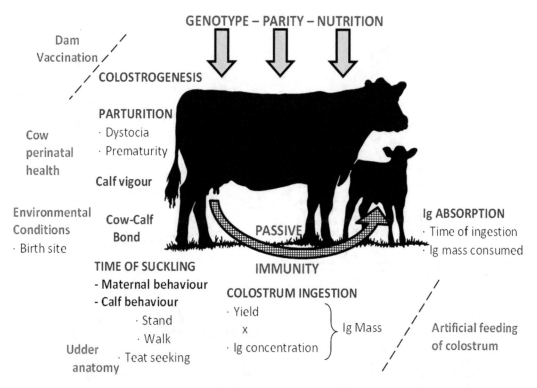

Fig. 5.24 Main factors affecting passive immunity in beef-suckler calves. (Adapted from McGee and Earley [13] with permission from Cambridge University Press)

After colostrum ingestion, IgG serum concentration should be above 10 mg/mL. In field conditions, serum total proteins in calves between 1 and 7 days of life can indirectly appraise insufficient or absence of IgG absorption (total proteins <5.2 g/dL). Values below these thresholds are considered a sign of failure in passive transfer (FPT).

Many factors can lead to a FPT, and these are summarized here and in **Fig. 5.24** for suckler cow-calf. The factors may have dam, calf or human origin:

– Calf-derived FPT – probably related with reduce vigour and ability to ingest enough colostrum
  – Foetal stress and trauma during the calving.
  – Premature and twin calves.
  – Very large calves.
  – Calves showing head/tongue oedema (i.e. incomplete tongue withdrawal) or abnormal mucous membrane colour, still present 10 min. after birth.
  – All these calves are more likely to show inadequate FPT compared with calves from those normal and unassisted calving (odds ratio = 26.1; $P = 0.005$) or even those born from difficult calving but with immediate assistance (odds ratio = 7.04; $P = 0.02$) [18].
– Cow-derived FPT
  – Very high colostrum yields after calving (<8.5 kg).
  – Age and parity – not too old nor too young.
  – Body condition – cows too fat or too thin will have low protein and energy resources, and this will correlate with colostrum quality.
  – Too long or too short dry period (dairy cows).
  – Mastitis or blood in the milk – very low quality colostrum.

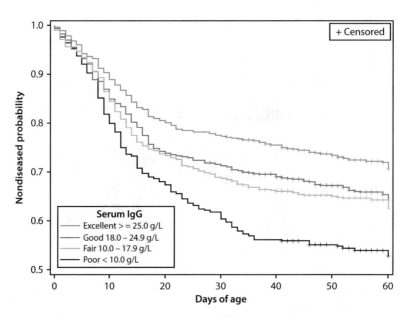

◧ Fig. 5.25 No disease probability for preweaned heifer calves by days of age and serum IgG concentration categories. Legend: Serum IgG levels of greater than or equal to 25.0 g/L, 18.0–24.9 g/L, 10–17.9 g/L, and less than 10.0 g/L corresponded to serum total protein categories of greater than or equal to 6.2 g/dL, 5.8–6.1 g/dL, 5.1–5.7 g/dL, and less than 5.1 g/dL, respectively and Brix score categories of greater than or equal to 9.4%, 8.9–9.3%, 8.1–8.8%, and less than 8.1%, respectively. See Hernandez et al. [5] for further explanation. (Adapted from Godden et al. [4] with permission from Elsevier)

- "Leakers" that are cows that start dripping the milk days before calving.
- Very stressed cows.
- Cows with very large udders – difficulty in getting to the teats.
- Human derived
  - Poor hygiene at milking, preserving or delivering the colostrum.
  - Poor freezing and thawing technique.
  - Trying to speed ingestion.
  - Delaying time for colostrum administration.
  - Pooling colostrum – bad quality colostrum will affect the lot.

Adequate ingestion of colostrum ensures a systemic immunologic protection up to four weeks (◧ Fig. 5.25). However, the negative impact of FPT has been demonstrated in calves and even in heifers, months or years after birth. Many studies have shown increase prevalence in diarrhoea and respiratory disease and a delay in first heat and pregnancy, in animals that were diagnosed with FPT soon after birth. For example, higher prevalence of respiratory disease and of antimicrobial treatments and mortality was shown in calves that showed low IgG concentration in an enzyme-linked immunosorbent assay (ELISA) test performed at 7–10 days of age [25].

5.7.4 Soft Tissue and Bone Injuries

Although it may be a complex procedure, in a knowledgeable and experienced assistance, no trauma or injury is expected to either the dam or the foetus. However, the possibility exists, and all care should be taken to prevent damage to the animals involved.

Although rare, inadequate foetal manipulation can lead to foetal soft tissue compression or distension and bone fractures. The main bone fractures usually involve the metacarpus, the metatarsus or the mandibula. Nevertheless, other bones such as the femur, tibia, humerus, radius-ulna and ribs can be fractured due to excessive foetal compression or traction. Also nerves in the calf can be damaged by forceful traction (e.g. paralysis of the sciatic nerve after traction in posterior presentations, ◧ Fig. 5.26).

■ **Fig. 5.26** Forceful
traction in a calving with
posterior presentation
resulted in nerve damage
and flaccid paralysis (left
hindlimb dragging).
(Original from George
Stilwell)

External reduction and immobilization of
limb closed fractures can be easily made using
PVC or aluminium bandage splinting (e.g.
Thomas splint). Also, external fixation of
long bones can be done using transfixation
pinning, and casting can be used in closed or
open fractures. For dislocated and fragmented
fractures, the internal fixation using some
techniques such as screw pinning, intramedul-
lary cerclage wiring, interlocking pin or
dynamic compression plate is more appropriate. Depending on fracture type (i.e. closed vs.
open, oblique vs. transversal and fragmented
vs. non-fragmented) and method used to
reduce and immobilize the fracture and evolution of the osteosynthesis evaluated by radi-
ography, more than six weeks can be required
to remove the external fixators.

Costal fractures (■ Fig. 5.27) are probably more common than is usually thought.
This condition is further addressed in the
Dystocia chapter, as it usually occurs after
complicated parturition.

5.7.5 Maladjustment Syndrome

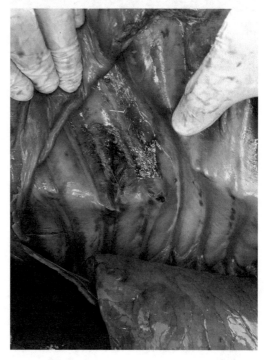

■ **Fig. 5.27** Rib fractures on a calf born after severe
dystocia. This animal was lethargic and had to be intu-
bated to receive colostrum. It died 48 h after delivery.
(Original from George Stilwell)

The natural transition from a state of neuroinhibition to one of neurostimulation that
occurs during parturition has been presented
in ▶ Chap. 3. However, in some rare cases,

this transition does not seem to through all its
stages resulting in calves showing signs of
neonatal maladjustment syndrome (NMS) –
indifference to environmental stimuli, lack of

◘ Fig. 5.28 Thoracic squeeze on a very large Blonde d'Aquitaine that showed signs of neonatal maladjustment syndrome after an uncomplicated calving. After 20 min., the calf woke up, nursed and stranded of with the dam. (Original from George Stilwell)

affinity for the dam, failure to find the udder, refusal to suck even when helped by putting the teat in their mouths, aimless wandering or motionless standing. In practice these animals are sometimes referred to as dummy calves.

The "Madigan squeeze technique" is a procedure that induces a slow wave sleep and hormone changes similar to what occurs during delivery, by physical compression of the chest with a rope. This sleep-like unconsciousness persists until the squeeze is discontinued. The compression simulates the forces exerted on the foetus during parturition apparently replicating the neonatal transition from neuroinhibition to neuroactivation. This technique has been used for years in dummy foals and has now been tried with great success in many calves showing NMS signs [26] (George Stilwell, personal communication).

The postnatal thoracic squeeze is a slight modification to the original Reuff's method of casting cattle. A soft cotton rope, approximately 3 cm in diameter, is looped around the thorax of the calf three times (starting on the neck to shoulder transition area), and pressure is applied to the loose end of the rope (◘ Fig. 5.28). The calf is then lowered into lateral recumbency while the loops of rope around the thorax are gently tightened. Some struggling will occur at first but is easily managed by one person at the head. Calves will then go into a deep sleep manifested by closed eyes, absence of leg or other body movements, slow breathing, reduced heart rate and lack of response to noise or human presence. The squeeze should be maintained (unless signs of distress occur), for 20 min., after which the ropes are loosened, and the calf made to stand.

Usually, after removal of the rope, the calves exhibit alert consciousness and maternally directed behaviours that were not originally present. Most calves will look for the udder and start sucking, in the same way a newborn does.

It should be recalled that other conditions can produce similar dummy signs in the calf – congenital defects or illness, pain, dehydration, hypoxia/acidaemia, etc. This means that the thoracic squeeze *should only be performed after a thorough clinical examination by a veterinarian*. For example, squeezing a calf with broken ribs is unacceptable.

> **Tips**
>
> A U-shaped aluminium splinting fixed by bandage is very useful to immobilize injured fetlock joint by fracture and/or tissue distension due to excessive traction during assistance to calving.

> **Tips**
>
> Limb bandage up to its proximal edge is required to prevent distal oedema and necrosis due to blood flow impairment and consequent amputation of the distal part of the limb.

Case Study 5.1 Vaginal Delivery: Foetal Posterior Presentation

A group of nine Holstein-Frisian heifers at pregnancy term (270 days) was previously moved to the calving area according to the calving protocol of the farm. This morning (6 am), during the first daily observation of the group, two newborn calves were found close to their dams, and one more heifer showed restlessness and discomfort behaviour, as well as other periparturient signs, such as udder oedema and vulvar relaxation. At 8 am, the heifer was observed lying down and presented regular abdominal contractions. The water sac was not observed at this moment. Twenty minutes later, no calving progress had occurred yet, and an obstetrical intervention was planned. After cleaning, washing and disinfecting the perineal and vulvar area with povidone-iodine 1%, the obstetrician inserted his hand, previously lubricated, into the vagina to explore the birth canal. The amniotic sac was intact, and two hooves were identified with the soles facing up. The hoof was pinched with immediate retraction, indicating high foetal vitality. However, the cervix had not fully dilated. In order to differentiate posterior presentation with dorsal position from anterior presentation with ventral position and deviation of head, the tail was identified as well as opposite joint flexion (pastern and fetlock joints vs. hock joint) of both hindlimbs. Thus, an insufficient cervical dilatation was diagnosed, probably originated by the posterior presentation of the foetus. Two options were considered: (a) delaying the intervention for 20–30 min. or more, once the amniotic sac was intact and the foetus showed high vitality signs; in these cases, it is expected that foetal pressure and hormonal activity will eventually lead to full dilatation of the cervix; (b) applying a mechanical force to help fully dilate the cervix. In this case, the vulva was relatively well dilated (adequate hormonal balance); also, the conic shape formed by both hind limbs may have been the limiting factor regarding the full cervical dilatation. So, a mechanical cervical dilation, using both hands for 5 min., was eventually successful. Meanwhile, the amniotic membrane ruptured due to the obstetrical manipulations. One rope was attached to each foot, and the foetal hips were rotated 45°. Each foetal hindlimb was alternately pulled when abdominal contractions occurred. As the hips entered the birth canal, quick traction was exerted to prevent foetal fluid aspiration. Some caution was required at the vulvar level to avoid trauma and lacerations at hips passage. The calf vitality was evaluated using the Apgar score. At the end, the reproductive tract of the dam was examined. No haemorrhages, trauma or lacerations were observed. Nevertheless, 40 IU of oxytocin were administered (i.m.) and 1 g. of oxytetracycline was introduced into the uterus between the endometrium and the placental membranes. A postpartum close monitorization was recommended, regarding the retained placenta and the possibility of puerperal metritis.

Key Points

- The periparturient should be moved to the maternity pen or to a clean and safe facility, before Stage I starts.
- Timing of obstetrical intervention needs to be appropriate, avoiding early or late assistance and, as a consequence, adverse effects for both the calf and the dam.
- The design of calving protocol for farmers is crucial to maximize animal health, welfare and productivity and needs the close collaboration of all intervenients.

- Adequate restraint of the cow is essential although the type can vary accordingly to the obstetrical approach anticipated.
- Systematic obstetrical examination of the birth canal starts with morphological inspection of the vulva and eventual identification of placental membranes and/or foetal body parts, followed by vaginal palpation of the birth canal and its contents.
- Dilatation of the soft birth canal, absence of obstruction or occlusion as well as

5

appropriate foetal disposition and viability/vitality needs to be ensured before initiating any obstetrical procedure.

— Compliance of foetal extraction guidelines, such as alternate traction and hips rotation, avoids or mitigates trauma and disease in both dam and calf, including the occurrence of lacerations and stillbirth.

— Primary post-calving care includes resuscitation and calf vitality assessment and dam's reproductive tract evaluation.

— The evaluation of colostrum and its timeline administration to the calf is mandatory to reach adequate immunity protection during the next 4–6 weeks of life.

? Questions

1. Describe the main reference signs to decide for an obstetrical intervention.
2. What is the "Utrecht method of delivery" and what are its advantages?
3. How can you assess the calf vitality before and after birth?

✓ Answers

1. The onset of abdominal contractions and the appearance of the placental membranes (intact chorioallantoic or amniotic sacs) are two main signs which serve as markers. Once abdominal contraction starts, the dam needs to be observed at least every 30 min., and intervention is needed if no progress occurs in the following 2 h. In normal calving, the amniotic sac (or hooves, if amniotic sac has already ruptured) appears approximately 60 min after the rupture of the chorioallantoic membrane and 70 min. before birth. However, normal calving progress should be evaluated every 15–20 min.

2. The "Utrecht method of delivery" follows the steps of a natural calf delivery: (1) the birth canal is confirmed entirely dilated without obstructions or occlusions; (2) the foetus is checked to be in longitudinal presentation and dorsal position with the extension of the head and forelimbs (or just hind limbs for posterior presentations); (3) the dam is kept in lateral recumbency; and (4) the abdominal contractions exert most of the force to expel the foetus. In consequence, the obstetrician should mimic these different steps to improve the

chance of obtaining a live calf and mitigate adverse effects on the health and welfare of both calf and dam. A methodical obstetrical examination is needed to diagnose and solve potential dystocia due to foetal or maternal origin, as well as to evaluate foetal vitality. The dam should be lying in its right side. Mechanical traction is alternatively exerted on the feet of the foetus at the same time as abdominal contractions occur. Once the head and chest are outside the vulva, a 45-degree rotation of the foetal hips allows for its adaptation to the largest diameter of the dam's bone pelvis. Now, both foetal limbs can be pulled simultaneously and downwards until the delivery of the foetus is completed.

3. Foetal vitality can be evaluated using the biophysical parameters of the foetus. During obstetrical examination, the foetus is assessed by vaginal or transrectal palpation so that its accessible parts can be touched or pinched. This procedure stimulates a reflex reaction of the live foetus: withdrawal reflex (limbs), suckle reflex (tongue), palpebral reflex (eyes) or rectal sphincter reflex (anus). Furthermore, the degree of movement, i.e. time to react and the reaction intensity, can be objectively evaluated. A delay in the reaction and a decrease in its intensity occur as the degree of foetal hypoxia increases. These reflexes can be absent in live foetuses presenting deep depression. The heartbeat or umbilical pulse can also be used to evaluate foetal via-

bility. In the newborn calf, the Apgar score can be used to evaluate heart rate, respiratory rate and effort, irritably reflex and mucous membrane colour. Also, the meconium staining extended over the body is a sign of foetal stress. The newborn should be evaluated approximately 1 and again 5 min. after birth.

References

1. Abrahamsen EJ. Ruminant field anesthesia. Vet Clin North Am Food Anim Pract. 2008;24(3):429–41., v. https://doi.org/10.1016/j.cvfa.2008.07.001.
2. Ames NK. Noordsy's food animal surgery. 5th ed. Iowa: Wiley; 2014.
3. Becker M, Tsousis G, Lüpke M, Goblet F, Heun C, Seifert H, Bollwein H. Extraction forces in bovine obstetrics: an in vitro study investigating alternate and simultaneous traction modes. Theriogenology. 2010;73(8):1044–50. https://doi.org/10.1016/j.theriogenology.2009.11.031.
4. Godden SM, Lombard JE, Woolums AR. Colostrum management for dairy calves. Vet Clin North Am Food Anim Pract. 2019;35(3):535–56. https://doi.org/10.1016/j.cvfa.2019.07.005.
5. Hernandez D, Nydam DV, Godden SM, Bristol LS, Kryzer A, Ranum J, Schaefer D. Brix refractometry in serum as a measure of failure of passive transfer compared to measured immunoglobulin G and total protein by refractometry in serum from dairy calves. Vet J. 2016;211:82–7. https://doi.org/10.1016/j.tvjl.2015.11.004.
6. Hiew MW, Constable PD. The usage of pelvimetry to predict dystocia in cattle. J Vet Malaysia. 2015;27(2):1–4.
7. Hiew MW, Megahed AA, Townsend JR, Singleton WL, Constable PD. Clinical utility of calf front hoof circumference and maternal intrapelvic area in predicting dystocia in 103 late gestation Holstein-Friesian heifers and cows. Theriogenology. 2016;85(3):384–95. https://doi.org/10.1016/j.theriogenology.2015.08.017.
8. Hirsbrunner G, Zanolari P, Althaus H, Hüsler J, Steiner A. Influence of prostaglandin E2 on parturition in cattle. Vet Rec. 2007;161(12):414–7. https://doi.org/10.1136/vr.161.12.414.
9. Hodgson DS, Dunlop CI, Chapman PL, Smith JA. Cardiopulmonary effects of xylazine and acepromazine in pregnant cows in late gestation. Am J Vet Res. 2002;63(12):1695–9. https://doi.org/10.2460/ajvr.2002.63.1695.
10. Jackson P. Chapter 4 – dystocia in the cow. In: Handbook of veterinary obstetrics. 2nd ed. London: Saunders Ltd; 2004. p. 37–80. https://doi.org/10.1016/B978-0-7020-2740-6.50009-2.
11. Ko J, Ruble M. Using maternal pelvis size and fetal hoof circumference to predict calving difficulty in beef cattle. Veterinary Medicine (USA); 1990: 1030–6.
12. Kumbhar GB. Physio-pathological studies and efficacy of different therapeutic approaches for cervical dilatation in large ruminants. Master thesis of Veterinary Science. MAFSU, Nagpur, India; 2015.
13. McGee M, Earley B. Review: passive immunity in beef-suckler calves. Animal. 2019;13(4):810–25. https://doi.org/10.1017/S1751731118003026.
14. Mee JF. Newborn dairy calf management. Vet Clin North Am Food Anim Pract. 2008;24(1):1–17. https://doi.org/10.1016/j.cvfa.2007.10.002.
15. Murray C. Characteristics, risk factors and management programs for vitality of newborn dairy calves. PhD Thesis. Univ. of Guelph, Guelph, ON, Canada; 2014.
16. Murray CF, Veira DM, Nadalin AL, Haines DM, Jackson ML, Pearl DL, Leslie KE. The effect of dystocia on physiological and behavioral characteristics related to vitality and passive transfer of immunoglobulins in newborn Holstein calves. Can J Vet Res. 2015;79(2):109–19.
17. Norman S, Youngquist RS. Chapter 42 – parturition and dystocia. In: Youngquist RS, Threlfal WR, editors. Current therapy in large animal theriogenology. 2nd ed. St. Luis: Saunderds; 2007. p. 310–35. https://doi.org/10.1016/B978-072169323-1.50045-3.
18. Pearson JM, Homerosky ER, Caulkett NA, Campbell JR, Levy M, Pajor EA, Windeyer MC. Quantifying subclinical trauma associated with calving difficulty, vigour, and passive immunity in newborn beef calves. Vet Rec Open. 2019a;6(1):e000325. https://doi.org/10.1136/vetreco-2018-000325.
19. Pearson JM, Pajor EA, Campbell JR, Caulkett NA, Levy M, Dorin C, Windeyer MC. Clinical impacts of administering a nonsteroidal anti-inflammatory drug to beef calves after assisted calving on pain and inflammation, passive immunity, health, and growth. J Anim Sci. 2019b;97:1996–2008. https://doi.org/10.1093/jas/skz094.
20. Proudfoot KL, Jensen MB, Heegaard PM, von Keyserlingk MA. Effect of moving dairy cows at different stages of labor on behavior during parturition. J Dairy Sci. 2013;96(3):1638–46. https://doi.org/10.3168/jds.2012-6000.
21. Ravary-Plumioën B. Resuscitation procedures and life support of the newborn calf. Rev Méd Vét. 2009;160(8–9):410–9.
22. Robinson AL, Timms LL, Stalder KJ, Tyler HD. Short communication: the effect of 4 antiseptic compounds on umbilical cord healing and infection rates in the first 24 hours in dairy calves from a commercial herd. J Dairy Sci. 2015;98(8):5726–8. https://doi.org/10.3168/jds.2014-9235.
23. Schuenemann GM, Bas S, Gordon E, Workman JD. Dairy calving management: description and assessment of a training program for dairy personnel. J Dairy Sci. 2013a;96(4):2671–80. https://doi.org/10.3168/jds.2012-5976.

5

24. Schuenemann GM, Bas S, Workman JD. Calving management: the first step in a successful reproductive program. In: Stevenson JS, editor. Proceedings of 2013 Dairy Cattle Reproduction Council Conference Indianapolis, Indiana – November 7–8. Kansas State University; 2013b. p. 8–17.

25. Stilwell G, Carvalho RC. Clinical outcome of calves with failure of passive transfer as diagnosed by a commercially available IgG quick test kit. Can Vet J. 2011;52(5):524–6.

26. Stilwell G, Mellor DJ, Holdsworth SE. Potential benefit of a thoracic squeeze technique in two newborn calves delivered by caesarean section. N Z Vet J. 2020;68(1):65–8. https://doi.org/10.1080/0048016 9.2019.1670115.

27. Trefz FM, Lorch A, Feist M, Sauter-Louis C, Lorenz I. Construction and validation of a decision tree for treating metabolic acidosis in calves with neonatal diarrhea. BMC Vet Res. 2012;8:238. https://doi.org/10.1186/1746-6148-8-238.

28. Tsousis G, Becker M, Lüpke M, Goblet F, Heun C, Seifert H, Bollwein H. Extraction methods in bovine obstetrics: comparison of the demanded energy and importance of calf and traction method in the variance of force and energy. Theriogenology. 2011;75(3):495–9. https://doi.org/10.1016/j.theriogenology.2010.09.017.

29. Uystepruyst CH, Coghe J, Dorts T, Harmegnies N, Delsemme M-H, Art T, Lekeux P. Effect of three resuscitation procedures on respiratory and metabolic adaptation to extra uterine life in newborn calves. Vet J. 2002;163(1):30–44. https://doi.org/10.1053/tvjl.2001.0633.

30. Vannucchi CI, Rodrigues JA, Silva LC, Lúcio CF, Veiga GA. Effect of dystocia and treatment with oxytocin on neonatal calf vitality and acid-base, electrolyte and haematological status. Vet J. 2015;203(2):228–32. https://doi.org/10.1016/j.tvjl.2014.12.018.

31. Varga J, Mester L, Börzsönyi L, Erdész C, Vári A, Körmöczi P, Szenci O. Adaptation of respiration to extrauterine-life in healthy newborn calves. Reprod Domest Anim. 1999;34:377–9. https://doi.org/10.1111/j.1439-0531.1999.tb01268.x.

32. Villettaz Robichaud M, Pearl DL, Godden SM, LeBlanc SJ, Haley DB. Systematic early obstetrical assistance at calving: I. Effects on dairy calf stillbirth, vigor, and passive immunity transfer. J Dairy Sci. 2017;100(1):691–702. https://doi.org/10.3168/jds.2016-11213.

33. Vincze B, Gáspárdy A, Kézér FL, Pálffy M, Bangha Z, Szenci O, Kovács L. Fetal metacarpal/metatarsal bone thickness as possible predictor of dystocia in Holstein cows. J Dairy Sci. 2018;101(11):10283–9. https://doi.org/10.3168/jds.2018-14658.

34. Wieland M, Mann S, Guard CL, Nydam DV. The influence of 3 different navel dips on calf health, growth performance, and umbilical infection

assessed by clinical and ultrasonographic examination. J Dairy Sci. 2017;100(1):513–24. https://doi.org/10.3168/jds.2016-11654.

Suggested Reading

Drost M. Calving management: a team approach. In: Risco CA, Melendez P, editors. Dairy production medicine. West Sussex: Wiley; 2011. p. 19–26.

Funnell BJ, Hilton WM. Management and prevention of dystocia. Vet Clin North Am Food Anim Pract. 2016;32(2):511–22. https://doi.org/10.1016/j.cvfa.2016.01.016.

Murray CF, Haley DB, Duffield TF, Pearl DL, Deelan SM, Leslie KE. A field study to evaluate the effects of meloxicam NSAID therapy and calving assistance on newborn calf vigor, improvement of health and growth in pre-weaned Holstein calves. The Bovine Practitioner. 2015;49(1):1–12.

Murray CF, Leslie KE. Newborn calf vitality: risk factors, characteristics, assessment, resulting outcomes and strategies for improvement. Vet J. 2013;198(2):322–8. https://doi.org/10.1016/j.tvjl.2013.06.007.

Pearson JM, Pajor EA, Caulkett NA, Levy M, Campbell JR, Windeyer MC. Benchmarking calving management practices on western Canada cow–calf operations. Transl Anim Sci. 2019;3(4):1446–59. https://doi.org/10.1093/tas/txz107.

Vaginal Delivery Procedures

Whittier JC, Thorne JG. Assisting the Beef Cow at Calving Time. University of Missouri (Extension), USA; 2020. From: https://extension2.missouri.edu/g2007, accessed on September 18, 2020.

Sprott LR. Recognizing and Handling Calving Problems. Texas A&M AgriLife System: USA; 2020. From: https://agrilifeextension.tamu.edu/library/ranching/recognizing-and-handling-calving-problems/, accessed September 18, 2020.

The National Animal Disease Information Service (NADIS). Calving Part 1 The Basics UK; 2020. From: https://www.nadis.org.uk/disease-a-z/cattle/calving-module/calving-part-1-the-basics/, accessed September 18, 2020.

University of Wisconsin-Madison School of Veterinary Medicine (SVM) Dairy Apps

Murray-Kerr C, Leslie K, Godden S, McGuirk S. Calf Vigor SCORER app. University of Wisconsin, USA, 2020. From: https://www.vetmed.wisc.edu/fapm/svm-dairy-apps/calf-vigor-scorer/ accessed on November 5, 2020.

Other relevant SVM Dairy Apps on reproduction and reproductive management from University of Wisconsin can be accessed from https://www.vetmed.wisc.edu/fapm/svm-dairy-apps/

Obstetrical Manoeuvres

Contents

Electronic Supplementary Material The online version of this chapter (https://doi.
org/10.1007/978-3-030-68168-5_6) contains supplementary material, which is available to
authorized users. The videos can be accessed by scanning the related images with the SN
More Media App.

© Springer Nature Switzerland AG 2021
J. Simões, G. Stilwell, *Calving Management and Newborn Calf Care*,
https://doi.org/10.1007/978-3-030-68168-5_6

⚙ **Learning Objectives**

- To understand the fundamentals of the methodology aiming at improving foetal accessibility and minimizing trauma.
- To identify the obstetrical material and describe its manipulation.
- To classify and describe obstetrical manoeuvres.
- To define the sequential procedures to solve successfully the main foetal maldispositions.

6.1 Introduction

Obstetrical manoeuvres involve all mechanical forces exerted by the obstetrician on the foetus aiming to correct any foetal maldisposition, i.e. allows for foetal mutation. Mutation includes all movements and procedures aiming at returning the foetus to a natural orientation with respect to presentation, position and posture. The foetus is manually manipulated inside the uterus or in the birth canal. Most often the mutation of the foetus needs to be performed inside the uterus as the birth canal is relatively narrow, limiting or impeding major obstetrical manoeuvres of the foetal extremities (i.e. limbs and head). Specific obstetrical equipment such as chains, ropes, foetal extractor and a Krey hook serve as an additional tool to assess, move or fix foetal parts or to transmit physical forces from the obstetrician's hands to the foetus.

Foetal accessibility is a significant limitation at intervention time. It is more evident in dams presenting a deep abdomen, long birth canal and/or oversized foetus. In case of foetus found very deep in the abdomen, the obstetrician's hands/arms may have to exert force on the foetus without any leverage. However, most commonly, the obstetrician's forearms and elbows can rest on the pelvic floor or lateral wall serving as a lever. Caution is needed to prevent too much pressure, bruises, lacerations and other type of trauma to the cervix, vaginal wall or nerves (sciatic nerve and obturator nerve). These may lead to haemorrhage, inflammation and oedema, causing pain and discomfort and eventually leading to post-calving complications (e.g. downer cow). Strong continuous abdominal and myometrial contractions may also push the foetus into the pelvic inlet, decreasing the space of the uterine cavity momentarily and exerting opposite forces to the obstetrical procedure.

In addition to the different aspects above-mentioned, a sequential reasoning should be applied to choose the best obstetrical approach for each dystocia: (1) "is foetal mutation possible so that vaginal extraction of the foetus can occur without complications?"; (2) "is partial or total foetotomy of a dead foetus possible and helpful?"; or (3) "is the life and welfare of cow and foetus at severe risk so that a Ceasarean section (C-section) should be considered?".

Obstetrical manoeuvres should be selected when it is expected that the entire foetus can be delivered per vagina, without causing significant short- or long-term adverse health/welfare effects or death of the dam or foetus. When the general health status of the dam is comprised (e.g. intense weakness, infection and shock) and it cannot be stabilized, the vaginal delivery of the foetus may be the unique treatment option for reducible foetal dystocia. Nevertheless, it should be remembered that the manipulation of a live foetus presenting low vitality or that is under stress can cause its death. Unless dystocia is minor and vaginal delivery is expected to be quick, C-section is the best option to decrease the risk of stillborn. So, the general health status of the dam, foetal viability and vitality, degree of dystocia and the expected duration of foetal manipulation, and its consequences on the dam and foetus health, should be evaluated judiciously.

In the present chapter, several preparatory aspects essential to manipulate the foetus inside the dam are reported, as well as the description of successive obstetrical manoeuvres required to successfully convert foetal maldispositions to the normal presentation, position and posture. Normal vaginal delivery was described in the previous chapter.

6.2 Foetal Accessibility and Ease of Handling

Usually, mutation of the foetus is easier when the cow is standing up. This position prevents additional pressure exerted from the viscera in the abdominal cavity, especially the rumen. If possible, the dam should only be lying down at final extraction (pulling) of a well-orientated foetus (◘ Fig. 6.1 Video Box). If the dam is lying down during obstetrical procedures, the pressure will be higher on the side closest to the floor. Thus, foetal manipulations will be facilitated if they take place in the upper hemisphere. This means that the cow may have to be rotated to right or left decubitus, according to the expected foetal manipulations.

> **Tip**
>
> More extensive obstetrical manoeuvres should only be performed between the abdominal contractions or may be unceasing if an epidural block has been given.

6.2.1 Blocking Cow's Straining and Myometrial Contractions

Pharmacological inhibition of abdominal and even myometrial contractions of the dam can

◘ **Fig. 6.1** Video box: Forced traction of a relatively large foetus in anterior presentation using a calving jack. A halter is applied due to head retention into the birth canal. (Courtesy of António Carlos Ribeiro) (▶ https://doi.org/10.1007/000-2qj)

facilitate the obstetrical manoeuvres, in moderate to severe dystocia.

The major limitation of these inhibitions is the lack of dam's contractions at expulsion time; i.e. the foetus will be pulled only by artificial traction. For moderate to severe dystocia, when complicated or lengthy obstetrical manoeuvres are expected, caudal epidural anaesthesia is strongly recommended. It is also appropriate for when the intrauterine space is relatively small regarding its content or for aggressive or very active cows as it will reduce resistance to manoeuvres. It is also commended on welfare grounds if severe pain is expected or if the foetus is showing stress signs. Nevertheless, it can be dispensable in mild or some moderate foetal dystocias, which can be easily corrected.

6.2.1.1 Caudal Epidural Anaesthesia

Abdominal contractions can decrease because of the elimination of stimuli from the distended vaginal wall and cervix. This indirect effect is obtained performing a low caudal epidural anaesthesia. The most commonly used drugs are local anaesthetics such as procaine or lidocaine (in countries where it is still registered for use in cattle). Other substances such as bupivacaine or even sedatives (e.g. xylazine) can be used at a low volume (◘ Table 6.1).

The technique consists of injection into the epidural space through the intervertebral space – most commonly between the last sacral (S5) and first coccygeal (Co1) or between the two first coccygeal vertebrae (Co1–Co2). This will cause caudal sacral nerve desensitization, namely, S3, S4 and S5. Several areas such as the posterior croup, anus, rectum, vulva, vagina and urethra are desensitized at least for 1 h. This regional anaesthesia also prevents defecation. The bladder relaxes, but the internal urethral sphincter remains tonic. The cow may continue standing up if the motor nerves of hindlimbs are not affected, contrary to the more cranial epidural anaesthesia (high caudal epidural anaesthesia; ◘ Fig. 6.2), but this mainly depends on the volume which diffuse cranially. Low caudal epidural anaesthesia

■ **Table 6.1** Onset and duration (mean ± standard deviation) of low caudal epidural anaesthesia effects according to the administered pharmacologic agents

Drug and dose	Anaesthesia		Reference
	Onset (min.)	Duration (min.)	
2% procaine (0.1–0.2 mg/kg; 5–15 mL)	*	*	Sato et al. [9]
0.5% bupivacaine (0.06 mg/kg)		247 ± 31	Vesal et al. [12]
5% ketamine (20 mL)[a]	5.0 ± 0.0	62.5 ± 8.7	Lee et al. [6]
2% xylazine (0.05 mg/kg)[b]	11.7 ± 1.0	252.9 ± 18.9	Grubb et al. [4]
Tramadol (1.0 mg/kg; 100 mg/mL)	14.10 ± 1.57	306.8 ± 8.58	Bigham et al. [2]
2% lidocaine (0.2 mg/kg)[c]	2.82 ± 0.33	59.8 ± 3.4 116 ± 11.52 127 ± 25	Dehghani and Bigham [3] Imani Rastabi et al. [5] Vesal et al. [12]
2% lidocaine (0.2 mg/kg)-epinephrine (5 µg/mL)		137.7 ± 9.98	Imani Rastabi et al. [5]
2% lidocaine (0.1 mg/kg)-tramadol (0.5 mg/kg)	4.84 ± 0.68	174 ± 4.84	Bigham et al. [2]

[a]At this dose, ketamine produces analgesic and ataxic (for 48.0 ± 17.9 min) effects without significant sedation. By norm, the dam remains standing. For low doses (5 or 10 mL), the duration of the analgesic effect is shortened (17.0 ± 6.7 min. and 34.0 ± 7.6 min., respectively). Ketamine acts in the spinal cord as a noncompetitive antagonist at N-methyl-D-aspartate (NMDA) receptors [6]
[b]Xylazine is an alpha-2-adrenergic agonist, producing selective sensory blockade in the spinal cord. Mild ataxia and sedation effects can be observed. Usually, the total volume varies between 5 and 7 mL
[c]The use of lidocaine in production animals was banned in several regions, including the European Union, due to the hepatic production of 2,6-dimethylaniline (2,6-xylidine) metabolite which is considered carcinogenic and mutagenic
*Following lumbar segmental epidural analgesia in cows using 5% procaine solution, the onset and duration of anaesthesia are 15.9 ± 3.8 min. and 53.7 ± 14.3 min., respectively [11]

will be achieved by low doses of local anaesthetics (e.g. 5–10 mL of procaine 2%), while high caudal epidural anaesthesia, which will cause the cow to lie down, will be attained by higher volumes (over 10 mL of procaine).

Epidural anaesthesia is contraindicated in dams suffering from hypotension.

> **Tip**
>
> Technique for caudal epidural anaesthesia (caudal to the sacrum): "pump" the basis of the tail to identify the sacrococcygeal (S5–Co1; preferred local for low caudal epidural anaesthesia) or the first intercoccygeal (Co1–Co2; preferred local for low caudal epidural anaesthesia) space. Shave and aseptically prepare the area. A sterile 18-gauge and 38 mm needle is the inserted at 45° directed cranially. If it is expected to repeat the injection in the epidural space, an epidural catheter should be considered. The needle placement is confirmed by a popping sensation when the flavum ligament is punctured and by hearing air hissing through the needle. Another way is to have the hub of needle full of fluid (local anaesthetic) that will be drawn in when the epidural space is reached. This can be further corroborated by low resistance during solution administration. If the needle is advanced too far and bone is reached, the needle should be withdrawn approximately 0.5 cm. An administration rate of 0.5–1.0 mL/sec. of local anaesthetic is appropriate. Pinprick the skin of the perineum or vulva to confirm desensitization.

6

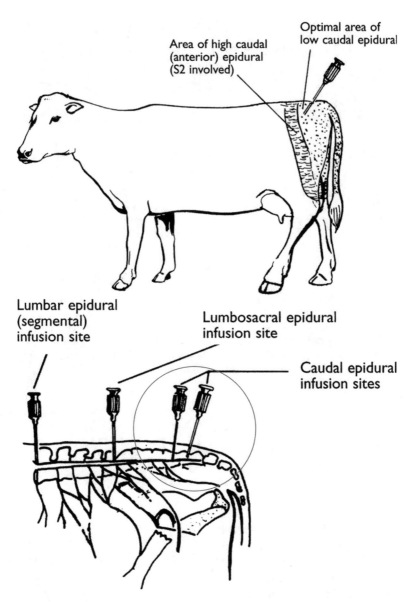

Area of high caudal
(anterior) epidural
(S2 involved)

Optimal area of
low caudal epidural

Lumbar epidural
(segmental)
infusion site

Lumbosacral epidural
infusion site

Caudal epidural
infusion sites

☐ **Fig. 6.2** Desensitized area caused by low and high caudal epidural anaesthesia and site of administration Legend: Top – The low caudal epidural anaesthesia desensitizes the sacral nerves S3, S4 and S5 and in consequence the area between the tail base and ventral perineal skin, approximately 25–30 cm to each side. The high caudal epidural anaesthesia is achieved with higher volumes of anaesthetics and will also desensitize sacral nerve S2. In this last case, the cow will lie down, so can

be used to keep it in recumbency or to prevent abrupt falls in weak animals. Bottom – Epidural anaesthesia infusion sites. Mainly the volume and concentration of the anaesthetic or analgesic substance and site of administration (S5–Co1 or Co1–Co2) are relevant for cranial progression of the solution in the epidural space and obtention of an optimal desensitization of sacral nerves. (Modified from Ames [1] with permission from John Wiley & Sons)

6.2.1.2 Blocking Myometrial Contractions

Beta-adrenergic agonists are pharmacological agents which can produce smooth muscle relaxation of the myometrium. Uterine wall

slackening facilitates obstetrical foetal manipulations. Severe foetal dystocia can be easily and quickly solved when the uterine wall is not tense and when uterine space is not too limited. Clenbuterol is a classical beta-adrenergic agonist presenting high utility to correct bovine dystocias. The myometrial relaxation starts approximately 3 min. after the intravenous administration of a dose of 0.6–0.8 µg/kg [7]. However, clenbuterol is a growth-promoting substance when used at low and repeated doses, and so, similarly to lidocaine, its use has been banned for food-producing animals. As an alternative, 10 mL of 1:1000 epinephrine solution can be used.

6.2.2 Lubrification of the Birth Canal and Foetal Fluid Replacement

After rupture of the second water sac, the amniotic fluid lubricates the birth canal. Amniotic fluid is progressively expelled by successive myometrial and abdominal contractions of the dam. However, as a result of foetal fluid losses, the birth canal quickly dries, increasing the friction between the mucosa and the skin/hair of the foetus or the hands and arms of the obstetrician. Dryness of the birth canal will start at approximately 15–20 min. after the rupture of the amniotic membrane. Consequently, the foetus' hair inside the birth canal will also become dry.

In these cases, a great amount of obstetrical gel (2–4 L) should be introduced into the birth canal to lubricate and facilitate any movement, arm introduction or the final passage of the foetus, preventing lacerations or bruising of the maternal mucosa.

Methylcellulose-based lubes are considered one of the best choices for this purpose. A liquid Vaseline is a secondary but also a viable option. The lubricant should be repeatedly introduced at different intervals according to the duration and intensity of the obstetrical manoeuvres.

The intrauterine cavity also progressively dries due to foetal fluid losses, increasing the risk of uterine lacerations and foetal death during the obstetrical manoeuvres. For extensive or difficult manipulations, the replacement of foetal fluid is strongly recommended. The intrauterine fluid replacer can be obtained adding 1 kg of methylcellulose to 45 L of warm water. Magisterial formulations using Vaseline or linseed tea (250 g/10 L of warm water) can also be efficient. Soap (warm) water can be an alternative in an emergency situation although some mucosal irritation will occur. The water solution is pumped deep into the uterus once or more times. After foetal delivery, both the fluid replacer and the lubricant should be pumped out as much as possible in order to prevent mucosal inflammation and the establishment of a good environment for bacteria growth.

> **Tip**
>
> In delayed Stage II after the amniotic sac breaking, complicated obstetrical manoeuvres inside the uterine cavity are more efficiently and safely performed using foetal fluid replacer.

6.2.3 Obstetrical Equipment and Its Uses

There is some equipment that most obstetricians would designate as essential, while others will only be used in very complicated dystocias. Basic obstetric instrument includes at least two obstetric chains, two handles, a foetal extractor, a Krey hook, and a basic surgery pack (▶ Box 6.1; ◼ Fig. 6.3). Obstetrical chains are preferred over ropes because they are more easily cleaned and because they are less likely to cause damage to the foetus or dam.

All obstetrical equipment (chains, obstetrical forks/rods, hooks, etc.) need to be adequately disinfected or sterilized.

6

> **Box 6.1   Main Equipment and Material Used for Obstetrical Manoeuvres**
> - Chains or ropes.
> - Snares (80–100 cm rope with loop).
> - Snare introducer – Lindhorst and Schriever models.
> - Cämmerer's torsion fork (with canvas cuffs).
> - Kühn's obstetrical crutch (a snare can be adapted).
> - GYN-Stick® torsion fork.
> - Krey-Schöttler double hook (applied to the joints and bones).
> - Two Ostertag's blunt eye hooks (to fix on the orbital bones).
> - Obermeyer's anal hook (used in the pelvis/anus).
> - Blanchard's hook (flexible and long – applied to the joints and bones).

In foetal longitudinal presentations, it is common to detect, partially or completely, one or both limbs (forelimbs or hindlimbs) into the birth canal or at pelvic inlet. In anterior presentations, the foetal head also can be detected. Usually, a chain or rope needs to be attached to the extended limb(s). This procedure allows one or both limbs to be kept readily accessible during and after performing the obstetrical manoeuvres. In a significant part of foetal maldispositions, the foetus and consequently its limbs are pushed inside the uterus to gain sufficient space for mutation. Snares are sometimes needed to fix the head or the lower jaw in lateral deviation of the head too.

Most obstetrical manoeuvres are done inside the uterus, as the pelvis is frequently too narrow to contain both foetal body parts and the obstetrician's arms. In these conditions, an obstetrical fork or rod may be used as arm extensions. Attached to the foetal body, they allow to push, fix or rotate the foetus. The obstetrical forks must be well fixed and used carefully with adequate pressure, to prevent foetal injuries or uterine lacerations if they escape and slide.

For dead foetuses, the use of obstetrical hooks (e.g. Krey hook with attached chain)

allows for quick fixation of foetal parts, reducing the time of intervention. The placement of Ostertag's blunt eye hooks on each orbital bone is one of the best choices to extend or pull the foetal head, in alternative to the use of a long snare. The tear of foetal soft tissues or even bones followed by uterine or vaginal lacerations is the main limitation for hook use.

6.3   Classification of the Obstetrical Manoeuvres

The forces used in the obstetrical manoeuvres can be classified according to their relative direction within different parts of the foetus or between the foetus and the longitudinal axis of the birth canal. In addition to forced traction, four distinct forces can be defined: retropulsion, rotation, version and extension. These forces are used by the obstetrician to correct the foetal maldisposition and should be gentle but consistently exerted on the foetus. This procedure mitigates soft tissue injuries and can improve the survival and welfare of the dam and calf.

6.3.1   Retropulsion or Repulsion

The retropulsion is defined as the force exerted on foetal parts to push the foetus from the birth canal back into the uterine cavity. This manoeuvre is necessary to unblock the pelvis and correct any foetal maldisposition inside the uterine cavity. The Benesch or Kühn crutch is a useful obstetrical tool which can work as an arm extension of the obstetrician or when the introduction of two hands is not possible or advantageous.

6.3.2   Rotation

The rotation is defined as the force exerted upon the foetal longitudinal axis. It is mainly used up to the right or left 180° to correct lateral or ventral foetal positions inside the uterine cavity. Also, a 45° rotation is normally exerted to the foetal hips during the vaginal delivery of the foetus. A higher rotation can be applied to

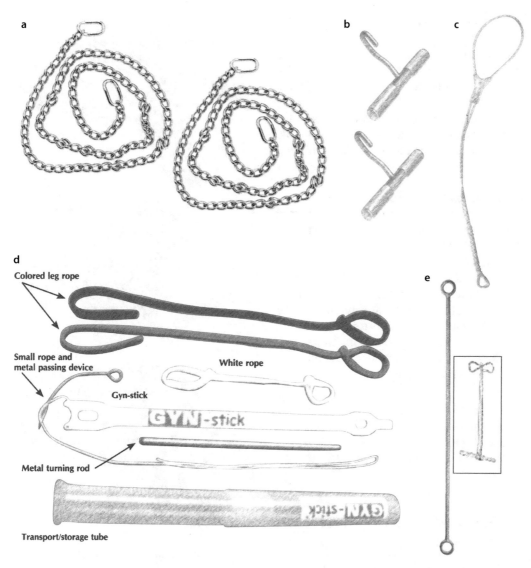

■ **Fig. 6.3** Instruments for obstetrical manoeuvres Legend: **a** A pair of obstetrical chains, 30″–60″ long (76.2–152.4 cm). **b** A pair of obstetrical chain handles, 4″ (10.2 cm). **c** Calf snare (foetal head fixation), 26″ (66 cm). **d** GYN-Stick® (like Kühn's crutch); see ▶ http:// www.gynstick.de/ (accessed on 25-08-2020). **e** Cornell detorsion rod (uterine torsion), 31½″ (82.6 cm). (Modified (FotoSketcher 3.60) from Jorgensen Laboratories, Inc. (USA) with permission)

the limbs present in the vagina or outside the vulva, forcing the rotation of the foetus body until the desired orientation.

6.3.3 Version

The version is defined as a horizontal, vertical or oblique transverse presentation con-

verted to an anterior or posterior longitudinal presentation. It is the most difficult manoeuvre to perform due to the weight and size of the foetus and the limited size of the uterine cavity. Usually, opposite forces are exerted in the different foetal extremities to move the body. The reactive movements of the live foetus may facilitate its displacement.

6.3.4 Extension

The extension is defined as the force used to the extent a flexed joint of foetal extremities such as the limbs, neck and head. This force allows the correction of malpostures. Usually, a simultaneous or sometimes previous retropulsion, creating sufficient space in the uterine cavity, is necessary to extend the flexed joint(s).

6.3.5 Forced Traction

The traction is defined as the force used to pull or extract the foetus or its parts. This force replicates the dam's expulsion forces. The traction is the main force used to pull the foetus, or its parts, towards the birth canal and for its vaginal delivery. It is also extensively used during some procedures such as foetotomy to create tension in the portion being cut. In general, veterinarians do not consider traction as a proper obstetrical manoeuvre.

> **Important**
> During obstetrical manoeuvres, be gentle but effective, be as quick as possible but give time, and use obstetrical lubricant abundantly.

6.4 Foetal Mutation: Anterior Presentation

6.4.1 Carpal Flexion (Knee Joint Flexion)

The carpal flexion can be presented unilaterally or bilaterally. For unilateral flexion, a chain or rope is attached to the extended contralateral forelimb. One hand is introduced into the uterus, and the hoof is grabbed and wrapped in a cupped hand (◘ Fig. 6.4). A loop can be previously applied on the carpal level to slightly pull the forelimb if the hoof is not accessible. A retropulsion of the foetal body or head is made using the second hand of the obstetrician unlocking the brim of the pelvis. A Benesch/Kühn crutch (e.g. GYN-Stick®) should be used for this purpose if the

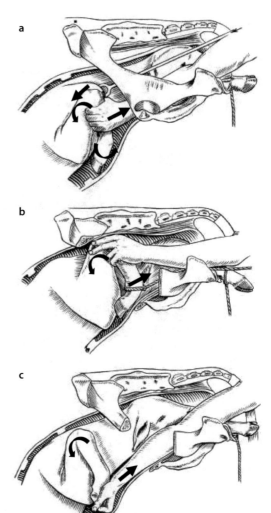

◘ **Fig. 6.4** Obstetrical procedure to solve a unilateral carpal flexion
Legend: The arrows illustrate the direction of the forces. **a** Foetal retropulsion using a Kühn's obstetrical crutch is simultaneously performed with the medial and lateral movements of the hoof and knee, respectively. **b** An obstetrical rope can be attached immediately below to the fetlock joint to help the traction and extension of the limb. **c** The hoof should be covered by the obstetrician's hand to prevent its retention below the pelvic rim and to avoid laceration of the uterine wall. (Modified from Parkinson et al. [8] with permission from Elsevier)

introduction of the second hand is not possible. Simultaneous to the foetal retropulsion, the hoof is moved towards the midline and the knee laterally. Finally, the forelimb is extended. The same procedure is applied to the other forelimb in bilateral carpal flexion.

6.4.2 Shoulder-Elbow Flexion or Elbow Lock Posture

Shoulder-elbow flexion can be presented unilaterally or bilaterally. The elbow remains locked on the brim of the pelvis, and the foetus cannot progress. After placing a chain in at least the contralateral or in both limbs, a retropulsion of the foetal body is made. Simultaneously, traction in the forelimb is exerted to extend it. One limb is extended in bilateral shoulder-elbow flexions at a time.

6.4.3 Shoulder Flexion

The shoulder flexion can be presented unilaterally or bilaterally. Two steps are necessary to solve this forelimb retention. The shoulder flexion should be converted into a carpal flexion which is solved as above-mentioned. The obstetrician secures and pulls the forelimb below the carpal joint until a carpal flexion is obtained (sometimes retropulsion force needs to be applied to the shoulder to increase the space needed for carpal flexion). If necessary, a loop can be applied using a snare introducer.

6.4.4 Deviation of the Head

The deviation of the head can be presented in right or left lateral and ventral postures. The dorsal deviation of the head is rare in bovine foetus, due to the relatively short length of the neck when compared with the equine foetus. Mainly in big dams, the lateral and dorsal deviations of the head can originate additional difficulty to access the foetal head, which is deeply located inside the pregnant uterine horn. Usually, the obstetrician grabs the lip commissure, the nostrils or the lower jaw with his fingers and pulls the head simultaneously with a slight retropulsion of the foetal body, if necessary (🔲 Fig. 6.5). Nevertheless, a loop can be placed between the foetal nape (under the ears) and the mouth. This loop facilitates the traction of the head and the extension of the neck and can also be useful to prevent posterior retention of the head, during the vagi-

nal extraction of the foetus. As an alternative, a snare can be applied to the lower jaw. In the latter case, a smaller force should be used in order to avoid trauma.

The ventral deviation ("vertex posture") of the head is easily corrected if it occurs because of non-extension of the muzzle. The bovine muzzle foetus is anatomically long, and it is sometimes retained under the anterior brim of the dam's pelvis. A slight retropulsion of the front head simultaneous to the muzzle extension is enough to solve this dystocia type. In some cases, the neck may be trapped underneath the basin floor (complete ventral deviation of the head). In dead foetuses, the use of an Ostertag's blunt eye hook in each orbital arch is a quick and useful technique to grab and extend the head. The foetotomy of the neck also quickly solves the more problematic cases.

6.5 Foetal Mutation: Posterior Presentation

> **Important**
> Never convert a posterior foetal presentation into an anterior foetal presentation. It requires a lot of time and much intrauterine space. The foetus can die due to prolonged anoxia or placenta displacement, and the uterine wall can be significantly injured due to the obstetrical manoeuvres. Also, the retention of the head on the pelvis inlet can occur, and additional time is needed to manage it.

6.5.1 Hock Flexion

Hock flexion can be presented unilaterally or bilaterally. The resolution of this postural dystocia is similar to the carpal flexion. A lateral-forward retropulsion of the hock is made simultaneously with the hoof moving towards the midline, followed by complete extension of the limb (🔲 Fig. 6.6). The attachment of an obstetrical rope to the limb passing through the hoof interdigital space is a practical procedure which facilitates the

6

a

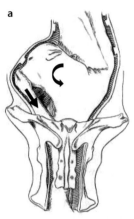

b

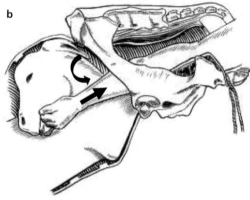

■ **Fig. 6.5** Obstetrical procedure to solve the right lateral deviation of the head
Legend: The arrows illustrate the direction of the forces. Left: The muzzle **a** or the labial commissure **b** of the foetus is grabbed by the obstetrician's hand. The muzzle is pulled with a slight medial rotation of the head. In some cases, a simultaneous retropulsion of the chest is needed to obtain more space for the neck extension. (Modified from Parkinson et al. [8] with permission from Elsevier). Right: Demonstration of the application of an obstetrical snare to fix the head of the foetus. (Courtesy of Carlos Cabral)

extension of the hindlimb, preventing bruises or other trauma to the soft birth canal. These manoeuvres are repeated in the contralateral hindlimb (bilateral flexions), and the foetus is then pulled in posterior presentation.

6.5.2 Hip Flexion or "Breech Presentation"

The hip flexion is usually presented as bilateral flexion (only the tail is palpated in the birth canal) which needs firstly to be converted into a hock flexion (■ Fig. 6.7). A retropulsion of the foetal perineum forward and upward exerted previously or simultaneously to the traction forces applied above the hock joint allows for this conversion. A loop at the hock joint level can also be helpful here. The hock flexion is solved according to what was previously described.

6.6 Foetal Mutation: Transverse Presentation

The transverse presentation requires complex foetal manipulation using the version manoeuvre. These malpresentations and oblique-vertical presentations are rare in cattle being more common in the mare. For live foetuses, a C-section should be considered as prolonged manipulations are to be expected. For dead foetuses, partial foetotomy removing one or two limbs or even the destruncation is recommended.

a

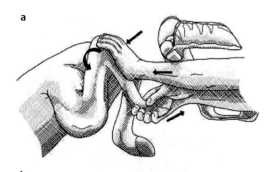

b

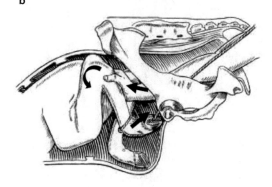

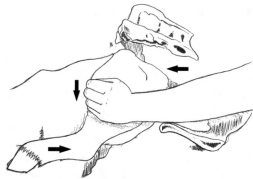

Fig. 6.7 Obstetrical procedure to solve bilateral hip flexion
Legend: The obstetrician slides his hand from the hip towards the hock joint and pulls the member reducing the hip flexion to a hock flexion. A previous or simultaneous retropulsion of the foetus is made. The same procedure is applied in the other member. The hock flexion is solved as reported in the previous figure. (Original from Soraia Marques)

Fig. 6.6 Obstetrical procedure to solve bilateral hock flexion
Legend: The arrows illustrate the direction of the movements. **a** The hock is pushed laterally (retropulsion). Simultaneously, the hoof is moved medially and pulled until the extension of the hindlimb. **b** A rope can be attached immediately above the leg fetlock joint and passing between the hooves. The joint flexes and the limb can be pulled simultaneously with the retropulsion of the hock. (Modified from Parkinson et al. [8] with permission from Elsevier)

6.6.1 Ventro-transverse Presentation

Ventro-transverse presentation can be presented in a right or left cephalo-iliac position. The version manoeuvre aims to convert this malpresentation into an anterior or posterior longitudinal presentation. Both forelimbs and hindlimbs need to be previously identified and followed by the retropulsion of one pair of limbs, including the head in the case of forelimbs. The foetal extremity closer to the pelvic inlet can be chosen and pulled (■ Fig. 6.8). If possible, the opposite foetal extremity should be pushed simultaneously. The fixation of a snare in a limb using a Kühn's obstetrical crutch can be useful. At the same time, a rota-

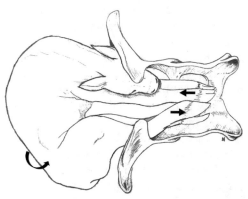

Fig. 6.8 Obstetrical procedure to solve a ventro-transverse presentation
Legend: Once the foetal extremities are identified, the forepart (preferred) or hindpart of the foetus is pushed, and a version manoeuvre is made to put the foetus in a posterior presentation and dorsal position. The retropulsion of the hindpart of the foetus is recommended when this part is more cranial than his forepart. In this last case, the foetus mutates to an anterior presentation. (Original from Soraia Marques)

tion of the foetal body should be tried. Sometimes the foetus remains in the lateral position. In that case, a rotation is performed, such as described below. Foetal extraction in a posterior presentation prevents potential head retention.

6.6.2 Dorso-transverse Presentation

Overall, a version manoeuvre similar to the previously described must be performed to solve the dorso-transverse presentation. Due to the flexion of the vertebral column, foetal extremities may not be entirely accessible. The obstetrician needs to identify specific foetal points to fix the hands or make a snare loop around one limb or even the trunk towards one of the foetal extremities.

6.7 Foetal Mutation: Ventral and Lateral Positions

The ventral (dorso-pubic) and right or left lateral (dorso-ilial) positions mainly occur in singletons which have sufficient space to slightly change their arciform posture. A 45-degree rotation can be considered normal, so that abnormal lateral position is only above this limit. Both ventral and lateral positions can occur in anterior or posterior presentation. These malpositions are less common in cows than in mares in which the foetus needs to rotate from a ventral to a dorsal position during the Stage I or earlier.

In an anterior foetal presentation, these dystocias can be solved rotating the foetus in the opposite direction, up to 180°, until a dorsal position is accomplished. Sometimes a simultaneous slight retropulsion of the foetus is required. This goal is usually achieved by crossing and rotating the two hindlimbs. The use of a Cämmerer's torsion fork is helpful. Both limbs and head need to be extended, and the feet manipulation can be done inside the birth canal or just outside the vulva. Raising the backquarters of downer cows can facilitate foetal rotation.

It should be noted that an anterior or posterior presentation in which the foetus is rotated more than 45° should lead to a suspicion of uterine torsion. Diagnosis and obstetric management of uterine torsion is discussed in ▶ Chap. 2.

> **Tip**
>
> In live foetuses presenting a lateral or dorsal position and normal vitality, the obstetrician can redirect the foetal movements into a rotational force: pinch the foetus to stimulate its movements while crossing and rotating the limbs until reaching the dorsal position.

Rolling the cow can be useful to decrease the degree of rotation. In some cases, it is possible to convert the lateral or ventral position into a foetal dorsal position. Nevertheless, special attention should be given to prevent iatrogenic uterine torsion. Since the cow is rolled up to 180°, both hindlimbs should be fixed or rotated in the opposite direction by the obstetrician, limiting a uterine torsion around the uterine axis.

In a posterior presentation, similar techniques can be used to rotate the foetus. In cows, and contrarily to mares, the risk of rectovaginal fistula as a complication of the dorsal position of the foetus is low. Nevertheless, it can provoke second-degree perineal lacerations [10].

6.8 Foetal Mutation: Twins

The extremities of both foetuses may simultaneously enter into the birth canal impeding the progress of any of them. Most commonly the first foetus will be in posterior presentation and the second in anterior presentation. Usually, three limbs, one head or one tail can be detected in the birth canal. In that case, the retropulsion of the farthest away foetus unblocks the birth canal (◻ Fig. 6.9). The identification of the foetal parts belonging to each one of them is a crucial step to solve this dystocia – we suggest that a string of different colour be secured to the limbs of each calf so that instant recognition is possible when starting traction. Despite the evaluation of joint flexion and relative anatomy of body parts, the slight traction or repulsion of an extremity (limb, head or tail) causes perceptible movements of adjacent portions of the same foetus. This perception is a useful procedure facilitating the identification of each foetus.

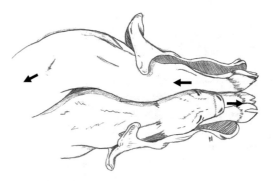

◻ **Fig. 6.9** Obstetrical procedure to solve a dystocia caused by twins
Legend: Once the foetal extremities are identified, the retropulsion of one the foetuses is made unblocking the birth canal. The other foetus is adjusted to a normal foetal disposition, and his extraction takes place. Similar foetal manipulation is made to the second foetus. (Original from Soraia Marques)

Case Study 6.1: Obstetrical Manoeuvres – Dystocia Caused by Twins

A Charolais breed dam weighing around 700 kg, at third pregnancy, presents a delay of Stage II of labour. After sedation (xylazine 2%; 0.04 mg/kg, i.m.), the dam keeps standing up, and a vaginal palpation is performed. Three not crossed legs and one head block the pelvic inlet. Three differential diagnoses are taken into consideration: dystocia due to twins (the most likely), a foetal ventro-transverse presentation or a schistosomus reflexus (extremities presentation). The monstrosity was discarded due to the absence of evident deformities and arthrogryposis. Two forelimbs and one hindlimb were identified comparing the joints' flexion of each one. To differentiate between twins and a single foetus, a slight retropulsion of the head was made dragging both forelimbs without evident movement of the hindlimb. A hand was introduced into the uterus to confirm the parallelism between the forepart and hindpart. The dam was laid down according to Reuff's method. A caudal epidural anaesthesia was performed using 8 mL of procaine 2%. Because the hindlimb was located on the left side of the birth canal, right recumbency was initially chosen to mitigate the pressure. One forelimb was fixed by an obstetrician's hand, and the retropulsion of the forelimb was made. The hindlimb was then fully reintroduced into the uterus. Also, the respective foetus was pushed as deeply as possible to make more free space close to the pelvic inlet. An obstetrical chain was inserted on each forelimb remaining inside the birth canal. Lubricate fluid was poured into the birth canal to facilitate the passage of the calf. Both forelimbs and the head were extended, and the foetus was extracted according to the calving guidelines. Each hindlimb of the second foetus was fixed with chains and extended. Dorsal position was confirmed and the foetus extracted in posterior presentation. At the end, the uterus was inspected for a possible third foetus. Aftercare of the dam and calves was ensured.

Key Points

- Obstetrician's hands and arms are the most important tool to achieve foetal mutation. Other instruments, such as obstetrical forks, are useful in increasing length and strength.
- Foetal mutation is usually performed inside the uterus or through the birth canal.
- Caudal epidural anaesthesia prevents abdominal contractions and defecation of the dam. It improves welfare as it reduces pain and stress in prolonged interventions.
- Plenty lubrification of the birth canal and foetal fluid replacement can determine the success of a foetal mutation.
- Five different obstetrical manoeuvres can be used to manipulate the foetus.
- The accessibility of foetal parts, free intrauterine space and friability of the uterus are significant factors influencing foetal mutation.
- The appropriate foetal point fixation by the obstetrician's hands, snares, forks or chains is crucial to moving the foetus.
- The movements of the live foetus can be redirected helping its mutation.
- The version manoeuvre requires the application of forces on the forepart in the opposite direction to those applied on the hindparts.

❓ Questions

1. When is the foetal fluid replacer required during foetal manipulation?
2. Describe the obstetrical manoeuvres used for foetal mutation and give examples.
3. What are the hardest foetal dystocias to solve using obstetrical manoeuvres? Justify.

✅ Answers

1. In prolonged Stage II, after amniotic sac breaks, the surfaces of the birth canal, foetus, placenta and endometrium become too dry. In consequence, the increasing friction between the foetus and the maternal surfaces makes the movement of both the foetus and the obstetrician's hands during obstetrical manoeuvres difficult. The introduction of foetal fluid replacer in the uterus allows the obstetrician to lubricate these surfaces, and it temporarily expands the intrauterine cavity. Warm foetal fluid replacer should be pumped inside the uterus as soon as desiccation is detected. Usually, up to 4 L at a time is pumped. This procedure is repeated whenever friction increases: three or four times for prolonged intrauterine foetal manipulation (>30 min.). For more effortless obstetrical manoeuvres performed close to the pelvic inlet, or inside the birth canal, the use of plenty lubricant is enough. A uterine lavage to remove the foetal fluid replacer and lubricant is needed at the end.

2. Five different obstetrical manoeuvres can be used for foetal mutation. They are classified according to the relative direction of the forces exerted on the foetus. The retropulsion or repulsion is the act of cranially pushing one foetal body part into the uterus. It is mainly used to obtain enough space in the birth canal or at the pelvic inlet, allowing for other obstetrical manoeuvres (e.g. to extend a limb or the head). It can be used previously or simultaneously with other manoeuvres. The rotation aims to switch the longitudinal axe of the body foetus up to 180°. This force is used to solve ventral and lateral positions. The force (more than 180°) is usually applied to the limbs, by force crossing them, and transmitted to the foetal body. The extension is the act of extending a flexed joint of one foetal extremity. All foetal malpostures are due to the flexed joints of the limbs (e.g. tarsal and hock flexions) or cervical vertebrae (head deviation). The version is defined as the act of transforming a transverse or vertical/oblique presentation into an anterior or posterior longitudinal presentation. In practice, this involves the use of repulsion, traction and rotation forces. Finally, traction is the act of pulling one part of the foetal body. It is mainly used to deliver

the foetus. However, several small forced tractions on the foetus are made to adjust the foetus inside the uterus during foetal manipulations.

3. Several main factors can affect the ease of the obstetrical manoeuvres and foetal mutation: foetal size, weight, conformation, number and status (dead or alive), size of the dam, intrauterine free space, diameter of the birth canal, straining of the dam and myometrial contractions, friability of the uterine wall, use of lubricant and foetal fluid replacer and the ability and strength of the obstetrician. Related to these factors, some foetal maldispositions require more intensive foetal manipulations. The transverse presentations are usually the hardest maldispositions to solve. The version manoeuvre requires the intrauterine displacement of the entire foetus from a perpendicular to a longitudinal axis. The dorso-transverse presentation is more challenging than the ventro-transverse presentation due to the less accessibility of the limbs or head and neck. In most cases, C-section seems to be better or the unique option to solve this dystocia. Also, the bilateral hip flexion can be problematic mainly in foetus with a large hip. Sometimes it is not possible to transform a hip flexion into a hock flexion in both limbs. In addition, sometimes the lateral and complete ventral deviation of the head is irreducible. The head remains inaccessible, becoming impossible to fix the obstetrician's hand, a snare or even a pair of Ostertag's blunt eye hook in dead foetuses. In this last case, a partial foetotomy of the head quickly solves the dystocia.

References

1. Ames NK. Chapter 4: Epidural anesthesia. In: Noordsy's food animal surgery. 5th ed. Ames: John Wiley & Sons, Inc; 2014. p. 39–50. https://doi.org/10.1002/9781118770344.ch4.

2. Bigham AS, Habibian S, Ghasemian F, Layeghi S. Caudal epidural injection of lidocaine, tramadol, and lidocaine-tramadol for epidural anesthesia in cattle. J Vet Pharmacol Ther. 2010;33(5):439–43. https://doi.org/10.1111/j.1365-2885.2010.01158.x.

3. Dehghani SN, Bigham AS. Comparison of caudal epidural anesthesia by use of lidocaine versus a lidocaine-magnesium sulfate combination in cattle. Am J Vet Res. 2009;70(2):194–7. https://doi.org/10.2460/ajvr.70.2.194.

4. Grubb TL, Riebold TW, Crisman RO, Lamb LD. Comparison of lidocaine, xylazine, and lidocaine-xylazine for caudal epidural analgesia in cattle. Vet Anaesth Analg. 2002;29(2):64–8. https://doi.org/10.1046/j.1467-2995.2001.00068.x.

5. Imani Rastabi H, Guraninejad S, Naddaf H, Hasani A. Comparison of the application of lidocaine, lidocaine-dexamethasone and lidocaine-epinephrine for caudal epidural anesthesia in cows. Iran J Vet Res. 2018;19(3):172–7.

6. Lee I, Yoshiuchi T, Yamagishi N, Oboshi K, Ayukawa Y, Sasaki N, Yamada H. Analgesic effect of caudal epidural ketamine in cattle. J Vet Sci. 2003;4(3):261–4.

7. Ménard L. The use of clenbuterol in large animal obstetrics: manual correction of bovine dystocias. Can Vet J. 1994;35(5):289–92.

8. Parkinson TJ, Vermunt JJ, et al. Dystocia in livestock: delivery per vaginam. In: Noakes DE, Parkinson TJ, England GCW, editors. Veterinary reproduction and obstetrics. 10th ed. Edinburgh: Elsevier; 2019. p. 250–76. https://doi.org/10.1016/B978-0-7020-7233-8.00014-8.

9. Sato RE, Kanai E, Tsukamoto A. Comparison of lidocaine-xylazine and procaine-xylazine for lumbar epidural anesthesia in cattle. J Hell Vet Med Soc. 2019;70(3):1727–32.

10. Sato R, Kamimura N, Kaneko K. Surgical repair of third-degree perineal lacerations with rectovestibular fistulae in dairy cattle: a series of four cases (2010–2018). J Vet Med Sci. 2019;81(5):703–6. https://doi.org/10.1292/jvms.19-0004.

11. Skarda RT, Muir WW, Hubbell JA. Comparative study of continuous lumbar segmental epidural and subarachnoid analgesia in Holstein cows. Am J Vet Res. 1989;50(1):39–44.

12. Vesal N, Ahmadi M, Foroud M, Imani H. Caudal epidural anti-nociception using lidocaine, bupivacaine or their combination in cows undergoing reproductive procedures. Vet Anaesth Analg. 2013;40(3):328–32. https://doi.org/10.1111/vaa.12000. Epub 2012 Oct 20.

Obstetrical Manoeuvres Online

Drost M, Samper J, Larkin PM, Gwen Cornwell D. Obstetrics: calving problems. The visual guide to bovine reproduction. Gainesville: UF College of Veterinary Medicine; 2019. From: https://visgar.vetmed.ufl.edu/en_bovrep/calving-problems/calving-problems.html. Accessed on 3 Mar 2020.

Suggested Reading

Mortimer RG, Toombs RE. Abnormal bovine parturition. Obstetrics and fetotomy. Vet Clin North Am Food Anim Pract. 1993;9(2):323–41. https://doi.org/10.1016/S0749-0720(15)30650-2.

Foetotomy

Contents

© Springer Nature Switzerland AG 2021
J. Simões, G. Stilwell, *Calving Management and Newborn Calf Care*,
https://doi.org/10.1007/978-3-030-68168-5_7

☞ Learning Objectives

- To identify the indications, contraindications and complications of foetotomy.
- To prepare the dam for a foetotomy.
- To manipulate adequately the foetotomy material.
- To define the appropriate cut sections and procedures for total or partial percutaneous foetotomy.
- To take decisions regarding the postoperative care of the dam.

7.1 Introduction

The foetotomy or embryotomy is an obstetrical technique which reduces foetal volume. As a result, a dead foetus is severed into two or more parts. It is an alternative to caesarean section when the vaginal delivery of the entire dead foetus is not possible or recommended.

The development of the foetotomy technique started in the nineteenth century using the subcutaneous technique. In the early twentieth century, the surge of the obstetrical foetatome using a saw wire allowed for a safer method to cut the foetus inside the uterine cavity in large animals. This percutaneous technique replaced the subcutaneous foetotomy [1], and it is currently the most widely used method in cows.

The foetotomy requires a sufficiently large birth canal to ensure good access to the foetus. The integrity of the uterus and birth canal is also a key factor ensuring a successful foetotomy: ruptured uterine or vaginal wall may result in haemorrhages, peritonitis, sepsis, shock and death of the dam. Also, further fertility can be adversely influenced due to uterine adhesions, fibrosis and infection. Total foetotomy commonly involves oversized foetuses and requires significant physical effort and experience of the obstetrician to complete the whole cut sequence. Partial foetotomy is useful to solve dystocia due to irreducible foetal malposture or skeletal malformations. Adequate restraint of the dam, safety of the dam and obstetrician and preoperative care, including epidural anaesthesia, are essential for a successful foetotomy. A hygienic and aseptic procedure, abundant use of obstetrical lubricant during foetal manipulation and appropriate aftercare, reduces postoperative complications.

In the present chapter, we mainly describe the use of the percutaneous foetotomy.

7.2 Indications and Contraindications of the Foetotomy

Foetotomy may be the only way to achieve vaginal delivery of dead foetuses when wishing to avoid the use of caesarean section (C-section). The indications for foetotomy are related to the incapability to remove an entire foetus despite having accessibility to manipulate it. So, foetal dystocias blocking the pelvic inlet of the birth canal are the main indication to perform a foetotomy (▶ Box 7.1). The most frequent reasons are foetopelvic disproportion and foetal maldispositions. The emphysematous foetus is also a particular indication since the survival rate of the dam is low when a C-section is performed.

Foetal death is the first premise for performing foetotomy. In the presence of schistosomus reflexus and other monsters, the foetus can still be alive at intervention time. Uncertainty as to foetus death can occur, for example, in severe foetal hypoxia, in which no foetal movements or pulse is detected.

> **Box 7.1   Indications of Foetotomy**
> - Foetal death.
> - Irreducible foetal maldisposition.
> - Oversized foetus and foetopelvic disproportion (anterior presentation).
> - Alterations of foetal conformation (e.g. teratology foetal monsters).
> - Emphysematous foetus.

Unless the foetus is blocked in the birth canal, the main cuts must be performed inside the uterine cavity. The birth canal must be sufficiently broad to allow easy access to the foetus. Uterine and vaginal lacerations are a crucial aspect to take into consideration when deciding for a foetotomy. Uterine haemorrhages and bacterial contamination of the pelvic and peritoneal cavities decrease survival rate or the dam's further fertility. In fact, trauma of the soft birth canal and the duration of the labour have been reported as major risk factors of pelvic phlegmons [2]. Particular attention should also be given when the uterine wall is friable.

All these points are or can be contraindications for foetotomy (▸ Box 7.2). However, if the C-section is not recommended (see indications, contraindications and complications of C-section in ▸ Chap. 8) and a dead foetus is accessible, total foetotomy is an option to consider. Nevertheless, recumbent and deeply depressed dams in septic shock carrying an emphysematous foetus are not candidates to either foetotomy or C-sections. Even if pre-stabilization using fluids, antimicrobial and analgesic therapies and intensive post-intervention care are ensured, the expected survival rate of the dams is very low. In this condition, euthanasia of the dam is usually the best option.

Box 7.2   Contraindications of Foetotomy
- A live foetus.
- Narrow birth canal.
- Haemorrhage.
- Uterine or vaginal rupture.
- Significant vulvar lacerations.
- Friable uterus.
- Dams in shock without previous stabilization.
- Oversized foetuses in posterior presentation (enlarged thorax).

7.3 Foetotomy Techniques

7.3.1 Percutaneous Versus Subcutaneous Foetotomy

In the percutaneous foetotomy, the friction and abrasion of the saw wire will cut through soft tissues and skeletal bones. The tissues need to be under tension, limiting its movement. This aspect is important to cut soft tissue, such as skin and subcutaneous tissue, viscera and muscles. To achieve quick and complete cutting, external traction and tension should be concomitantly exerted on the body part being severed.

One of the major limitations of the percutaneous technique is the lack of tension points on the portion to be cut. However, the length of the foetatome (approximately 90 cm) usually is enough for a safe and single cut in any accessible part of the foetus. In oblique cranial cuts, the foetatome applies pressure on the foetus in the opposite direction to the foetal traction. This opposite pressure allows for an additional tension on the cut.

❯ Important

Usually, the bone cut leaves sharp edges, which should be covered by the obstetrician's hand at the time of extracting the respective foetal portion, preventing uterine, vaginal or vulvar lacerations.

Subcutaneous foetotomy is performed using an obstetrical knife attached to the obstetrician's indicator finger (e.g. Guenther's foetotomy finger knife). The cuts are limited to the skin, subcutis and other soft tissues which are easily accessible. It can also be used to disarticulate joints. The use of subcutaneous foetotomy requires special care to prevent injuries to the dams' soft tissues along the birth canal.

7.3.2 Total Versus Partial Foetotomy

Foetotomy can be classified as total or partial. Total foetotomy is defined as a complete cut sequence of the foetus and remotion of its several portions by vaginal extraction. Usually, the Utrecht technique is used and involves up to six or seven cuts ([3]; ▸ Box 7.3; ◼ Fig. 7.1). The complete cut sequence is adequate for oversized foetuses. However, because some body parts may still be quite large making it difficult to pass through a relatively narrow birth canal, additional cuts can be made. Similarly, additional cuts may also be needed in posterior presentations. The Hannover technique, according to Götze, involves up to nine cuts ([4]; ◘ Fig. 7.2). Additional cuts can be performed on carpal/tarsal joints and thorax (two hemi-transverse and one longitudinal cuts).

Box 7.3 Complete Cut Sequence for Total Foetotomy, Based in the Utrecht Technique

Anterior presentation
1. Amputation of the head (caudal to the ears).
2. First forelimb cut (including the scapulae).
3. Second forelimb cut.
4. Thoracic detruncation at the middle of the sternum (removal of neck and the cranial portion of the chest).
5. Thoracic detruncation immediately after the last rib followed by abdominal evisceration.
6. Longitudinal cut of the rib in very large foetuses (optional).
7. Bisection of the pelvis.

Posterior presentation
1. First hindlimb cut.
2. Second hindlimb cut (optional).
3. Abdominal detruncation (before the last rib) followed by evisceration.
4. Thoracic detruncation (middle of the sternum).
5. Longitudinal cut of the rib in large foetuses (optional).
6. First forelimb cut or diagonal forepart cut.
7. Second forelimb cut (optional).

According to the Schätz technique, the number of cuts can be reduced down to four sections in normal or small-sized foetuses (◘ Fig. 7.3). In anterior foetal presentation, the cuts are (1) one diagonal cut that simultaneously removes the head, neck, one forelimb and a cranial portion of thorax; (2) a second diagonal cut that is made at the thorax level, performing the thoracic detruncation with thoracic and abdominal evisceration; (3) a third cut that involves a deep abdominal cut; (4) and, finally, a diagonal longitudinal bisection of the pelvis. In the posterior presentation, a diagonal section of the hindpart removing one limb is followed by two diagonal cuts of the trunk; finally, a diagonal cut is similar to the first one of the anterior presentation. A modification of the Utrecht technique to reduce the number of cuts was also described by Mortimer et al. [6].

In foetal ventro-transverse presentation, the head, neck and limbs are the most accessible part of the foetus. These foetal extremities are the first ones to be removed. Both carpal and tarsal joints are sectioned. In each hindlimb, a second cut must be performed at stifle joint level.

In dorso-transverse presentation, the first cut aims to achieve detruncation and to split the foetus into two parts, improving the accessibility to each one in the following cuts.

Partial foetotomy can be defined as the cut of one or more (usually two) foetal portions, without completing the whole cut sequence. It is very useful for those cases where the foetal obstruction is caused by an irreducible foetal presentation, position or postural defect or for some anchylosed or deformed monsters as well (e.g. schistosomus reflexus, perosomus elumbis, hydrocephalus, dicephalus and spina bifida). The cut of the foetal portion which blocks the pelvic inlet can be removed, and the dystocia is usually solved. Partial foetotomy is also commonly used for postural defects such as carpal, hock or hip flexions and lateral deviation of the head. The cut of these extremities should be just cranial to the pelvic inlet. Flexion usually provides adequate tension during the cut, and no evisceration is needed. The limb extremities can be quickly cut and removed without significant contamination of the uterus by foetal blood and tissues. In consequence, uterine lavage after partial foetotomy is generally unnecessary.

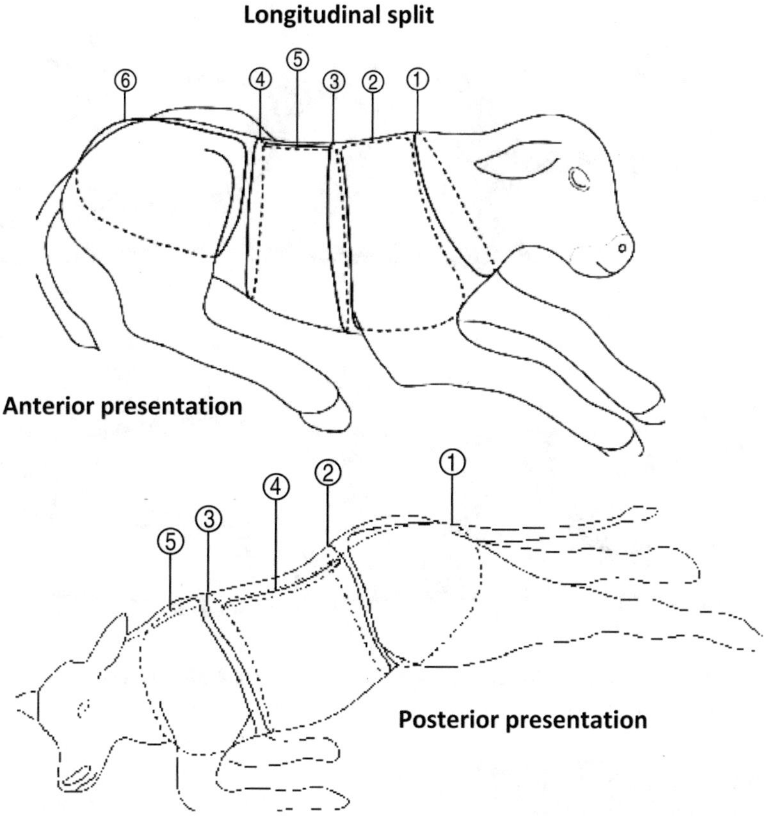

Longitudinal split

⑥   ④⑤ ③ ② ①

Anterior presentation

⑤ ③ ④ ② ①

Posterior presentation

◘ **Fig. 7.1** Schematic representation of sequential cuts required for a total percutaneous foetotomy. The longitudinal cut of the thoracic/lumbar ring is optional (also see ▸ Box 7.3 for the Utrecht technique). (Modified from Vermunt and Parkinson [5] with permission from Elsevier)

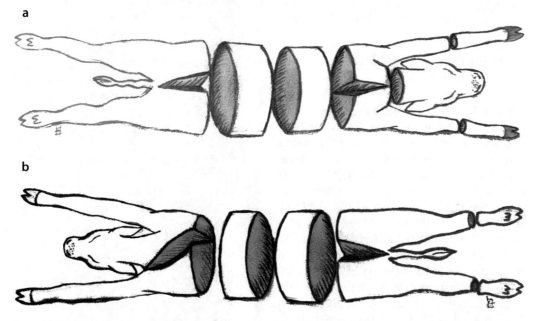

a

b

◘ **Fig. 7.2** The Hannover technique according to Götze. **a** Anterior presentation (9 cuts). **b** Posterior presentation (8 cuts). (Original from Sorraia Marques)

7

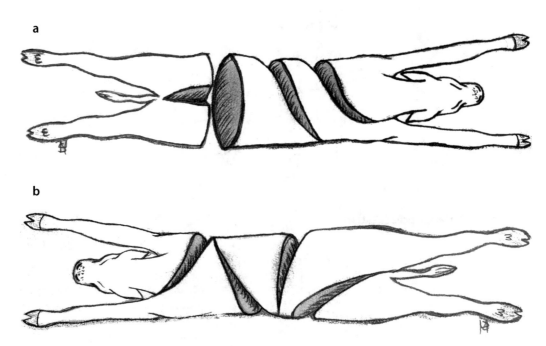

◘ Fig. 7.3   Total foetotomy according to Schätz. **a** Anterior presentation (4 cuts). **b** Posterior presentation (4 cuts). (Original from Sorraia Marques)

> **Tip**
>
> Partial foetotomy can also be used during the C-section. It is done through the uterine incision and aims to reduce the foetal diameter (e.g. schistosomus reflexus) preventing tears of the uterine wall.

7.3.3 Equipment and Other Material

The basic material involves the general obstetrical equipment used in an assisted calving and specific tools to perform the foetotomy. The foetatome, saw wire and its accessories as well as obstetrical hooks to fix foetal parts are the main tools needed to perform a percutaneous foetotomy (▶ Box 7.4; ◘ Fig. 7.4). All the material should be sterilized or at least washed and disinfected in their first use. The equipment cleanliness and disinfection between cuts can be necessary in heavily contaminated environments. In addition to this, tissue debris originating from the previous cut need to be removed from the saw wire to facilitate the next sawing.

> **Box 7.4   Basic Equipment for Foetotomy Procedure**
> - Thygesen's foetatome (Swedish modification) or Utrecht foetatome.
> - Obstetrical saw wire according to Liess: 5 to 6 metres (at least five times the length of the foetatome); the availability of one extra complete coil (12 metres), for broken wire substitution, is recommended.
> - Obstetrical saw wire handles (a pair).
> - Wire threader.
> - Wire introducer (e.g. Schriever model).
> - Wire brush (for cleanliness).
> - Plier (to cut the wire).
> - Krey hook with attached chain or rope.
> - Obstetrical chains or ropes.
> - Obstetrical handles (a pair; e.g. Moore's or T-bar models).
> - For subcutaneous foetotomy: Guenther's or Linde's foetotomy finger knife and a long spatula.

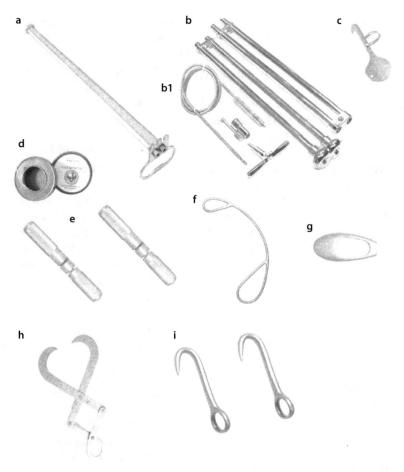

◻ Fig. 7.4 Obstetrical instruments for foetotomy. Legend: **a** Utrecht foetatome, 38″ (96.5 cm). **b** Thygesen's foetatome, 33″ (83.8 cm long). Two-tube model with adjustable handle for obstetrical chain. B1 Insertion coil and cleaning brush for foetatomes. **c** Linde's embryotomy knife, 5″ (12.7 cm). **d** A pair of obstetrical saw wire, 10 m. **e** A pair of obstetrical saw wire handles (Sutter style), 4″ (10.2 cm). **f** Obstetrical wire guide, 7″ (17.8 cm). **g** Obstetrical wire drop guide, 5 ½" (14.0 cm). **h** Obstetrical (double action) Krey-Schöttler hook, 8″ (20.3 cm). **i** A pair of Ostertag's blunt eye hooks, 2″ (5.1 cm). (Modified (FotoSketcher 3.60) from Jorgensen Laboratories, Inc. (USA) with permission)

Obstetrical gel and foetal fluids' replacer should be used in vast quantities. Lubrication of the birth canal avoids inflammation and lacerations due to successive re-introduction of the obstetrician's hands and arms and the foetatome and the retrieval of foetus parts. For total foetotomy, large quantities of foetal fluids' replacer (see ▸ Chap. 6) should be previously prepared. Five to ten litres at a time are needed to lubricate the uterine cavity mainly during foetal destruncation and whenever the birth canal is getting dry. Foetal fluids' replacer is introduced inside the uterus using a pump which also serves to siphon the fluids at the end of the foetotomy procedures or to perform a final uterine lavage.

7.4 Preparation of the Dam

The area of the farm where the foetotomy is performed needs to be safe and clean. In the end, the used area should be washed and disinfected. Also, the different parts of the foetus need to be appropriately packaged and removed from the farm according to the biosecurity norms.

7.4.1 Dam Position and Cleanliness

Foetotomy can be performed with the cow standing or lying. The standing position is preferable for total foetotomy as it minimizes the external pressure exerted on the uterine cavity and on the foetus. In lying cows, the hindquarters must be raised. The cow usually is first placed in left lateral decubitus to remove the forepart (anterior presentation) or backside (posterior presentation).

After the tail is tied, the perineal and vulvar regions must be cleaned and washed using neutral soap and disinfected. Warm water and antiseptic iodine or chlorhexidine solution should be available in two buckets to wash and soak the obstetric instrument and the obstetrician's hands/arms during the foetotomy. This procedure mitigates uterine contamination caused by repeated insertion of arms and instruments into the birth canal.

7.4.2 Premedication and Epidural Anaesthesia

The stabilization of depressed and hypovolaemic dams is described in ► Chap. 8. The advantages and limitations of antimicrobial, analgesic, sedative, uterine myorelaxant and caudal epidural anaesthesia are also reported in that chapter. A similar use of each drug may be recommended in the foetotomy manipulation.

The use of uterine myorelaxant is controversial. It is true that myometrial contractions are blocked and that iatrogenic uterine atonia may facilitate foetal manipulation during the foetotomy procedure. This myometrial relaxation is particularly beneficial in the presence of large foetuses, providing more intrauterine space to move it. However, uterine relaxation may produce folds of the wall which are very prone to laceration when arms and foetatome are introduced repeatedly. Particular caution is needed to prevent the inclusion of a uterine fold in the cut or during the traction of foetal body parts.

Caudal epidural anaesthesia is mandatory in total foetotomy. This regional anaesthesia reduces abdominal contractions originated by the sensory receptors present in the vulva, vagina and cervix. This decrease in abdominal contractions is helpful to successively manipulate the foetus. In recumbent dams, the restraint of hindlimbs and its elevation can also be facilitated by the cranial epidural anaesthesia.

7.4.3 Percutaneous Foetotomy Manipulation

Once the foetal portion to cut is identified, the saw wire must be placed around it. Two techniques can be used for this purpose:

1. The saw wire attached to the wire threader is introduced into one barrel of the foetatome. The half-threaded foetatome is inserted into the birth canal or uterus. At that time, a loop covering the foetal portion is made using a wire introducer. The wire is attached to the threader located in the other barrel and is pulled back. This method is useful to insert the saw wire around the trunk or around a flexed extremity.

2. A previous loop is made introducing the saw wire in both barrels of the foetatome. The foetatome is partially inserted into the birth canal, and the loop is passed around the portion to be cut passing using the free foetal extremity.

> **Tip**
>
> For the limbs, the saw wire can be temporally fixed in the interdigital cleft of the foetal hoof pushing the foetatome. Then, the loop wire is more easily placed on the cut region.

A handle is attached to each end of the wire to allow wire tension and back and forth sawing movements. After the placement of the saw wire around the foetal portion to be cut, the barrels of the foetatome are fixed against the foetus by a hand of the obstetrician (► Box 7.5).

Four cut sections can be used: longitudinal, transversal, oblique cranial and oblique caudal (◻ Fig. 7.5). This selection depends on the relative topography between the localization of the foetatome extremity and the direction of the wire loop around the foetal portion to be cut. The main vectorial forces are simultaneously exerted towards the centre of the foetal portion and the head of the foetatome. The direction of these vectorial forces allows a safe cut under the tension of the foetal portion to be cut and the saw wire. The tension of the foetal portion needs to be applied during the whole cut procedure by concomitant traction using obstetrical chains or a Krey hook with an attached rope. Tension on the wire is ensured by pulling alternatively each handle (and wire end).

Crossing the wire while sawing may cause it to break. A too rapid sawing movement is another factor leading to wire breakdown, due to the abrupt increase in heat. The sawing, when performed by an assistant, should be firm and consistent but always directed and coordinated by the obstetrician.

It should be remembered that the wire will cut or damage anything it rubs against. For this reason, care should always be taken to protect or keep it away from cows' mucosae and the obstetrician's hands/arms.

Box 7.5 Basic Procedures of Percutaneous Foetotomy
- Ensure that all foetotomy equipment, other material and assistances are available.
- Use epidural anaesthesia.
- Restrain the dam adequately.
- Make a work plan of successive cuts even if you do not complete it.
- Re-evaluate the foetal disposition before starting.
- Use plenty lubricant.
- Use foetal fluids' replacer.
- Place the saw wire and the head of the foetatome on the section to be cut.
- Cutting joints is preferable to bones.
- Fix the foetal part and perform the cut – swiftly and continuously.
- The foetal portion should be sufficiently small but avoid cutting irrelevant portions.

- Cover the sharp edges of the bones with the hand at removal time.
- Ensure aseptic conditions during the whole foetotomy procedure.
- Check uterine integrity and content at the end.
- Perform uterine lavage, if deemed necessary.

> **Important**
> One hand is placed close but outside of the head of the foetatome to fix it during the cut. The other hand firmly sustains the opposite extremity of the foetatome during sawing. An assistant is responsible for the wire back and forth movements.

7.5 Cut Sequence of Total Foetotomy: Anterior Presentation

The following main sequences and procedures are described considering normal foetal position posture. For lateral and ventral positions, the head of the foetatome is fixed to the foetus in the same superficial locals.

7.5.1 Amputation of the Head and Neck

The decapitation or neck amputation of the foetus is the first cut. It is performed in a single cut (▢ Fig. 7.6). A wire loop around the neck is made using one of the techniques mentioned above. The removal of the head and a great portion of the neck provides more space for the next cut. The cut can be transversal or oblique cranial. For a relatively or absolutely large head, the wire is placed in the cranial portion of the neck, immediately after the foetal nape according to the Utrecht technique. In that case, the cut is transversal. A Krey hook or two orbital hooks are used to apply tension on the head, and a transversal cut is performed. The cut of the neck close to the withers removes a great part of it, and it decreases the foetal diameter at the shoulder level. Sometimes

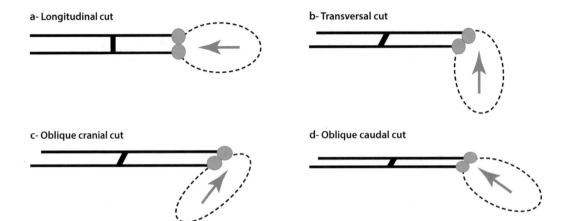

a- Longitudinal cut

b- Transversal cut

c- Oblique cranial cut

d- Oblique caudal cut

◻ **Fig. 7.5** Classification of the cuts. (Original from João Simões). Legend: Each cut is classified according to the direction of the cut towards the extremity of the foetatome: the cut can be parallel **a** and transversal **b** and oblique cranial **c** or caudal **d** to the long axis of the foetatome

7

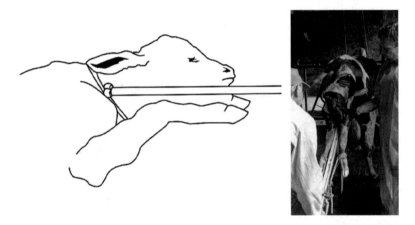

◻ **Fig. 7.6** Decapitation and neck amputation. Legend: Left A transversal cut of the neck is performed caudally, close to the chest, unless if a larger foetus is present. (Modified from Walters [7] with permission from John Wiley and Sons. Right Extraction of a foetus after decapitation. Courtesy of Carlos Cabral)

it is enough to remove the remaining part of the foetus. In small-sized foetuses, the head, neck and one forelimb may be severed using a single diagonal cut (◻ Fig. 7.7).

Tip

Sectioning the neck close to the withers circumscribes the movements of the exposed cervical vertebra with sharp edges, preventing injuries to the uterus wall and birth canal.

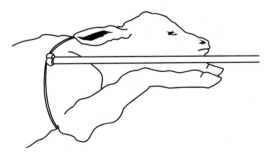

◻ **Fig. 7.7** Diagonal cut of the head, neck and one forelimb. (Modified from Walters [7] with permission from John Wiley and Sons)

Fig. 7.8 Cut of one forelimb. (Modified from Jackson [8] with permission from Elsevier)

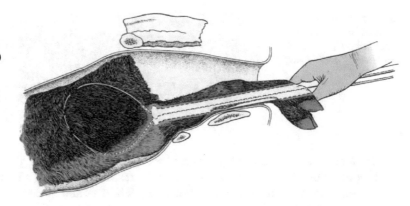

7.5.2 Cut of the Forelimb(s)

Once the head and neck are removed, one of the forelimbs must be severed. A wire loop is previously made and introduced into the birth canal with the foetatome. The loop is placed around the hoof. The foetatome is pushed laterally to the limb, dragging the wire. The saw wire and foetatome head are placed craniocaudally and dorsocaudally to the scapula, respectively. The cut is oblique cranial. A significant amount of tension on the wire and foetatome is needed to fix them in this position. The tension on the forelimb is ensured by a chain placed above the hoof. As an alternative, the foetatome can be introduced medially to the limb up to the foetal axilla. The wire needs to surround the scapula. In that case, the wire is fixed under the scapula by constant tension. The cut is longitudinal (**Fig. 7.8**). The saw wire sliding is a common problem when cutting this triangular bone. Remaining sharp fragments of the scapula should be removed or covered so that they do not scratch against the uterine wall.

Usually, the removal of one forelimb is enough, and the second forelimb can be used to pull the thorax. However, for oversized foetuses, the second forelimb may also need to be sectioned and removed according to what was described for the first one.

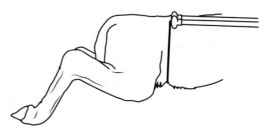

Fig. 7.9 Foetal destruncation. (Adapted from Walters [7] with permission from John Wiley and Sons)

7.5.3 Abdominal Detruncation and Evisceration

A wire loop is made around the thorax, usually using a wire introducer. Both the wire and the head of the foetatome are pushed caudally to the last rib in the lumbar region as deeply as possible. The foetatome head is placed dorsolaterally to the foetal trunk to obtain a transverse cut at mid-abdomen (**Fig. 7.9**). A Krey hook is applied onto a cervical vertebra. If only one forelimb was removed, the second forelimb can also be used to apply tension on the foetus. After being cut, the thorax is removed. At this time, most parts of the abdominal viscera are torn and removed (evisceration procedure). Special care is needed to prevent uterine injuries from sharp edges of

■ **Fig. 7.10** Bisection of the pelvis. (Adapted from Jackson [8] with permission from Elsevier)

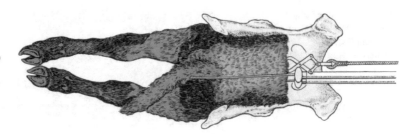

the sectioned vertebra. One useful procedure is the application of the Krey hook in the sectioned vertebra, exerting slight traction and covering the vertebra with a hand concomitantly to evisceration.

> **Tip**
>
> The diaphragm of an emphysematous foetus can be manually ruptured and the viscera removed reducing the thoracic volume.

> **Important**
>
> The use of the Krey hook to grab the sectioned vertebra is essential to keep tension and to more easily pull the foetus during foetotomy. The sectioned vertebra will also be straightened, preventing uterine and birth canal trauma from its sharp edges. All care should be taken to prevent the bonny sharp edges getting stuck on the uterine and vaginal walls during the traction procedure.

In oversized foetuses, a second abdominal cut may be necessary. The loop of the saw wire is inserted ventrally to the wings of the foetal ilium. The head of the foetatome is placed dorsally to the trunk, and a transverse cut is made.

7.5.4 Bisection of the Pelvis

The saw wire attached to the wire introducer is inserted sagittally between both hindlimbs around the foetal pelvis (■ Fig. 7.10). The head of the foetatome is placed cranially to

the pelvis, and a longitudinal cut is made. The remaining trunk is slipped laterally to the vertebral column. Once the bisection is completed, each limb is removed one at a time. The Krey hook is used to pull the member. Each member can be carefully turned inside the uterus and removed.

7.6 Cut Sequence of Total Foetotomy: Posterior Presentation

Overall, a similar number of cuts can be inflicted to complete the foetotomy of foetuses presenting a posterior presentation with the hindlimbs extended.

The foetatome is introduced laterally to the first hindlimb. An oblique cranial cut is made using a wire loop between the foetal groin and the great trochanter (■ Fig. 7.11). For oversized foetuses, the other hindlimb must also be sectioned and removed. The following step is the destruncation at midline abdominal level, followed by evisceration. According to the size of the foetus and the different methods reported in ■ Figs. 7.1 and 7.2 or ■ Fig. 7.3, the bisection of pelvis and two more transversal or diagonal cuts can be performed to split the foetal trunk.

> **Tip**
>
> The destruncation just after the iliac tuberosities followed by the bisection of the pelvis of the oversized foetus reduces the duration of the total foetotomy.

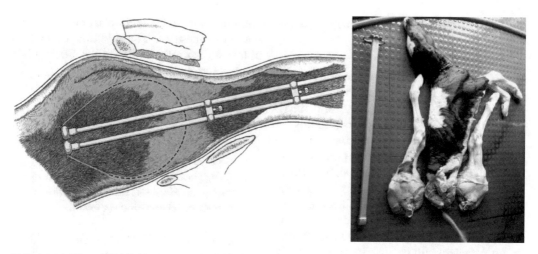

◻ Fig. 7.11 Cut of hindlimbs. Legends: Left Oblique cranial cut of one hindlimb (Thygesen's foetatome). (Adapted from Jackson [8] with permission from Else-vier). Right Cut of both hindlimbs (Utrecht foetatome) followed by foetus extraction. (Courtesy of Carlos Cabral)

7.7 Partial Foetotomy in Abnormal Postures, Hiplock and Foetal Abnormalities

Partial foetotomy is useful to solve dystocia caused by unilateral or bilateral abnormal posture in dead foetuses. Without destruncation, foetal manipulation is quick and relatively clean —nevertheless, the introduction of foetal fluids' replacer can be necessary, and a final uterine lavage is recommended. Sometimes, additional cuts are needed to ensure appropriate accessibility to the postural defect.

In most cases, the half-threaded foetatome is inserted in the birth canal, and a saw wire loop is passed around the flexed articulation using a wire introducer. Usually, the flexion of the extremity is in the opposite direction of the pelvic inlet. Once the foetatome is rethreaded and the saw wire is adjusted to the foetal extremity, the fixation and tension of the saw wire on the foetal section to be cut are guaranteed. No additional points of fixation are usually necessary:

Deviation of the head: the saw wire is placed around the flexed neck as close to the thorax as possible.

Carpal and hock *flexion*: the saw wire is placed immediately distal to the flexed joint. This loop placement prevents uterine or vaginal rupture during the extraction of the foetus.

Hip flexion: the loop is made passing the saw wire at groin level. The foetatome head is positioned against the foetal ischium, performing a longitudinal cut.

Hiplock: the destruncation is performed followed by the bisection of the pelvis or the diagonal cut of one hindlimb.

In the presence of monsters, the cuts aim to reduce the diameter of the foetus or overhaul lack of flexion of some joints. Usually, the abnormal foetal conformation is due to ankylosed joints and distortion, such as in schistosomus reflexus [9] and in perosomus elumbis [10]. The schistosomus reflexus is the most frequent foetal teratologic monster causing dystocia in cows. In visceral presentation, the trunk is severed near the vertebral angulation (point of deviation) after viscera removal. If necessary, the ankylosed members can also be severed to reduce the diameter of each foetal half. In the dorsal (extremities) presentation, the cuts are performed in each accessible member followed by the trunk split and visceral removal. Particular caution is needed as it is likely to have several sharp bones exposed. A C-section is the only option if vaginal delivery of the foetus remains impossible. The perosomus elumbis causes dystocia due to the hindpart anomalies, and it should be solved similar to the hiplock dystocia.

7.8 Postoperative Foetotomy Care and Complications

At the end of the foetotomy, the presence of foetal portions or additional foetuses, uterine lacerations, injuries of the soft birth canal and pelvic fractures or bleeding is checked by vaginal palpation. At this moment, the clinical health status of the dam must be evaluated. Supportive treatment with fluids and sometimes steroids must be immediately administered in depressed and hypovolaemic dams. Blood transfusion (4–5 L with sodium citrate) may be an excellent alternative to save severely debilitated and exhausted cows.

> **Tip**
>
> The intravenous administration of a hypertonic saline solution (2–3 L of NaCl 7%; 4–5 mL/kg) followed by oral water supplementation (20–25 L) is a useful and cheap rehydration technique in cows.

Uterine lavage may be required after total foetotomy. Uterine lavage is performed by pumping and siphoning 4–5 L of warm water (40–45 °C) until fluids coming out are clear. A total of four to six cycles of uterine lavage may be required. The objective of the lavage is the removal of tissue debris, meconium, blood and blood clots and lubricant. A small amount of non-irritating antiseptics may be added to the water. A solution of up to 0.5% iodopovidone (antibacterial and antifungal) or 1% chlorhexidine (antibacterial) seems to be adequate to kill bacteria, preventing chemical inflammation of the endometrium and intraluminal adhesions. The addition of salt to obtain an isotonic solution (NaCl 0.9%) is a viable alternative.

> **❯ Important**
>
> The uterine lavage is contraindicated in dams suffering uterine rupture. Uterine content may drain into the peritoneal cavity causing focal or diffuse peritonitis and its consequences (e.g. toxaemia, sepsis, visceral and abdominal adhesions or death).

A dose of 40–50 IU of oxytocin should be administered to increase myometrial contractions. Even if clinical signs of hypocalcaemia are not observed, the administration of calcium borogluconate 24% (up to 300 mL) is beneficial, but this can be dangerous in toxaemic cows, so a previous comprehensive clinical examination is paramount. Systemic NSAIDs and antimicrobial should be administered during the next two and four days, respectively. Antimicrobial therapy (e.g. oxytetracycline, ceftiofur or amoxicillin + clavulanic acid) should continue for longer in case of metritis or if the cow shows hyperthermia.

According to Wehrend et al. [2], the most common complications after foetotomy in 131 dams are retained placenta (37.4%; $n = 49$), lochiometra (16%; $n = 21$), pelvic phlegmons (12.2%; $n = 16$), vaginal wounds (12.2%; $n = 16$) and neurotripsy (12.2%; $n = 6$). These researchers also reported that the mean duration of postoperative treatment (antibiotherapy and analgesia) is approximately 4 days but in case of severe sequelae, such as the pelvic phlegmon, can go up to 14 days.

> **Case Study 7.1 Partial Foetotomy: Left Lateral Deviation of the Head**
>
> The calving of a Holstein-Friesian cow (4.5-years-old; third pregnancy; approximately 650 kg live weight) was unsuccessfully assisted by farm staff due to foetal dystocia caused by left lateral deviation of the head. The dam was found standing up and alert without clinical signs of septicaemia or hypovolaemia. Abdominal contractions were still recurrent. The birth canal was fully dilated without apparent tears or bleeding. Lateral deviation of the head and a dead foetus were confirmed. A caudal epidural anaesthesia was applied to reduce abdominal straining and discomfort. The head was fully projected cranially, and it was inac-

cessible to the obstetrician's hand, a halter or orbital hooks. A solution of epinephrine (1:1000; 10 mL) was administered intravenously to relax the uterine wall. Two attempts to reach and extend the head were unsuccessful. Since there was no evidence of trauma, haemorrhage or rupture of the uterus and birth canal, a partial foetotomy was decided. After plenty lubrification of the birth canal, a half-threaded foetatome was introduced. A saw wire was attached to the wire introducer, and a loop was made around the neck. The saw wire was introduced in the other barrel, and both extremities were connected to the saw wire handles. The wire loop was placed close to the withers; the foetatome was positioned to perform a longitudinal cut; and slight pressure was exerted on the saw wire to fix it to the neck. While the obstetrician held the foetatome, an assistant sawed the neck according to his instructions. Once the cut was completed, a Krey hook with an attached chain was placed on the cervical vertebra. The foetal body and both forelimbs (with chains previously placed) were repulsed to gain more free space. The head and the neck were carefully removed, with the obstetrician covering the sharp edges of the exposed vertebra with one hand. The Krey hook was cleaned, disinfected and fixed onto the exposed cervical vertebra of the foetal body. Both forelimbs were extended and the foetus was removed. No other foetus was found. After two cycles of uterine lavage, 40 IU of oxytocin were intramuscularly administered. Flunixin meglumine (1.1 mg/kg) and an association of penicillin (12,000 IU/kg) + streptomycin (10 mg/kg) were intramuscularly administered for 2 and 4 days, respectively. No placental retention was observed. The reproductive tract was examined two weeks later. No metritis or uterine adhesions were observed. Uterine involution was classified as normal.

Key Points

- Foetotomy allows for the removal of small portions of dead foetuses by vaginal delivery, avoiding a C-section.
- Intrauterine foetal accessibility through the birth canal and mucosal integrity are key factors to perform foetotomy.
- Foetotomy is performed in standing or in lying down cows, and in any case, epidural anaesthesia is required.
- Percutaneous foetotomy can be made by transverse, longitudinal or oblique cuts.
- A sequence of four to nine cuts including the foetal destruncation is used to complete a total foetotomy.

- Partial foetotomy usually involves one or two cuts and is very useful to solve irreducible foetal malpostures.
- Foetal fluid replacers and lubricant and appropriate use of the foetatome prevent uterine and vaginal lacerations.
- Uterine lavage, as well as systemic administration of oxytocin, analgesic and antimicrobials, prevents postoperative infections.
- After procedure discomfort may be severe so pain management is crucial. NSAID or other analgesic should be administered to ensure animal welfare.

7

? Questions
1. What are the advantages and disadvantages of foetotomy?
2. Describe the main premises for a foetotomy.
3. What is the better technique to perform a total foetotomy?

✓ Answers
1. Foetotomy can be more advantageous than other alternative procedures. Overall, the cost of the total foetotomy is lower than a C-section. On weak or depressed dams, foetotomy may be less aggressive than a surgery. However, this is not always true, and so the pros and cons of each procedure should be carefully accessed before a decision is taken. The aftercare can be less intense than in surgery, mainly in non-complicated total foetotomy or in partial foetotomy. In both these cases, if significant uterine or vaginal ruptures are prevented during the foetotomy, the rate of survival and fertility of the dam can be improved. Total or partial foetotomy can also avoid excessive obstetrical manipulation when severe foetal dystocia is present, even if the foetal mutation is possible. The main disadvantage of foetotomies is related to iatrogenic perforations of the uterus and birth canal, which can lead to complications such as peritonitis, septicaemia and pelvic phlegmons. This poses a threat to the dam's fertility and chances of survival. Prolonged interventions increase the risk of provoking the dam's exhaustion and uterine or vaginal tears. Also, the foetotomy cannot be used when the uterine wall is friable. Finally, there is a risk of injuries to the obstetrician due to the foetatome and hook manipulation.
2. Despite innumerous indications for foetotomy, this procedure can only be tried under certain circumstances. Firstly, the basic foetotomy material needs to be available: (1) a foetatome and its accessories (saw wire, wire introducer, wire threader and saw handles), (2) a Krey hook and (3) obstetrical chains or ropes. Additionally, foetal fluid replacer and a pump are strongly recommended. One or two people are required to restrain the dam and to help during the procedures. The birth canal needs to be sufficiently large for the body parts being removed, and foetal accessibility has to be ensured. Usually, the foetatome, one or two obstetrical chains or ropes and the obstetrician's hands/arms must be simultaneously introduced in order to successfully perform a cut.
3. There are at least three main techniques to perform a total foetotomy. The Utrecht and Hannover (according to Götze) techniques allow the foetus to be splitted up to seven or nine sequential cuts, respectively. Both methods are appropriate to be used in relatively and absolutely oversized foetuses. The purpose is to originate small foetal portions which can be easily removed by vaginal route. In small or normal-sized foetuses, the number of cuts can be reduced. Diagonal cuts of the forepart or hindpart can provide small foetal portions, and the total foetotomy, according to Schätz, is recommended. Nevertheless, the Utrecht and Hannover techniques can also be used: after one or two cuts, it is probable that the foetus can be removed without destruncation.

References

1. de Kruif A. History of veterinary obstetrics. Vlaams Diergeneeskundig Tijdschrift. 2011;80(5):367–71.
2. Wehrend A, Reinle T, Herfen K, Bostedt H. Foetotomy in cattle with special reference to postoperative complications-an evaluation of 131 cases. Dtsch Tierarztl Wochenschr. 2002;109(2):56–61.
3. Bierschwal CJ, DeBois CHW. The technique of foetotomy in large animals. Bonner Springs: VM Publishing Co.; 1972.
4. Grunert E. Operative verfahren zur entwicklung des fetus. In: Tiergeburtshilfe, Richter J, Götze R, Grunert E, Arbeiter K (Eds). Verlag Paul Parey; 1993. Berlin, Germany. pp. 301–51.

5. Vermunt JJ, Parkinson TJ. Defects of presentation, position and posture in livestock: delivery by foetotomy. In: Veterinary reproduction and obstetrics, Noakes DE, Parkinson TJ, England GCW (Eds), 10th ed. Elsevier: Amsterdam, Netherlands; 2019. pp. 277–290. doi: https://doi.org/10.1016/B978-0-7020-7233-8.00015-X.

6. Mortimer RG, Ball L, Olson JD. A modified method for complete bovine foetotomy. J Am Vet Med Assoc. 1984;185(5):524–6.

7. Walters K. Obstetrics: mutation, forced extraction, foetotomy. Chapter 47. In: Bovine reproduction, Hopper RM (Edt.) Wiley.: Iowa; 2015. pp.416–423. doi: https://doi.org/10.1002/9781118833971.ch47.

8. Jackson P. Chapter 12 – Foetotomy. In: Handbook of veterinary obstetrics. 2nd ed. London, UK: Saunders Ltd.; 2004. p. 199–207. https://doi.org/10.1016/B978-0-7020-2740-6.50017-1.

9. Laughton KW, Fisher KR, Halina WG, Partlow GD. Schistosomus reflexus syndrome: a heritable defect in ruminants. Anat Histol Embryol. 2005;34(5):312–8. https://doi.org/10.1111/j.1439-0264.2005.00624.x.

10. Agerholm JS, Holm W, Schmidt M, Hyttel P, Fredholm M, McEvoy FJ. Perosomus elumbis in Danish Holstein cattle. BMC Vet Res. 2014;10:227. https://doi.org/10.1186/s12917-014-0227-2.

Foetotomy Procedures

Drost M, Samper J, Larkin PM, Gwen Cornwell D. Obstetrics: foetotomy. The visual guide to bovine reproduction. Florida: UF College of Veterinary Medicine; 2019. From: https://visgar.vetmed.ufl.edu/en_bovrep/foetotomy/foetotomy.html accessed on March 13, 2020

Suggested Reading

Vermunt JJ, Parkinson TJ. Defects of presentation, position and posture in livestock: delivery by foetotomy. Chap. 15. In: Noakes DE, Parkinson TJ, England GCW, editors. Arthur's veterinary reproduction and obstetrics. 10th ed. Elsevier: Amsterdam, Netherlands; 2019. p. 277–90.

Caesarean Section

Contents

© Springer Nature Switzerland AG 2021
J. Simões, G. Stilwell, *Calving Management and Newborn Calf Care*,
https://doi.org/10.1007/978-3-030-68168-5_8

8

Learning Objectives

- To describe the indications, contraindications and complications of the caesarean section.
- To elucidate the preoperative options, including pre-anaesthesia and anaesthesia according to the obstetrical evaluation.
- To identify the major factors which can influence the surgical procedures in each step.
- To make a prognosis and a postoperative management.

8.1 Introduction

The caesarean section (C-section) is a surgical procedure to extract the foetus by laparotomy (celiotomy). It is performed in several situations where vaginal delivery is not possible or appropriate. Ultimately, the dam and calf survivals are the factors to choose this option. Nevertheless, the caesarean section (C-section) is an aggressive surgical procedure requiring the exteriorization of the uterus by an abdominal incision from 30 up to 70 cm of length. The dam needs to be able to support this surgery which is mainly made in field conditions.

The estimated survival rate of the dam at the 14-day post-surgery is 80.6%, and the adequate exteriorization of the uterus seems to be the most significant factor ($P < 0.001$) to contribute to dam's survival [16]. Another relevant factor is blood clot removal during surgery. At least in non-complicated C-sections, the appropriate exteriorization of the pregnant uterine horn is the key to minimize the contamination of the peritoneal cavity. In the presence of significant bacterial contamination, an adequate uterine exteriorization can prevent septicaemia and local or even diffuse peritonitis, which can lead to the dam's death. After the immediate- or short-term dam's prognosis, the reproductive prognosis of the dam is the most prominent aspect to consider. Uterine adhesion to abdominal viscera and wall or adhesion between the ovary and the ovarian bursa can lead to different degrees of infertility and consequent culling of the dam.

Several approaches of the C-section can be made [22] according to the obstetrical examination to mitigate the risk of infection and hypovolaemic or septic shock, as well as other adverse effects on the dam's survival and fertility. Also, the short duration of the surgery, less than 1 hour, is related to low calf perinatal mortality [16].

In the present chapter, the relevant aspects related to the C-section approach and procedures are described. Several technical options regarding complicated and non-complicated surgeries are also discussed.

8.2 Indications, Contraindications and Complications of the Caesarean Section

The main determining factor to take a decision for C-section achievement is the viability of the dam and the calf. The C-section is usually an emergency when the vaginal delivery is not possible or not recommended (► Box 8.1), including when the obstetrician shows an inability to perform some alternative techniques such as foetotomy. Early intervention can improve the chance of obtaining a live calf, mainly if it is under distress. Also, elective C-sections are performed in some situations largely related to some breeds' (e.g. Belgian Blue Breed cows) value of the offspring or dam or high economic mainly if irreducible dystocia is expected.

The contraindications of the C-section are mainly based on the condition of the dam's health status. Usually, dams strongly weakened due to acute or chronic disease and starvation are not recommended for surgery. Also, depressed dam due to pyrexia, septicaemia or toxaemia and hypotensive shock are significant risk factors. These last occurrences, including the friability of the uterine wall and the bacterial contamination of the peritoneal cavity, are frequently associated with the emphysematous foetus [1]. In this specific case, a C-section is elected only if it is not possible to perform a foetotomy since the probability of the dam's survival dramatically decreases. In some extreme situations such as the presence of an emphysematous foetus with uterine rupture, the dam euthanasia is the best option.

The decision to perform a C-section also depends on many factors related to the surgery and farm management (▶ Box 8.2), as well as some post-surgery aspects, namely, the occurrence of potential complications (◘ Fig. 8.1) [7]. It is also evident that the C-section avoids the risk of dam and foetus's injuries which can be caused by forced vaginal traction.

Box 8.1   Indications of the Caesarean Section
Foetal dystocia
- Absolute and relative foetal oversize.
- Irreducible foetal maldisposition.
- Foetal distress (foetal hyperactivity followed by depression, due to hypoxia).
- Foetal deformities (e.g. schistosomus reflexus, ankylosed foetus, conjoined twins and hydrocephalus).
- Foetal anasarca.
- Foetal death (oedematous and emphysematous foetus).
  Maternal dystocia
- Pelvic immaturity (immaturity of the heifer), narrow or abnormal pelvic conformation.
- Pelvic fractures.
- Incomplete cervical dilatation.
- Uterine inertia not responsive to pharmacological treatment.
- Irreducible uterine torsion.
- Vaginal or uterine rupture.
  Elective caesarean
- Predictable foetopelvic disproportion (e.g. breed, heifers and prolonged pregnancy).

Box 8.2   Main Risk Factors for a Successful Caesarean Section
- Health status of the dam.
- Hygiene of the surgical environment, including the surgeon.
- Bacterial contamination during the surgery.

- Presence of an oedematous or emphysematous foetus.
- Duration of dystocia and foetal manipulation.
- Condition of the uterus: oedema and friability.
- Adequate people assistance.
- Adequate facilities (e.g. environmental cleanliness and light).

8.3   Surgical Approaches

Different surgical approaches can be used for the C-section procedure (▶ Box 8.3). However, the left paralumbar laparotomy with the dam in the standing position is the most used procedure in field conditions. This approach is useful for non-complicated cases of dystocia and when the health status of the dam is not compromised or when deep sedation is unnecessary, such as in non-aggressive dams. The dam's recumbency is used for the ventral approach, but it is also viable for every other approach (◘ Fig. 8.2). In recumbent dams, an appropriate (deep) sedation is recommended, the legs should be tied with ropes, and two people are necessary to restrain the cow in dorsal or right lateral decubitus.

Box 8.3   Laparotomy Approaches
- Standing left paralumbar.
- Standing right paralumbar.
- Standing left oblique.
- Recumbent left oblique.
- Recumbent left paralumbar.
- Recumbent right paralumbar.
- Recumbent ventral midline.
- Recumbent right ventral paramedian.
- Recumbent right ventrolateral.

8.3.1   Left and Right Paralumbar Laparotomy

The left or right paralumbar laparotomy is performed in standing position. The legs of very aggressive cows should be hoppled

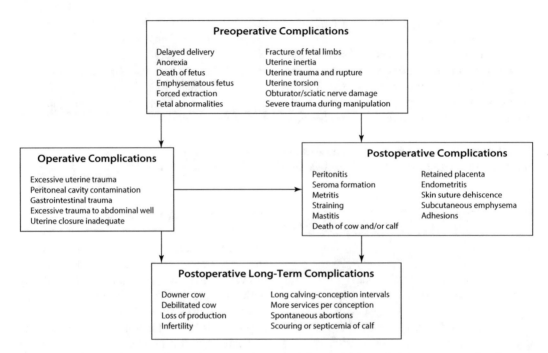

Preoperative Complications

Delayed delivery	Fracture of fetal limbs
Anorexia	Uterine inertia
Death of fetus	Uterine trauma and rupture
Emphysematous fetus	Uterine torsion
Forced extraction	Obturator/sciatic nerve damage
Fetal abnormalities	Severe trauma during manipulation

Operative Complications

Excessive uterine trauma
Peritoneal cavity contamination
Gastrointestinal trauma
Excessive trauma to abdominal well
Uterine closure inadequate

Postoperative Complications

Peritonitis	Retained placenta
Seroma formation	Endometritis
Metritis	Skin suture dehiscence
Straining	Subcutaneous emphysema
Mastitis	Adhesions
Death of cow and/or calf	

Postoperative Long-Term Complications

Downer cow	Long calving-conception intervals
Debilitated cow	More services per conception
Loss of production	Spontaneous abortions
Infertility	Scouring or septicemia of calf

◘ **Fig. 8.1** Preoperative, operative and postoperative potential complications of the caesarean section in cows. (Adapted from Newman [20] with permission from Elsevier)

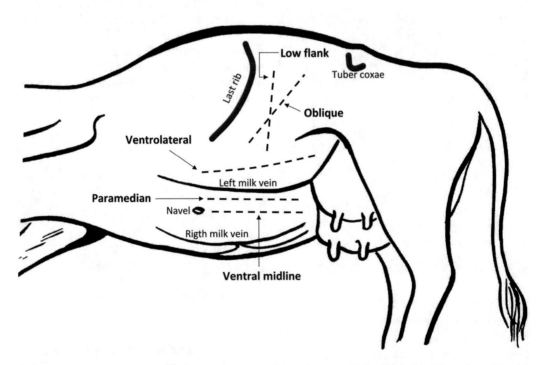

◘ **Fig. 8.2** Laparotomy approaches in the recumbent (right, left or dorsal) dam. (Modified from Ames [3], with permission from John Wiley and Sons)

◻ Fig. 8.3 The legs of a Salers heifer were hoppled to ensure surgeon security. Original George Stilwell

(◻ Fig. 8.3). However, these two approaches can be achieved in recumbent dams.

The surgical incision starts approximately 10 cm below the transverse process of lumbar vertebrae in the right or left flanks and progresses vertically for 30 to 50 cm according to the size of the dam. This choice of the left paralumbar fossa has significant advantages such as (1) the possibility of the dam standing up during the surgery and (2) only the dorsal sac of the rumen being able to inflate and exteriorize because of abdominal contractions. The biggest limitation of this approach is the accessibility of the uterine wall and foetal extremities when the pregnancy occurs in the right uterine horn. If that is the case, a rotation between 90° and 180° of the uterine axis is needed to exteriorize the distal part of the pregnant horn. The pregnant uterine horn can be previously and easily identified by transrectal palpation.

The access through the right flank (right paralumbar laparotomy) allows a greater uterine and foetal accessibility to the right uterine pregnant horn. Nevertheless, the biggest problem of this approach is the exteriorization of loops from the intestine during the surgery, even when the abdominal contractions are scarce. In case the right approach is

selected, it is crucial to have a cleaned and disinfected surgery assistant to prevent and reintroduce the intestinal loops. In some conditions, the right-flank approach is recommended [2]: the presence of significant scar tissues in the left paralumbar region due to previous laparotomy can difficult the incision and posterior suture of the abdominal wall; possible presence of adhesions because of previous laparotomy or localized peritonitis over the rumen; or the presence of a very full rumen that can difficult accessibility and exteriorization of the pregnant uterine horn.

8.3.2 Ventral, Ventrolateral and Left Oblique Laparotomy

The ventral approach can be performed on the midline or on a paramedian line. The dam is recumbent in dorsal decubitus with a temporary 45° inclination to facilitate the pregnant uterine horn exteriorization. The dam stays in recumbency during the whole surgery. Both approaches are usually reserved for emphysematous or very heavy foetuses, since the access to the uterus is straightforward. These approaches have the advantage of both right and left pregnant uterine horns being

8

easily exteriorized, limiting the bacterial contamination of the peritoneal cavity with uterine contents. However, ruminal tympany is very likely since eructation is not possible due to the cardia obliteration by ruminal fluids. Furthermore, pressure on internal structures, including the lungs, diaphragm, large abdominal blood vessels and muscles, can cause significant respiratory distress, thromboembolism or inflammation. A soft bedding should be provided to prevent these potential adverse effects. Additionally, there is a higher risk of dehiscence, herniation or evisceration through the sutured incision in the days following surgery. Some veterinarians state that the paramedian approach can reduce these risks.

In the ventrolateral laparotomy (Hannover approach), the dam is in right lateral recumbency. This position is more comfortable and causes less pressure on the dam's internal organs than the dorsal decubitus. The slightly curved incision is made from the groin and is cranioventrally extended up to 50 cm. This approach can also be used for emphysematous foetuses, and it is adequate for oversized foetuses. Nevertheless, because of abdominal tension, the risk of suture dehiscence after the surgery is not negligible.

In the left oblique laparotomy (Liverpool approach), surgery is performed either in a standing or in a lying animal (right lateral decubitus). A cranioventral incision line is done starting in the paralumbar fossa. The biggest advantage of this approach is that exteriorization of the distal portion of the pregnant uterine horn is easy while reducing the risk of suture dehiscence and herniation or evisceration of the other recumbent approaches. The Liverpool approach can be used for oversized or emphysematous foetuses facilitating uterine and foetal manipulation.

8.4 Facilities, People and Equipment Requirements

Usually, C-section is made under field conditions on the farm. This fact does not mean that important surgical and hygiene rules do not apply. Features such as the safety of animals and people, environmental hygiene and

sterilized surgery are crucial for the procedures' success. The surgeon should be able to perform all steps according to the best surgical norms. Farm people or assistants are usually needed to restrain the dam and to care for the newborn calf. The presence of at least one skilled and sanitized assistant is highly recommended, especially for complicated surgeries.

8.4.1 Facilities on the Farm

The dam should be moved to a clean and well-lit pen and safely restrained. A non-aggressive dam can be simply haltered to a fixed point without sedation. The tail should be tied to prevent bacterial contamination during the surgery. Alternatively, a caudal epidural anaesthesia (3–5 mL procaine 2%) will paralyse the tail without affecting standing. A calving gate or crush can also be used for this purpose, but clean and clear access to the incision needs to be guaranteed if the cow falls during surgery. It is preferable to force recumbency before starting the surgery, if the probability of falling during the procedure is high, such as when mental depression, muscle weakness or general frailty is evident. Dams falling at mid-surgery have a higher risk of developing wound infections or peritonitis or even dying. For recumbent dams, the flooring should be washed, and a soft and clean bedding should be provided (e.g. new straw without dust) to prevent infections and injuries (e.g. myopathies and peripheral neuropathies).

8.4.2 Material Requirement

C-section should always be considered a natural and potential option when dealing with a cow at calving. So, all material and conditions should be available for surgery whenever an obstetric intervention initiates. This includes foetus facilities and ropes to restrain the dam or to tie its feet, general material to prepare the surgical site and ensure the surgeon hygiene (e.g. clean buckets, water, soap and razors), disinfectants, sedatives, local anaesthetics, antimicrobials, nonsteroidal anti-inflammatory drugs (NSAIDs), NaCl 0.9%

solution bottles and oxytocin as well as sterile hypodermic needles and syringes. Surgical kits containing appropriate surgical material, including suture lines, and other essential sterilized equipment, should always be ready to use. The essential material for a normal C-section is listed in ► Box 8.4 and illustrated in ◻ Fig. 8.4.

> **Box 8.4 Sterilized Specific Material and Equipment for Caesarean Section in Cattle**
> - Scalpel handle (Number 4) and scalpel blades (Numbers 21 or 22).
> - Rat tooth (1) and tissue (2–3) forceps (15.2 cm).
> - Lane tissue forceps (4; 15.2 cm).
> - Kelly haemostat forceps (≥4).
> - Uterine forceps (2; 28 cm).
> - Mayo scissors (2).
> - Needle holder (1; Mathieu, Gillies or McPhail's).
> - Surgical drapes (disposable or reusable).
> - Swabs.
> - Needles (pairs): ½ curved cutting-edge (6.7 cm); 3/8 circle cutting-edge and taper-point (4.7 cm and 7.0 cm); double-curved.
> - Absorbable suture: one reel of chromic catgut or multifilament polyglactin (size metric 6 to 8 and USP 2 to 4).
> - Non-absorbable suture: one reel of pseudo monofilament polyamide polymer (polymerized caprolactam), nylon or silk (size metric 2 and USP 2 to 4).
> - Surgeon equipment: surgical gloves and clothes or gown (additional: surgical mask, cap and boots).

8.5 Preoperative Cares

Sequential preparatory procedures include dam restraint, antimicrobial and analgesia injection, sedation (optional) and regional anaesthesia and, finally, surgical site preparation (► Box 8.5).

> **Box 8.5 Preparatory Procedures for Surgery**
> - Antimicrobial therapeutics – the injection of an antimicrobial before surgery reduces significantly the probability of infection versus the injection at the end of surgery.
> - Analgesia (pain control) – likewise it is more efficient if applied before painful stimuli are induced.
> - Caudal epidural anaesthesia (optional) – very low dosage may be useful in keeping the tail still. However, care should be taken if recumbency is to be avoided.
> - Sedation – mainly depending on temperament and health status of the dam; the risk of cow lying down increases; also, foetal cardiorespiratory effect should be considered.
> - Administration of uterine relaxant (easier handling of the foetus and uterine wall).
> - Preparation of the surgical site – washing, trichotomy, again washing, disinfection (alcohol followed by 1% active iodine solution or 4% chlorhexidine).
> - Local or regional anaesthesia – procaine or, in countries where it is allowed, lignocaine.
> - Final preparation of the surgical site and the surgeon/assistant.

8.5.1 Premedication and Pre-anaesthesia Assessment

The administration of some drugs before the C-section procedures is an adequate option in many cases under field condition.

▪ **Antimicrobials**

The prior administration of a broad-spectrum antimicrobial is highly recommended. Procaine benzylpenicillin (12,000 IU/kg) and dihydrostreptomycin (10 mg/kg) association must be intramuscularly administered as soon as possible. These antimicrobials reach their plasmatic

8

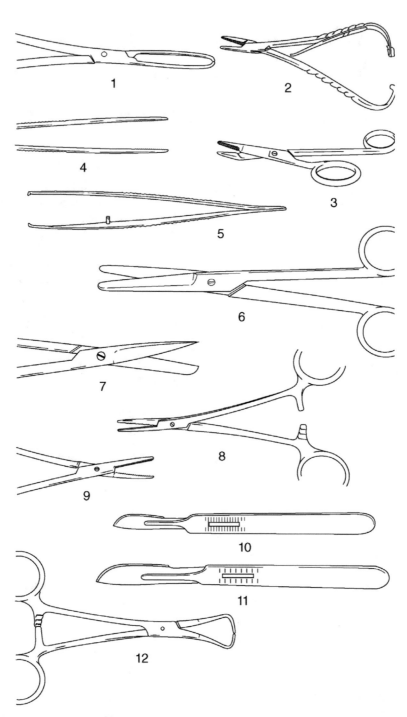

■ **Fig. 8.4** Basic instruments for laparotomy or caesarean section. Legend: 1 Allis tissue forceps; 2 McPhail's needle holder; 3 Gillies combined scissors and needle holder; 4 plain forceps; 5 rat tooth forceps; 6 Mayo scissors (blunt/blunt), slightly curved; 7 Mayo scissors (pointed/blunt), straight; 8 straight haemostatic forceps; 9 curved haemostatic forceps; 10 scalpel handle no. 4 and no. 22 blade; 11 scalpel handle no. 3 and no. 10 blade; 12 towel clip (Backhaus). (Adapted from Weaver et al. [26] with permission from John Wiley and Sons)

peak between 1 (dihydrostreptomycin) and 2 (penicillin G) hours after intramuscular administration. It is expected that they will act at this time to control peritoneal infection and bacteraemia. Another option is the intravenous administration of tetracycline (20 mg/kg body weight; BW). In cases where heavy contamination of the peritoneum occurs, the authors suggest the association of amoxicillin clavulanic with enrofloxacin (single dose, 7.5 mg/kg body weight; BW).

Likewise, we recommend intra-abdominal placing of 3 or 4 L of 1% iodine before final closure of the peritoneum, if contamination is likely.

Nevertheless, it is crucial to ensure asepsis during the whole surgery. Similar to human medicine and based on the Centers for Disease Control and Prevention (USA) definitions, surgical site cleanliness in cattle can be classified in four classes (▶ Box 8.6), with the risk of infection progressively increasing: clean (10.1% risk of infection), clean/contaminated (15.4% risk of infection), contaminated (26.7% risk of infection) and dirty (50% risk of infection) [8]. This aspect, as an exogenous risk factor for surgical site infections, is relevant to minimize the use of prophylactic antimicrobials. Other risk factors are related to the susceptibility of the cow and the bacteria species and its virulence.

> **Box 8.6 Classification of the Surgical Site Infections in Cows Based on Dumas et al. [8]**
> - Clean (class I) – Elective surgery where the surgical site is not inflamed or contaminated. An aseptic technique is ensured during the whole surgery.
> - Clean/contaminated (class II) – Elective surgery opening the genitourinary, gastrointestinal, respiratory or biliary tract with minimal leakage. Only minor break in aseptic technique occurs.
> - Contaminated (class III) – Gross contamination of the surgical site, but without active infection (e.g. gastrointestinal tract leakage and non-purulent

inflammation). A major break in aseptic technique occurs.
> - Dirty (class IV) – Active infection with purulent exudate at the surgical site. Ruptured viscus and presence of foreign bodies or faecal contamination.

■ **Nonsteroidal Anti-Inflammatory Drugs (NSAIDs)**

Preoperative use of analgesics is recommended to control secondary pain at peripheral and central nervous level. Pre-emptive use of analgesics will reduce inflammation and the wind-up effect that leads to neuroanatomic changes (plasticity) resulting in exaggerated pain states that are much more difficult to control. Several nonsteroidal anti-inflammatory drugs (NSAIDs) are available for cattle, although some can only be used off-label in some countries [2]: flunixin meglumine (2.2 mg/kg; i.m. or i.v.), meloxicam (0.5 mg/kg; s.c. or i.v.), ketoprofen (3 mg/kg; i.m. or i.v.), carprofen (1.4 mg/kg; s.c. or i.v.) and tolfenamic acid (4 mg/kg; i.v.) are the most frequently used worldwide. Intravenous administration of NSAIDs should be preferred to ensure immediate and full effect.

Therapeutic analgesia can also be provided in cows by alpha-2-adrenergic receptor agonists such as xylazine or detomidine sedatives. An association between NSAIDs and sedative improves the analgesic effect during the surgery, and it can reduce the sedative dose to obtain similar sedation levels. However, it should be recalled that xylazine is a strong muscle relaxant and hypotensive drug in cattle, so the risk of falling down during surgery increases significantly.

■ **Uterine Relaxants or Tocolytic Agents**

Beta-2-adrenergic agonists can provide the relaxation of the uterine wall. This effect facilitates the manipulation of the foetus as well as the uterus during surgery. Also, it has been reported that xylazine and detomidine can increase the myometrial tone as well as decrease the uterine blood flow in ruminants. Hence, the use of a beta-2-adrenergic agonist can mitigate these effects when sedation is necessary. Clenbuterol hydrochloride can be administered off-label by slow intravenous infusion at a dose of 10–15 mL (30 µg/mL). Isoxsuprine is a mixed beta-adrenoreceptor antagonist/agonist. It causes relaxation of vascular and uterine smooth muscle. Note that both these drugs are not registered for use in cattle in European countries, so the use is always off-label.

As an alternative, 10 mL of 1:1000 epinephrine solution can be administered slowly intravenously [19] or intramuscularly [9]. The myorelaxation effect starts within 2 min.

Meloxicam has shown tocolytic effect in vitro but in vivo studies are still lacking [6].

■ Sedative Drugs

In the first instance, the use of sedative medication is helpful to restrain the dam and to reduce anxiety and fear. They may be essential for a safe and swift C-section in aggressive or very frightened beef cattle. Sedatives can also promote additional analgesia, reducing the dose of other analgesic and anaesthetic drugs. Nevertheless, high doses of sedatives can cause foetal cardiorespiratory depression and increase the risk of stillborn. Similar effects can be found in depressed and toxaemic dams such as those carrying an emphysematous foetus. In these cases, use of very low doses (especially xylazine or detomidine) is recommended.

Depending on the degree of sedation, the dose of xylazine hydrochloride (2%) varies from 0.05 up to 0.3 mg/kg (i.m.), and the dose of detomidine hydrochloride varies from 10 up to 40 µg/kg (i.m.). The doses should be reduced in half when administered intravenously. Also, only half the dose of xylazine should be used if surgery is to be performed in a standing animal (► Box 8.7).

Box 8.7 Doses and Duration of Effect of Sedatives (Alpha-2-Adrenergic Receptor Agonists) for Caesarean Section

Sedation for standing dams

— Xylazine 2%: a single dose of 0.015–0.1 mg/kg (i.m.) or 0.0075–0.05 mg/kg (i.v.) for 30–45 min. of effect duration.

— Detomidine 1%: a single dose of 0.006–0.02 mg/kg (i.m.) or 0.002–0.015 mg/kg (i.v.) for 30–60 min. of effect duration.

Sedation for recumbent dams

— Xylazine 2%: a single dose of 0.2–0.4 mg/kg (i.m.) for 30–45 min. of effect duration or 0.1–0.2 mg/kg /kg (i.v.) for less than 30 min. of effect duration.

❯ Important

The use of sedatives depends on the dam's health (alert vs. depression: toxaemic, hypotensive or in shock), temperament (aggressive vs. docile) and position (recumbency vs. standing up), predicted duration of the C-section (more vs. less than 1 h) and availability of additional assistance (necessary/present vs. not necessary/absent).

Ketamine hypochlorite is a dissociative anaesthetic which was extensively used in the past in association with sedatives to obtain short-term (approximately 15 min.) anaesthesia by intravenous route. However, ketamine will only cause sedative effects at low doses. Nevertheless, no other alternative with similar anaesthetic effectiveness at reasonable cost is currently available. Some protocols using ketamine have been proposed to improve analgesia during C-section. These protocols involve association between a sedative, NSAIDs and ketamine (► Box 8.8) to promote analgesia but should never be thought to replace local/regional anaesthesia.

Box 8.8 Based Ketamine Protocols of Sedation and Analgesia

For standing dams

— Protocol 1: butorphanol (0.01 mg/kg) + xylazine (0.02 mg/kg) + ketamine (0.04 mg/kg) administered i.m. or s.c. for 60–90 min. of duration of effect. For a 500 kg dam, it is the denominated 5-10-20 technique which can be administered in the same syringe.

- Protocol 2: xylazine (0.04 mg/kg) + ketamine (0.2 mg/kg) administered i.m. or s.c. for 60–90 min. of duration of effect.

For recumbent dams

- Protocol 1: butorphanol (0.025 mg/kg) + xylazine (0.05 mg/kg) + ketamine (0.1 mg/kg) administered i.m. or s.c. for 30 to 45 min. of duration of effect.
- Protocol 2: xylazine (0.05–0.1 mg/kg) + ketamine (4 mg/kg) administered i.m. for 30–40 min. of duration of effect. Half the dose of both drugs can be intravenously administered for an additional 5- to 10-minute duration of effect.

■ **Caudal Epidural Anaesthesia**

The advantage of caudal epidural anaesthesia is decreasing the number of abdominal contractions of the dam during surgery. It is recommended for dams presenting frequent and strong abdominal contractions during obstetrical examination or parturition attempts. For exhausted or depressed dams, the caudal epidural anaesthesia is optional. The technical procedure of caudal epidural anaesthesia is described in ▶ Chap. 6. The use of adrenaline can increase the duration of the epidural anaesthesia, but it can cause ischaemic necrosis and so should be avoided, if possible.

> **Tip**
>
> Caudal epidural injection can be given to prevent straining during surgery.

8.5.2 Preparation of the Surgical Site

The surgical site needs to be prepared according to basic surgery norms. Trichotomy is performed by shaving the hair in the selected zone. For the flank approach, all area between the last rib and the ischial tuberosity, and between the transverse apophyses of the lumbar vertebrae and an imaginary line at udder level, should be shaved and prepared. It is important to ensure a larger area if surgical drapes are not to be used. This area should first be washed with hot water and soap. The skin is then disinfected with 70% ethanol solution, and a disinfectant solution such as iodopovidone 10% or chlorhexidine 4% in ethanol 70% is applied using swabs. The swab must be moved from top to bottom and inside out by a simple passage regarding the line of incision. It is recommended to repeat this procedure after anaesthesia infiltration. At this moment, the surgeon needs to prepare. Surgeons should wash hands and forearms with soap and hot water and disinfect them with 70% ethanol before dressing the sterilized surgical gloves and gown.

8.6 Local and Regional Anaesthesia

There are several methods to perform the anaesthesia of the flank: paravertebral, inverted-L and in the line of the incision. One of several local anaesthetics may be used depending on availability and desired duration. In the last decade, lidocaine's use in production animals was banned in several countries (e.g. European Union), but where it is available, it is the preferred drug. Procaine hydrochloride (ester local anaesthetic), with or without adrenaline, is the most common alternative as it is authorized for farm animals in many countries. However, its effectiveness and duration of effect are poorer than lidocaine. So, a high dose of procaine 2–4% varying from 60 up to 200 mL according to the selected method is used to mitigate this limitation. However, no more than 6.6 mg/kg of procaine should be administered as excessive absorption of local anaesthetics can block cardiac sodium channels leading to a depression of automaticity and conduction of cardiac impulses through the heart. Clinical signs of intoxication include twitching and convulsions to coma and death. Also, several anaphylactic reactions have been seen by the authors in cows with hypersensitivity to procaine (◘ Fig. 8.5). The signs include respira-

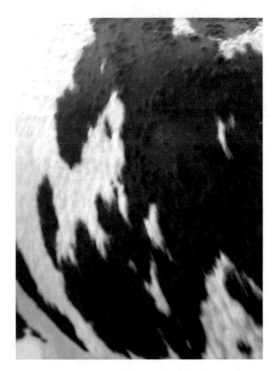

◘ Fig. 8.5 Urticaria in cow with hypersensitivity to procaine used as local anaesthetic. Original from George Stilwell

tory distress, salivation, urticaria, facial and perineal oedema, lacrimation, anxiety and agitation.

Several other local anaesthetics, such as bupivacaine and mepivacaine, or the more recent ropivacaine and levobupivacaine, still need to be adequately tested and become available at a reasonable cost to be used for local or regional anaesthesia in cows.

8.6.1 Paravertebral Anaesthesia

Paravertebral anaesthesia may be used for either the left or the right paralumbar nerve block. It consists of a perineural deposition of a local anaesthetic on the spinal nerves emerging from the vertebral canal through the intervertebral foramen space. It provides the anaesthesia of the whole flank/abdomen region covered by the segmental nerve block.

Proximal or distal paravertebral block techniques can be used (◘ Fig. 8.6). The proximal paravertebral anaesthesia is technically more difficult to perform than distal paravertebral anaesthesia, but it allows a better and uniform desensitization and muscle relaxation even with lower doses of local anaesthetic.

In the proximal paravertebral anaesthesia (Farquharson, Hall or Cambridge method), a 14-gauge 2.5 cm needle is used as a cannula to minimize skin resistance during insertion of an 18-gauge 10–15 cm spinal needle. To desensitize T13 nerves, the cannula needle is placed through the skin at the cranial edge of the transverse process of L1 at approximately 4 to 5 cm lateral to the midline. The 18-gauge 10–15 cm spinal needle is then passed ventrally until it contacts the bone of the transverse process of L1. The needle is then directed to the cranial edge of the transverse process of L1 and inserted deeper into the intertransverse ligament. When the ligament is passed, a "pop" feeling is evident. A total of 10 mL of local anaesthetic (e.g. procaine 2%) is injected to block the ventral branch of T13. The needle is then withdrawn 1–2.5 cm to the area just dorsal to the transverse process, and an additional 10 mL of local anaesthetic is infused to desensitize the dorsal branch of the nerve. To desensitize the nerves from L1 and L2, the needle is inserted just caudal to the respective transverse processes, and the same steps as for T13 are repeated. Both dorsal and ventral nerve roots emerging from the intervertebral foramina of T13, L1 and L2 are thus desensitized.

◘ Fig. 8.6 Proximal and distal paravertebral anaesthesia. Legend: Top: The Farquharson and Magda's methods are used for proximal and distal (ventral and dorsal to each transverse process), respectively. A 7 cm hypodermic needle is recommended to reach the nerve branches, mainly in dams with high body condition score. T thoracic. (Adapted from Mansour et al. [18] with permission from John Wiley and Sons). Bottom: Left – Regions of the flank involved after paravertebral block of the respective nerves. Right – Scoliosis caused by paravertebral block or lumbar epidural unilateral anaesthesia (spinal curvature towards the affected side). (Modified from Hall et al. [10] with permission from Elsevier)

P. Proximal (dorsal) paravertebral nerve block (Cambridge or Farquharson) at the intervertebral foramina

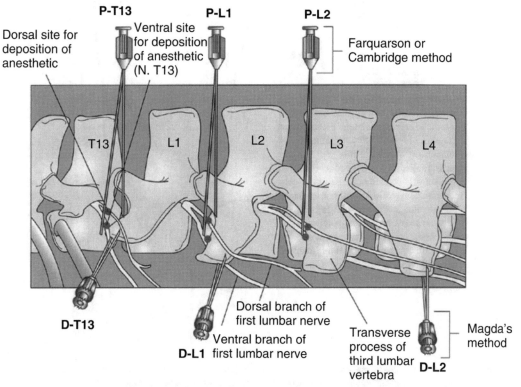

Dorsal site for deposition of anesthetic

P-T13

Ventral site for deposition of anesthetic (N. T13)

P-L1

P-L2

Farquarson or Cambridge method

T13   L1   L2   L3   L4

D-T13

D-L1 Ventral branch of first lumbar nerve

Dorsal branch of first lumbar nerve

Transverse process of third lumbar vertebra **D-L2**

Magda's method

D. Distal (lateral) paravertebral nerve bock (Magda or Cornell) at the tips of the transverse processes

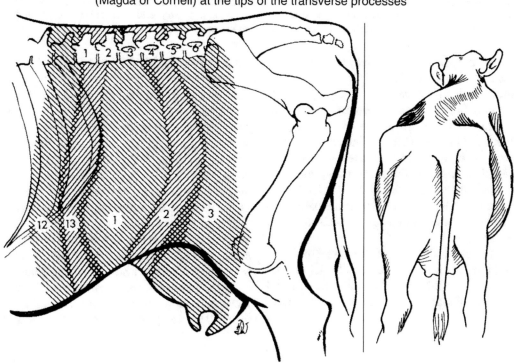

8

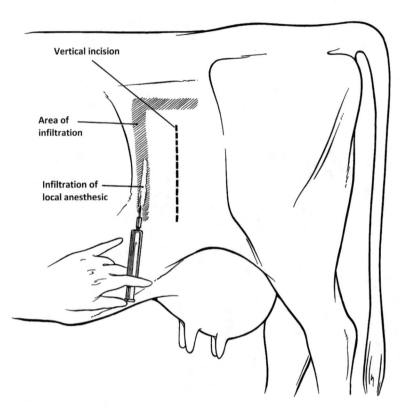

Vertical incision

Area of
infiltration

Infiltration of
local anesthesic

☐ **Fig. 8.7** Inverted-L anaesthesia and incision line for caesarean section. Legend: The local anaesthetic is infiltered in all muscle layers and subcutis approximately 10 cm below the transverse processes of lumbar vertebrae and 5–10 cm caudal to the last rib. A 30–40 cm vertical incision is made. In the C-section, the laparotomy is performed more ventrally than for other abdominal surgeries in order to facilitate the lifting and exteriorization of the pregnant uterine horn. A cranioventral oblique incision, less than 45°, using the inverted-L anaesthesia is also possible. (Modified from Mama [17] with permission from John Wiley and Sons)

In the distal paravertebral anaesthesia (Magda, Cakala or Cornell method), the needle is firstly placed ventrally to the tips of L1, L2 and L4 transverse processes, directed to the midline, and 10–20 ml of local anaesthetic are injected in a fan-shaped infiltration pattern. In a second step, the needle is reinserted dorsally to the same transverse processes, and an additional 5–10 ml of local anaesthetic is injected in the same pattern to desensitize the cutaneous branch of each dorsal rami.

The desensitization of the dorsal and ventral rami of the spinal nerves T13, L1 and L2 starts within 10 min. after the injection and lasts for about 90 min., in both techniques [15].

8.6.2 Inverted-L Anaesthesia

The inverted-L anaesthesia can also be used to desensitize the left or right flank. The trajectory of the nerve fibres in the paralumbar area is oblique, in a ventrocaudal direction. The deposition of the local anaesthetic is done subcutaneously, intramuscularly and over the peritoneum, horizontally at the top of the incision line and vertically all along the projected line of incision (inverted-L). This way the nerves are blocked at two sites establishing a very efficient block (☐ Fig. 8.7). This is an easy technique which allows satisfactory anaesthesia of the flank for more than 60 min. However, a large quantity of local anaesthetic, from 60 up to 200 mL of 2–4% procaine, is required. Note that deep injection is needed to adequately desensitize the peritoneum in beef cows that have thick abdominal walls.

8.6.3 Line Block

The line block technique blocks the nerve fibre along or just beside the surgical incision, i.e. it is proper local anaesthesia. This kind of anaesthesia can be used in the flank, ventral or ventrolateral approaches. The amount of local anaesthetic, duration of anaesthesia and not very satisfactory analgesia of the deep muscle layers are similar to the inverted-L anaesthesia. Additionally, oedema and haematoma can occur in the local, interfering with local anaesthesia efficacy (ideal pKa is slightly basic) and posterior healing. However, it is considered the easiest technique for flank desensitization.

8.7 The Caesarean Section Technique

Usually, if the surgical approach is properly chosen and all sequential steps are correctly performed, the duration of a non-complicated C-section takes from 30 to 60 min. Nevertheless, some more or less expected difficulties may be found during surgery (▶ Box 8.9) and need to be prevented or solved quickly. These complications can prolong the surgery for up to two hours or more.

> **Box 8.9 Expected and Unexpected Difficulties Found during Surgery [14]**
> - Recumbency during the surgery.
> - Abdominal wall muscle haemorrhage.
> - Overfilled and/or tympanic rumen.
> - Accidental incision of the rumen (left approach).
> - Tenesmus with or without the eventration of intestines or uterus.
> - Adhesion between the uterus, abdominal viscera and wall consecutive to previous surgery or peritonitis.
> - Difficulty in exteriorization of the pregnant uterine horn.

> - Rupture and haemorrhages of the uterine wall before, during or after foetus exteriorization.
> - Rupture of the broad ligament, with or without haemorrhages.
> - Laceration of the uterine suture.
> - Necrosis of the uterine wall in case of pronged cases of dystocia.

8.7.1 Left Paralumbar Laparotomy

A >30 cm (up to 50–70 cm) vertical or caudocranial continuous incision of the skin is made, starting >10 cm below the transverse processes of the lumbar vertebrae and 15 cm caudal to the last rib. Once the subcutaneous tissue is incised, the craniocaudal fibres of the external abdominal oblique muscle and fascia are cut in the same direction of the skin incision (◘ Fig. 8.8). In large incisions, the rectus abdominis muscle appears downwards, ventrally to the incision, and it can also be partially nicked. The internal abdominal oblique muscle is then identified (caudocranial fibres) and also incised, revelling the glistening aponeurosis of the transverse abdominal muscle. The peritoneum is closely attached to this muscle. At this moment, the muscle plus peritoneum are held up using a tissue forceps and are nicked in its dorsal part to prevent the incision of the rumen wall. Due to negative abdominal pressure, air entering the abdominal cavity is usually audible. Complete incision of the transverse abdominal muscle and peritoneum can then be made using a scalpel or a scissor.

▪ Exteriorization of the Uterine Wall

This is the most important step to prevent bacterial contaminations of the peritoneal cavity. A portion of the pregnant uterine horn needs to be exteriorized. Normally, in the left flank approach, the rumen presses the uterus to the right side of the abdominal cavity. The surgeon needs to insert his hand dorsally and below the ruminal dorsal sac to identify the pregnant uterine horn and foetal parts. One foetal limb should be identified to serve as a handle to move the uterine horn outside

8

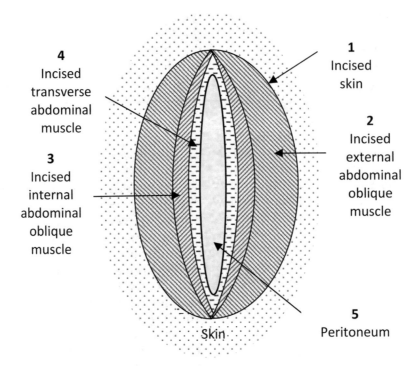

4
Incised
transverse
abdominal
muscle

3
Incised
internal
abdominal
oblique
muscle

1
Incised
skin

2
Incised
external
abdominal
oblique
muscle

Skin

5
Peritoneum

◻ **Fig. 8.8** Cutting sequence (1 to 5) of the skin, abdominal muscle layers and peritoneum in laparotomy by flank approach

through the abdominal incision. The rotation of the right pregnant uterine horn, up to 180°, is usually necessary to elevate the uterus to the abdominal incision. In the ventrolateral approach, the exteriorization of the pregnant uterine horn is easier due to its topographic proximity with the incision line of the cow's abdomen.

■ **Uterine Incision and Foetal Extraction**

The incision of the uterine wall should be made over a foetal limb (preferably over the hooves), as close as possible to the distal part of the exteriorized uterine horn. This procedure ensures that this portion of the uterine wall stays easily exteriorized for suturing. A > 30 cm incision is made as linearly as possible but avoiding the placentomes (◻ Fig. 8.9). The length of the incision needs to be large enough to prevent uterine lacerations in different directions caused by the foetal passage. Some care is also needed to avoid cutting the foetus. Sometimes, due to foetal malpresentations or to the presence of an oversized foetus, it is not possible to adequately exteriorize the uterus using a flank approach. In these cases, the ventrolateral approach is recommended, since the exteriorization of the uterus containing a portion of the foetal body or its extremities is more easily achieved. Partial or complete incision of the uterine wall can be made inside the abdominal cavity. There is no real danger if no or little vaginal handling of foetus was done or if the water bags are still intact. On the other hand, if there is a dead foetus or obstetrical procedures were going for long, it should be considered the last option, since it may be a major cause of contamination, peritonitis and posterior uterine adhesions.

After the uterus has been cut open, a uterine forceps should be placed at each extremity of the incision.

❯ **Important**

The uterine wall should be incised along its great curvature, where large vessels are scarce, and parallel to the long axis of the pregnant uterine horn.

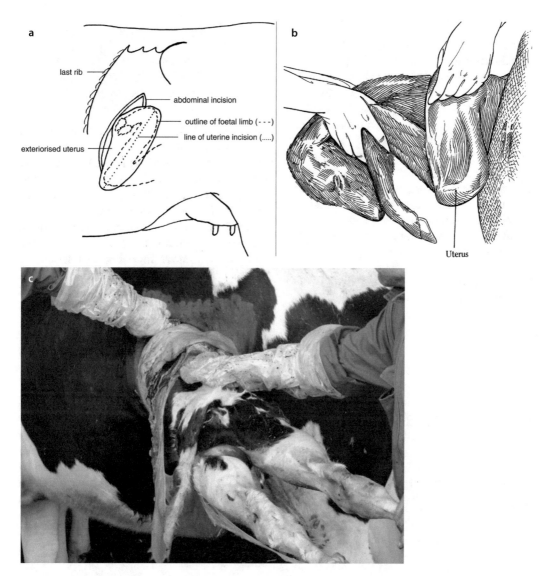

Fig. 8.9 Line for uterine incision and foetus extraction. Legend: **a** Line for uterine incision between the foetal hock and hoof. Placentome incision should be avoided. (Adapted from Weaver et al. [23] with permission from John Wiley & Sons, Inc.). **b** Extraction of the foetus. Adapted from Baird [4] with permission from John Wiley and Sons. **c** A 45–90° foetal rotation in his longitudinal axis can be useful to extract the foetus preventing uterine wall tears. Original from George Stilwell

After opening the placental membranes, the exteriorized limb and its contralateral pair are held up by the surgeon's hand. Sterilized obstetrical chains or ropes are then applied to each limb. The limbs should be pulled simultaneously by an assistant applying a rotation movement to the foetus if necessary. If the foetus is held by the hindlimbs, the foetus should pulled caudally. The hips or shoulders should be placed close to each top of the incision, preventing uterine lacerations. If the foetus is being pulled by the forelimbs, the head must be grasped together with the limbs. As soon as the foetus relieves tension over the incision area, the surgeon or the sanitized assistant should hold both uterine forceps to keep the uterus outside of the abdomen.

■ Suture and Accommodation of the Uterus

Once the calf is drawn out, the uterine incision is inspected to confirm the presence of lacerations or important bleeding. Small lacerations (1–2 cm) can be included in the main suture. A second suture pattern layer can be used. However, large lacerations, usually oblique or perpendicular to the uterine incision, need to be sutured individually. Placental tissues are introduced into the uterus cavity, and only free tissue or parts that may interfere with uterine closure should be removed using a scissor. Furthermore, the uterine wall should be washed using a sterile saline solution (NaCl 0.9%).

> **Important**
> For complicated cases such as when the uterus is friable or there was an emphysematous foetus, a second suture pattern layer 1–2 cm distal to the first one should be added to completely close the uterine incision.

> **Tip**
> Uterine forceps (■ Fig. 8.10) can be placed approximately 4–5 cm distal to each incision top and sustained by an assistant to avoid the uterus from falling back into the abdominal cavity once the foetus is removed.

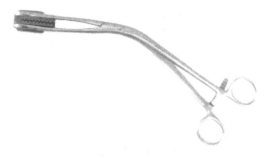

■ Fig. 8.10 Obstetrical caesarean forceps (12″) with removable soft rubber jaws to grabbing the uterus minimizing trauma of the uterine wall. (Modified (FotoSketcher 3.60) from Jorgensen Laboratories, Inc. (USA) with permission)

> **Tip**
> Injecting oxytocin (3–5 ml) into the myometrium close to the incision will promote contractions and aid closure.

An absorbable suture such as chromic catgut or multifilament polyglactin (United States Pharmacopeia; USP 2–4) is used for suturing the uterus incision. Usually, 1.5–2 meters' length as well as USP 2 in diameter is adequate for the uterine wall to make a horizontal mattress inverted suture (e.g. Cushing suture). The uterus is sutured downwards using a continuous inverting pattern suture without penetrating the lumen and modified by Utrecht method (■ Fig. 8.11). In this method, oblique stitches from 2 up to 0.5 cm of the incision edge are used to minimize the exposure of the suture along the whole trajectory including the initial and final knots (■ Fig. 8.12). The Utrecht method can prevent adhesions between the uterine wall and peritoneum or visceral organs. The initial and final knots should be placed approximately 2 cm distal to each end of the incision. All placental tissue should be removed from near the suture as it can facilitate leakage and lead to fistula formation after it suffers necrosis. Once suture is completed, the exteriorized portion of the uterine wall must be washed with saline solution, to remove all blood clots and foreign bodies. This procedure prevents uterine adhesions.

> **Tip**
> To wash the uterine wall, use 1 L of NaCl 0.9% solution containing heparin (25.000 IU) to prevent blood coagulation.

The uterus is now re-introduced into the abdominal cavity according to its normal topography. Special care is needed to avoid any significant uterine torsion (over 45°). This can be ensured by direct palpation of the uterine body as it enters the pelvic inlet.

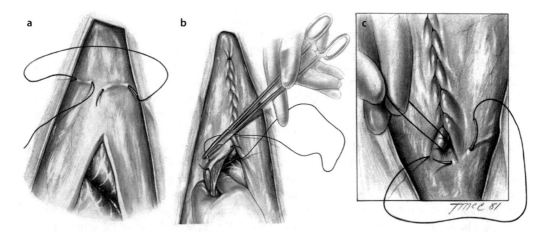

Fig. 8.11 Cushing suture pattern modified by the Utrecht method. Legend: The modified Utrecht method consists of making distal to proximal oblique bites using an inverting pattern without penetrating the lumen (Cushing suture). The initial knot starts about 2 cm dis- tal the top of the incision **a** and progresses with a con- tinuous pattern **b** up to the opposite top, where a final knot is also covered by the uterine wall **c**. (Adapted from Baird [4] with permission from John Wiley and Sons)

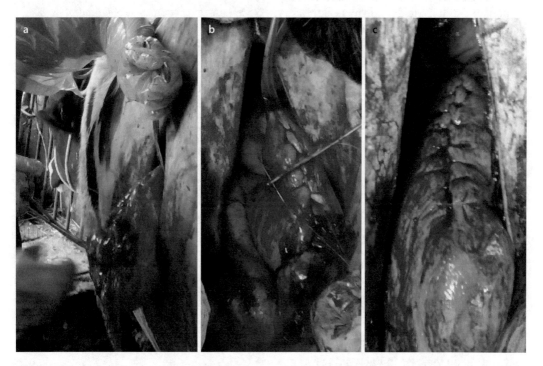

Fig. 8.12 Closure of the uterine incision. Legend: **a** Both extremities of the uterine incision are fixed by an assistant using two uterine forceps 2–3 cm away from each end. A slight tension is exerted by stretching both edges of the incision. **b** The obstetrician closes the inci- sion using a Cushing suture modified by the Utrecht method. The foetal membranes are pending and need be to replaced inside the uterus as their inclusion in the suture will prevent healing, contaminate the peritoneum and eventually produce a fistula. **c** External view of the compete uterine closure before washing the uterine sur- face with saline solution to remove blood clots. (Cour- tesy of Carlos Cabral)

8

■ **Laparotomy Incision Closure**

In the flank approach, the laparotomy incision is closed in three or four layers using simple single sutures. The first layer is the peritoneum and transverse abdominal muscle, which are sutured together using an USP 2 absorbable suture line in the upward direction, preventing any abdominal viscera protrusion. The second layer involves the suture of both internal and external abdominal oblique muscles using an USP 3 or 4 absorbable suture line (◘ Fig. 8.13). For thick muscles (mostly beef cattle), two independent layers can be made. Also, the rectus abdominis needs to be sutured if it was cut. Some stitches should involve the previous muscle layer at regular intervals to obliterate the dead space between these two layers.

The final layer corresponds to the skin suture. A USP 2 non-absorbable suture line (e.g. monofilament polyamide polymer, Supramid or silk) is adequate to perform a Ford interlocking pattern. Simple interrupted sutures or metallic skin staples can also be used.

Tip

The application of one anchored stitch every five to six single stitches of the continuous suture in each muscle layer is very helpful to maintain tension while the suture is performed.

For a contaminated surgery, it has been usual to administrate antimicrobial inside the uterine and peritoneal cavities as well as between muscle layers and subcutaneously once each incision is closed [5, 11]. Nevertheless, there is growing evidence of several disadvantages of these administration routes (▶ Box 8.10). Moreover, the topical use of antimicrobials may lead to an increase in bacterial resistance [21, 24, 25]. This is particularly important when this route has not been studied or approved for the drug and formulation being used. So, the use of injectable or intramammary antimicrobials on the abdominal incision should be considered extra-label, and adding it to the evidence-based antimicrobial therapy should be unequivocally justified. As alternative, the intraperitoneal administration of iodopovidone 1% solution (1–4 L, up to 4 g of active iodine) is recommended for this purpose. Favourable clinical evolution to prevent and treat diffuse peritonitis was obtained by the authors (Stilwell, personal experience).

Box 8.10 Intraperitoneal and Topical Administration of Antimicrobials During Caesarean Section

Although the antimicrobial administration by intraperitoneal (IP) route is generally considered non-irritating, there are a few issues that should be considered before its use in a cow during surgery:

1. Is the efficacy of the drug being used been confirmed when used IP? The fact that it is efficient by other routes does not mean that it works or is absorbed in the same way when put inside the peritoneum. The formulation of the antimicrobial, including excipients, may play an important role. For example, there is evidence of peritonitis in cows after IP infusions of an ampicillin anhydrate formulation [13]. Very few antimicrobials and formulations have been evaluated for peritoneal inflammation or safety.

2. There has been very little investigation of IP antimicrobials used for surgical prophylaxis in cattle. So, questions on

---

◘ **Fig. 8.13** Closure of the abdominal wall. Legend: **a** The first layer involves the peritoneum and transverse abdominal muscle using a simple continuous suture from bottom to top. Some intermittent individual anchored stiches can be applied to keep the tension on the suture. **b, c** The second layer involves the oblique muscles. As alternative to one single continuous suture, two simple continuous sutures can be used starting at midline of the incision towards each extremity. **d** Complete suture of the second muscle layer. **e, f** Ford interlocking suture pattern of the skin. **g** Application of iodopovidone 10% solution in the surgical wound. (Courtesy of Carlos Cabral **a, d–g** and António Carlos Ribeiro **b, c**)

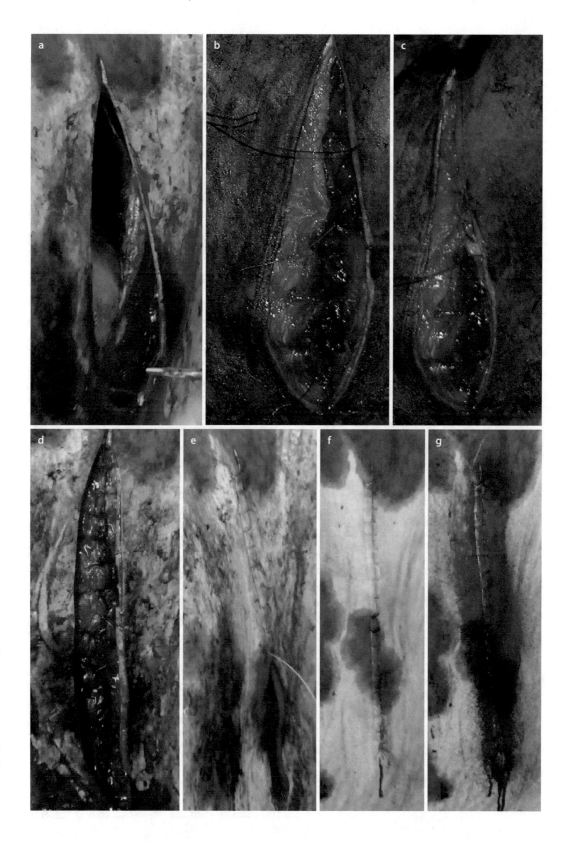

8

the appropriateness of drug doses, timing and withdrawal times after IP administration have not been answered. The impact of the drug on the immune response is also not fully known. Use by IP route should be considered as extra-label and so should meet the corresponding criteria and justifications.

3. There are no studies on the impact of IP use on the establishment of bacterial resistance. At a time when all care should be put on the fight against antimicrobial resistance, the use without scientific evidence should be avoided, especially with critical and priority antimicrobials (e.g. cephalosporins and others).

4. The question as to what is gained by the simultaneous use of IP and other conventional parenteral routes is pertinent, as it will increase the cost of surgery. It has been demonstrated that antimicrobials used by the conventional routes will reach very high concentrations in the peritoneum.

8.7.2   Ventral Midline and Paramedian Approaches

In ventral laparotomy, the skin, the subcutis and the *linea alba* are incised with the dam in dorsal decubitus. Once the pregnant uterine horn is identified, a 45° movement of the dam towards the surgeon is recommended for uterus exteriorization. The procedure, regarding uterine and foetus handling, is similar to the described for the flank approach foetus. Lastly the cow is again placed in dorsal decubitus to suture the abdominal incision. The linea alba is sutured in a single layer using an everting interrupted horizontal mattress pattern. This pattern ensures a safer healing compared with a simple continuous suture, as the suture will be subjected to quite high tension. An absorbable USP 4 suture line is recommended.

The incision in the paramedian laparotomy approach is made approximately 5 cm lateral to the *linea alba*. In this approach, the rectus abdominis muscle and its internal sheath are sutured in one layer.

Sometimes a large udder or a severe ventral oedema may limit the incision length. Also, an extensive ventral vasculature can pose some constraint, and so the trajectory of both cranial superficial epigastric veins should be carefully identified.

> **Tips**
>
> The greater omentum may have to be pulled cranially as it usually partially covers the uterus.

8.7.3   Ventrolateral Approach

The incision starts approximately 5 cm laterally to the umbilicus and runs dorsally to the left cranial superficial epigastric vein. A curvilinear trajectory is made approximately parallel to the last rib. Here muscles layers, tendinous intersections, aponeurosis and sheath can be found, as there is a confluence of the four skeletal abdominal muscles layers, i.e. external and internal abdominal oblique muscles, transverse abdominal muscle and rectus abdominis muscle (◻ Fig. 8.14). Two to three distinct layers are usually used to suture these muscles and their fibrotic extensions.

8.7.4   Left Oblique Approach

The incision line is made on the left flank, approximately 5–10 cm cranioventral to the *tuber coxae*, and runs downwards obliquely to the ventral area of the last rib. In this approach, four skeletal abdominal muscle layers are usually incised. The fibres of the internal abdominal oblique muscle show the same cranioventral direction and can be separated instead of being cut. A similar procedure is sometimes possible for the transverse abdominal muscle. Similar to what was described

Internal abdominal oblique muscle
(fleshy and aponeurotic parts)

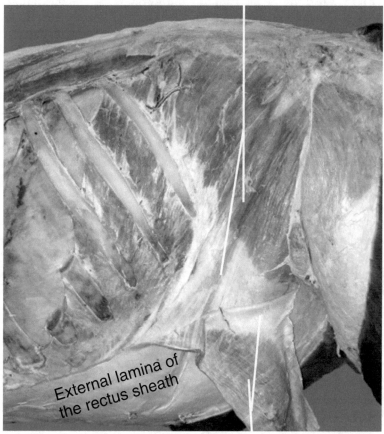

External lamina of
the rectus sheath

External abdominal oblique muscle
(fleshy and aponeurotic parts)
(reflected down)

◘ Fig. 8.14 Topography of the abdominal oblique muscles and their aponeurosis and external lamina of the rectus sheath. Legend: Dissection of the bovine internal abdominal oblique muscle by transection and reflection of the external abdominal oblique muscle ventrally. Distally, the aponeurosis of the external and internal abdominal muscles is fused forming the external lamina of the rectus sheath that inserts on the *linea alba* on the mid-ventral line (left lateral view). (Adapted from Mansour et al. [18] with permission from John Wiley and Sons)

above, the suture can be made using two to three distinct muscle layers.

8.8 Postoperative Care

At the end of surgery, a thorough physical examination of the dam is recommended so as to evaluate general health, pain level and hydration status. It is recommended the administration of 40 IU of oxytocin (i.m.), which increases myometrial contractions and expulsion of uterine contents.

Antibiotherapy should be performed for 3–4 days. The previously described antimicrobial combination or other beta-lactam antimicrobials (e.g. ampicillin or cephalosporin) can be administered. Other antimicrobial such as tetracyclines can also be used if beta-lactam antimicrobials have not been previously administered. For contaminated surgeries, the duration of the antibiotherapy may have to be extended after a re-evaluation of the dam in the next 24–48 h. For non-contaminated uncomplicated C-sections, the reduction or even the abolition of preventive antimicrobial treatment is possible [12].

NSAIDs should be concomitantly administered. The daily use of flunixin meglumine (1.1–2.2 mg/kg; i.m. or i.v.) up to four days is effective in controlling pain and also toxaemia/septicaemia in complicated cases such as those with a dead emphysematous foetus, in which endotoxic shock can occur. In the latter circumstance, fluid therapy prior, during or post-surgery is crucial, mainly in moderate (8–10% dam BW) to severe (11–12% of the dam BW) dehydration. For severe cases, stabilization before the surgery is crucial as can be achieved by intravenous infusion of 5–10 L of a sterile NaCl 0.9% solution. After surgery, oral administra-tion of 50 L of isotonic fluids is recom-mended. A practical and cheap alternative is the intravenous administration of approxi-mately 3 L of 7.2% hypertonic saline solution followed by oral administration of 20 to 40 L of water (not needed if the cow is willing to drink by itself). Re-evaluation of these cases should be done frequently, and, if deemed necessary, fluid therapy should be repeated during the next 12–24 h.

An antiseptic solution (iodopovidone), or a spray containing antimicrobial, is locally applied on the suture twice a day for at least 4 days. The stitches can be removed from the tenth day after the C-section.

The dam needs to be surveyed at 24 h after C-section, to access the presence of placental retention, and during the next two weeks for early detection of puerperal metritis. A full reproductive examination should be per-formed between the second and third week to assess uterine involution.

Case Study 8.1 Complicated Caesarean Section with an Oedematous Dead Foetus

A slightly exhausted Holstein heifer weighing approximately 500 kg was observed in Stage II of labour in the early morning. According to the farmer, a dead foetus with lateral head devi-ation had been identified and correction suc-cessfully performed using a halter. After this, the farmer had evaluated the possibility of foe-topelvic disproportion putting one hand in the pelvic inlet where both forelimbs and head were located. The result was inconclusive, and forced traction was exerted on the foetus by two peo-ple, but the shoulders got stuck on the pelvic inlet. After the successful retropulsion of the foetus into the uterine cavity, the veterinarian was called.

During physical examination, the dam was found recumbent, depressed with moderate dehydration (10% deficit estimation) and rectal temperature of 37.5 °C. A 4-second capillary refill time and heart rate of 86 heartbeats per minute were observed. Abdominal contrac-tions were scarce and weak. During obstetrical examination, the cervix was found to be fully dilated without any evidence of lacerations or haemorrhages, but the birth canal was dry. The dam showed great discomfort during vaginal palpation, even when a copious amount of gel was used, and some degree of vaginal inflam-mation consequent to manipulation was per-ceived. Placental membranes and fluids exhaled a foul but not intense odour. The dead foetus was palpated inside the uterus. Subcutaneous oedema (anasarca) of the foetal forepart was also diagnosed. In accordance with the clinical history and obstetrical evaluation, it was sus-pected that the foetal dystocia was primarily due to the lateral head deviation followed by early foetal death, at more than six hours. The foetal oedema/emphysema contributed to the birth canal obstruction of a relatively large foe-tus. Once the exhausted dam only showed an early phase of hypovolaemic shock and it was subjected to previous significant vaginal manipulation and extended Stage II of labour, a C-section was considered to prevent addi-tional injuries to the uterus and birth canal. The relative foetal oversize and enlargement due to oedema also contributed to this decision. Moreover, the physical evidence of emphysematous foetus can be only detected >24 h after foetal dead. In the current case, the alternative to C-section would be total foetotomy. Nevertheless, total foetotomy choice will increase foetal manipulation through the birth canal and will probably be very stressful for the heifer which has been previous subjected to intensive foetal manipulation causing some severe birth canal inflammation.

Two litres of hypertonic (7.2%) saline solution were intravenously administered followed by the oral administration of approximately 15 L of water, to stabilize the animal. Penicillin (12,000 IU/kg) and streptomycin (10 mg/kg) antimicrobial association was administered intramuscularly to prevent septicaemia as well as flunixin-meglumine (1.5 mg/kg). Since the cow was recumbent, and the foetus dead, showing an enlargement due to oedema with probable early generalized bacterial development, a ventrolateral approach was chosen. The 5–10-20 protocol (5 mg butorphanol + 10 mg xylazine + 20 mg ketamine) was used as pre-anaesthesia. Ten millilitres of 1:1000 epinephrine solution was intramuscularly administered to relax the uterine wall. Caudal

epidural anaesthesia was not considered necessary because of the dam's exhaustion.

The dam was placed in right lateral decubitus, the surgical site was prepared, and the abdominal incision line (45 cm) was infiltrated with 100 ml of 2% procaine. A slightly abnormal amount of peritoneal fluid was present, probably due to uterine inflammatory process. The pregnant uterine horn was effortlessly exteriorized, and an incision along the large curvature was done. The foetus was removed apparently without significant peritoneal contamination. Nevertheless, the uterus was carefully washed with saline solution after the closure of the uterine incision. The uterine wall was sutured using two layers, and 3 litres of a 1% povidone solution was poured into the abdominal cavity. At the end, 35 UI of oxytocin was administered intramuscularly, and the dam was re-evaluated and periodically monitored. No additional postoperative care was necessary. The heifer got up in the following two hours. Analgesia (flunixin meglumine, 1.1 mg/kg/day) and antibiotics were given for five days, also due to the retained placenta. No other complications occurred, and the cow became pregnant at 75 days postpartum.

Key Points

- The indications of the C-section are primarily related to the viability, health and fertility of dam as well as the survival of the foetus.
- C-section is usually (1) an emergency foetus especially for stressed but alive foetus or (2) an elective surgery when dystocia is probable and (3) when the presence of an emphysematous foetus poses severe challenges to the cow.
- Each one of the nine laparotomy approaches presents advantages and limitations. Nevertheless, left paralumbar approach is the preferable method in field conditions.
- Cow's pre-medication includes antimicrobials, analgesics, sedatives (optional), caudal

epidural (optional) and uterine relaxants (optional).
- Local anaesthetics are used to ensure nerve block. Paravertebral anaesthesia and inverted-L anaesthesia, with procaine, are the most commonly used. Nerve block lasts for 1–2 h.
- Appropriate uterine exteriorization is crucial to prevent peritoneal contamination.
- Uterine adhesions will impact future fertility of the dam and can be prevented by following proper surgical rules, including a clean environment and highly aseptic surgical conditions.
- Postoperative care includes clinical re-evaluation of the cow every 24–48 h.

❓ Questions

1. Report and elucidate the main indications for a C-section.
2. How can we prevent the contamination of the peritoneal cavity throughout surgery?
3. Describe the good practices of antimicrobial therapy for the C-section.

✔ Answers

1. The indications to perform a C-section are mainly related to foetal and/or maternal reasons. Any irreducible occlusion or obstruction of the birth canal is a potential justification for a C-section. If foetotomy is not possible or recommended in case of dead emphysematous foetuses, C-section is usually the only alternative. However, before the decision is made, prognosis for the dam and the foetus' viability, health, welfare and fertility should be considered. For live foetuses, the main indications of C-section are the foetal irrevocable maldisposition, oversized foetus or severe foetal distress. In this last case, it should be remembered that the death of the foetus is very likely during forceful and/or prolonged foetal manipulation. Usually, all motives for C-section are emergencies except in pre-emptive surgeries when foetal oversize is expected (e.g. Belgian Blue Breed calvings) or in some breeding programs or in prolonged pregnancies. Foetal deformities, such as teratological monsters, abnormal conformation of the foetus due to foetal death (e.g. emphysematous foetus) or foetal anasarca, are also indications for C-section. In these last cases, the partial or total foetotomy should be considered as a viable alternative. Underdeveloped heifers; deformity or fracture of the pelvic girdle; insufficient or non-dilatation of the cervix; vaginal, cervical or uterine lacerations; and irreducible uterine torsion are maternal factors impeding vaginal delivery of a complete foetus or foetotomy procedures. Several risk factors compromising the success of the C-section should also be carefully evaluated. Depressed and weak cows, whose general health status may be compromised, need to be previously stabilized. Special attention is required for a lengthy duration of Stage II of labour; a previously intense foetal manipulation and the presence of an emphysematous foetus can lead to vaginal and uterine wall inflammation and increase uterine friability or the occurrence of small lacerations. All procedures to prevent bacterial contamination of the dam should be ensured.

2. Contamination of the peritoneal cavity by environmental bacteria or originating from the uterine contents is one of the major hazards to be taken into consideration during a C-section procedure. Prevention of external contamination is achieved by following appropriate hygienic and aseptic guidelines for routine surgery. All surgical equipment and material need to be sterilized, and the surgeon needs to be adequately dressed. On farm, a clean, safe and well-lit location, with clean water available, is essential. Also a sufficient number of people are required to ensure dam's restraint, foetus extraction and calf care. These aspects are primordial for the surgeon to be able to focus on the successive steps of the surgical procedure. An adequate laparotomy approach, according to the clinical and obstetrical evaluation of the dam and foetus, is a significant measure to prevent peritoneal contamination from the environment or the uterus. If there is a risk of the cow falling during surgery, an approach for a recumbent animal should be selected. In the presence of an emphysematous foetus, the ventral or ventrolateral approach is the best choice to mitigate abdominal contaminations because uterine horn is more easily accessed and exteriorized. The key point to prevent bacterial contamination from uterine contents is adequate uterine exteriorization. The exteriorized uterine horn can only be

re-introduced into the abdominal cavity after incision closure and blood clot removal by washing it with a sterile saline solution. The presence of a second foetus or of uterine tears should be checked and solved. A double layer of the uterine suture is recommended for friable uterus or in complicated C-sections such as uterine lacerations (e.g. retrieval of foetus with foetal deformities or other abnormal conformations) or emphysematous foetus. Appropriate closure of the abdominal wall is the final step to prevent suture dehiscence or abdominal fistula.

3. By norm, and mainly in field conditions, prophylactic antibiotherapy is required since some degree of peritoneal contamination always occurs, even in non-complicated C-sections. In hospital environment, where surgical facilities and a surgical team are available, the use of antimicrobials can be avoided in uncomplicated cases. Antimicrobials with broad spectrum of activity should be selected since both Gram-positive and Gram-negative bacteria can contaminate the abdominal cavity and the uterus. Usually, beta-lactams (e.g. ampicillin or amoxicillin), penicillin + dihydrostreptomycin (*aminoglycoside*) or tetracycline antimicrobial groups are used. All these drugs can prevent peritonitis and septicaemia caused by susceptible bacteria. The World Health Organization's list of critically important antimicrobials should be respected to mitigate antimicrobial resistance in the human population. Therefore, the use of (fluoro)quinolones and 3rd- and fourth-generation cephalosporins should be avoided, if possible. The duration of systemic antibiotherapy should be as shortest as possible. Usually, the antimicrobials are administered until two days after the remission of clinical signs of infection. Nevertheless, the minimum duration should be three days to ensure a minimum inhibitory/bactericidal concentration during the prophylactic treatment.

Also, the concomitant administration of NSAIDs modelling the pain and the inflammatory process can help to reduce the antibiotherapy needed.

References

1. Alexander D. Bovine caesarean section 1. On-farm operations. In Pract. 2013;35:574–88. https://doi.org/10.1136/inp.f6679.
2. Alexander D. Bovine caesarean section 2. Difficult caesareans (potential pitfalls and how to overcome them). In Pract. 2014;36(1):1–4. https://doi.org/10.1136/inp.g128.
3. Ames NK. Noordsy's food animal surgery. 5th ed. Iowa: Wiley; 2014.
4. Baird AN. Bovine urogenital surgery (Chapter 14). In: Turner and McIlwraith's techniques in large animal surgery, 4th ed, Hendrickson DA, Baird AN (Edts); Wiley Blackwell: Iowa; 2013. pp. 235–271.
5. Chicoine AL, Dowling PM, Boison JO, Parker S. A survey of antimicrobial use during bovine abdominal surgery by western Canadian veterinarians. Can Vet J. 2008;49(11):1105–9.
6. Das YK, Aksoy A, Yavuz O, Guvenc D, Atmaca E. Tocolytic effects of meloxicam on isolated cattle myometrium. Kafkas Univ Vet Fak Derg. 2012;18(6):1043–8. https://doi.org/10.9775/kvfd.2012.6970.
7. Dehghani SN, Ferguson JG. Cesarean section in cattle: complications. Comp Contin Educ Vet Prac. 1982;4:s387–92.
8. Dumas SE, French HM, Lavergne SN, Ramirez CR, Brown LJ, Bromfield CR, Garrett EF, French DD, Aldridge BM. Judicious use of prophylactic antimicrobials to reduce abdominal surgical site infections in periparturient cows: part 1 – a risk factor review. Vet Rec. 2016;178(26):654–60. https://doi.org/10.1136/vr.i103677.
9. Funnell BJ, Hilton WM. Management and prevention of dystocia. Vet Clin North Am Food Anim Pract. 2016;32(2):511–22. https://doi.org/10.1016/j.cvfa.2016.01.016.
10. Hall LW, Clarke KW, Trim CM. Chapter 12 – anaesthesia of cattle. In: Veterinary anaesthesia. London, UK: WB Saunders; 2001. p. 315–39. https://doi.org/10.1016/B978-070202035-3.50013-9.
11. Hardefeldt LY, Browning GF, Thursky KA, Gilkerson JR, Billman-Jacobe H, Stevenson MA, Bailey KE. Cross-sectional study of antimicrobials used for surgical prophylaxis by bovine veterinary practitioners in Australia. Vet Rec. 2017;181(16):426. https://doi.org/10.1136/vr.104375.
12. Jorritsma R, van Geijlswijk IM, Nielen M. Randomized prospective trials to study effects of reduced antibiotic usage in abdominal surgery in cows. J Dairy Sci. 2018;101(9):8217–23. https://doi.org/10.3168/jds.2017-14158.

13. Klein WR, Firth EC, Kievits JMCA, De Jager JC. Intra-abdominal versus intramuscular application of two ampicillin preparations in cows. J Vet Pharmacol Ther. 1989; 12(2):141–6. https://doi.org/10.1111/j.1365-2885.1989.tb00655.x.

14. Kolkman I, Opsomer G, Lips D, Lindenbergh B, De Kruif A, De Vliegher S. Pre-operative and operative difficulties during bovine caesarean section in Belgium and associated risk factors. Reprod Domest Anim. 2010;45(6):1020–7. https://doi.org/10.1111/j.1439-0531.2009.01479.x.

15. Lee L. Local anesthesia and analgesia. Veterinary anesthesia and analgesia, 2003. From: https://www.westernu.edu/veterinary/anesthesia/. Accessed on 14 July 2021.

16. Lyons NA, Karvountzis S, Knight-Jones TJ. Aspects of bovine caesarean section associated with calf mortality, dam survival and subsequent fertility. Vet J. 2013;197(2):342–50. https://doi.org/10.1016/j.tvjl.2013.01.010.

17. Mama K. Anaesthesia and fluid therapy (Chapter 2). In: Turner and McIlwraith's techniques in large animal surgery, 4th ed, Hendrickson DA, Baird AN (Edts); Wiley Blackwell: Iowa, 2013. pp. 7–31.

18. Mansour M, Wilhite R, Rowe J. Chapter 3: the abdomen. In: Guide to ruminant anatomy: dissection and clinical aspects. New Jersey: Wiley; 2018. p. 91–138. https://doi.org/10.1002/9781119379157.ch3.

19. Newman KD, Anderson DE. Cesarean section in cows. Vet Clin North Am Food Anim Pract. 2005;21(1):73–100.

20. Newman KD. Bovine Cesarean sections: risk factors and outcomes. In: Anderson D, Rings M, editors. Current veterinary therapy: food animal practice. 5th ed. St. Louis: Saunders; 2009. p. 379–82. https://doi.org/10.1016/B978-141603591-6.10079-X.

21. Punjataewakupt A, Napavichayanun S, Aramwit P. The downside of antimicrobial agents for wound healing. Eur J Clin Microbiol Infect Dis. 2019;38(1):39–54. https://doi.org/10.1007/s10096-018-3393-5.

22. Schultz LG, Tyler JW, Moll HD, Constantinescu GM. Surgical approaches for cesarean section in cattle. Can Vet J. 2008;49(6):565–8.

23. Weaver AD, Owen AO, St. Jean G, Steiner A. Chapter 6: female urinogenital surgery. In: Bovine surgery and lameness. 3rd ed. New Jersey: Wiley; 2018a. p. 187–210.

24. Williamson D, Ritchie SR, Best E, Upton A, Leversha A, Smith A, Thomas MG. A bug in the ointment: topical antimicrobial usage and resistance in New Zealand. N Z Med J. 2015;128(1426):103–9.

25. Williamson DA, Carter GP, Howden BP. Current and emerging topical antibacterials and antiseptics: agents, action, and resistance patterns. Clin Microbiol Rev. 2017;30(3):827–60. https://doi.org/10.1128/CMR.00112-16.

26. Weaver AD, Atkinson O, St. Jean G, Steiner A. Chapter 1: general considerations and anaesthesia. In: Bovine surgery and lameness. 3rd ed. Oxford, UK: Wiley; 2018b. p. 1–52.

Caesarean Section Procedures

Drost M, Samper J, Larkin PM, Gwen Cornwell D. Cesarean Section. The Visual Guide to Bovine Reproduction. UF College of Veterinary Medicine: Florida, USA; 2019. From: https://visgar.vetmed.ufl.edu/en_bovrep/cesarean-section/cesarean-section.html, accessed on February 21, 2020.

Suggested Reading

Kelly E, Ryan E. Bovine caesareans: alternative approaches to difficult cases. Veter Ireland J. 2018;8(10):600–8.

Newman KD. Bovine cesarean section in the field. Vet Clin North Am Food Anim Pract. 2008;24(2):273–93, vi. https://doi.org/10.1016/j.cvfa.2008.02.009.

Hendrickson DA, Baird AN, editors. Turner and McIlwraith's techniques in large animal surgery. 4th ed. Iowa: Wiley Blackwell; 2013.

Puerperal Complications in the Dam

Contents

Electronic Supplementary Material The online version of this chapter (https://doi.org/10.1007/978-3-030-68168-5_9) contains supplementary material, which is available to authorized users. The videos can be accessed by scanning the related images with the SN More Media App.

9

Learning Objectives
- To identify the causes and predisposing factors of the main disturbs consequent to calving.
- To recognize the relationships between puerperal diseases, uterine involution and reproductive outcomes.
- To describe the pathogenesis, clinical signs and diagnosis of puerperal diseases.
- To manage downer cows, consecutive to maternal obstetrical paralysis.
- To recommend a clinical approach to retained placenta, regarding it as a risk factor for subsequent metabolic and infectious diseases.

9.1 Introduction

Several complications related to the calving process usually occur or are evaluated after the calf delivery. Parturition is a complex stressor event involving hormonal, immune and metabolic changes as well as an anatomical adaptation of the birth canal and mechanical straining to expel both the foetus and the placenta. Excessive or mispositioned pressure of the foetus in the birth canal can cause injuries in the soft tissues of the birth canal, such as bruises, peripheral nerve compressions and tears of the cervix, vagina and/or vulva. In some less common cases, significant uterine ruptures (8–10 cm or more), usually involving a friable uterine wall, can be detected near the pelvic inlet. Additionally, uterine prolapse can occur immediately after calf delivery or, less commonly, during the following few days and should be considered a reproductive emergency endangering the dam's life.

The puerperium (immediate postpartum period or puerperal period) is defined as the interval between birth and complete uterine involution. In cows, the normal length of this period is 40–42 days, simultaneous to the resumption of ovarian activity. The most significant morphological uterine changes occur during the first 30 days, during the which the uterus decreases 80% in length [15], reaching a size close to what it was before pregnancy. Lochia and uterine fluid secretions result from

tissue (caruncle) dehiscence and sloughing leading to endometrial regeneration. It is estimated that around half of the cows may develop abnormal inflammatory processes of the reproductive tract during the first 7 weeks postpartum. These disorders are mainly related to clinical metritis (up to 21 days) and endometritis (after 21 days), which originated from bacterial growth due to exposition to the environment. The retention of foetal membranes and subclinical hypocalcaemia represent a significant risk factor for uterine infections or metabolic diseases during the puerperium, with a significant negative impact on the cow's fertility, increasing the interval between calvings. Dairy cows presenting retained placenta are also more likely to develop metabolic diseases and reduced milk yield.

This chapter intends to describe and suggest a clinical approach to diagnose, treat and prevent the most relevant complications of the cow's puerperal period.

9.2 Uterine Prolapse

9.2.1 Aetiopathogenesis and Diagnosis

Uterine prolapse occurs in less than 0.5% of calvings, mainly within the first 2 h after calf delivery, but can occur up to 24 h. One of the authors has treated several mid gestation prolapse of the nongravid uterine horn. It consists of the eversion of the uterus through the birth canal. Commonly, it occurs during Stage III of labour, when the cervix is still sufficiently dilated. When the uterus is wholly everted, the pregnant uterine horn and body of the uterus and cervix are prolapsed outside the vulva. The uterine intercornual ligament usually prevents the eversion of the contralateral uterine horn. Sometimes the bladder and/or intestinal loops may be incarcerated in the prolapsed mass. The endometrium surface, including the maternal caruncles or placentomes, is exposed to the environment, and the vascular system of the uterus may be totally or partially compromised (◻ Fig. 9.1).

☐ Fig. 9.1 Uterine prolapse
Legend: Left picture – Posterior view of prolapsed mass consist on the everted previously pregnant uterine horn attached to the everted cervical canal with the cervical fold observed near the vulva. Maternal caruncles are well evident, once the foetal cotyledons and placenta were removed. (Courtesy of João Fagundes (Ilha Terceira, Azores)). Right picture – The oedema of the uterine wall quickly develops due to vascular compromise. (Courtesy of Carlos Pinto (University of the Azores))

Uterine prolapse is a reproductive emergency that can quickly lead to death if the dam develops hypovolaemic shock caused by haemorrhages from rupture or avulsion of the large uterine vessels. The prolapse should be reduced within the first two hours to mitigate progressive oedema, friability of the uterine wall, trauma or rupture of main blood vessels.

Several predisposing factors have been reported for uterine prolapse (▶ Box 9.1), and some of them are the same as for vaginal prolapse.

A uterine prolapse can be easily differentiated from a vaginal prolapse due to the volume exteriorized and the presence of maternal caruncles.

9.2.2 Treatment

The treatment of the uterine prolapse aims at quickly replacing the uterus back to its normal position and to prevent relapses. Prognosis is relatively good (over 80% survival rate) when there is timely veterinary intervention. In the case of extensive uterine damage, euthanasia or the amputation of the uterus (hysterectomy) may be the only viable solutions (▶ Box 9.2).

Box 9.1 Predisposing Factors for Uterine Prolapse
- Hypocalcaemia.
- Prolonged dystocia.
- Uterine inertia.
- Foetal oversize.
- Excessive foetal traction.
- Retained foetal membranes.
- Malnutrition and chronic disease.
- Tenesmus consecutive to severe dystocia.

Box 9.2 Hysterectomy
Although it comprises a poor prognosis, the amputation of the uterus presenting severe necrosis or gangrene is the only alternative before deciding for euthanasia. The technique is here summarized, but reading a detailed surgical description is strongly recommended [16].

Circumferential or transfixing non-absorbable strong sutures (e.g. umbilical tape) are applied just caudal to the cervix to ligate each half of the uterus. The resection of the prolapsed uterus is made approximately 5 cm caudal to the suture. Haemostatic absorbable sutures should be placed throughout the margin. A second technique involves an incision close to the body of the prolapsed uterus. The blood vessels of each broad ligament are ligated individually or in groups according to their size using catgut USP 4 or even umbilical tape. The broad ligament is sectioned around 1 cm distal to the sutures. In a second step, a mattress suture is applied just caudal to the cervix, and the uterus is amputated. In both techniques, the remaining uterine tissues stay located in the pelvic cavity.

It should be remembered that abdominal viscera and urinary bladder may be enclosed in the prolapsed uterus. These should be identified by making an incision on the uterus, and all viscera should be replaced into the abdominal cavity before amputation.

Tip

Any of these procedures should be done under epidural anaesthesia with procaine or lignocaine (lidocaine).

An analgesic drug should also be administered before replacing the uterus to provide analgesia quickly and reduce postoperative discomfort.

The prolapsed mass should be cleaned and washed with warm water. The foetal cotyledons are carefully detached from the maternal caruncles to avoid bleeding. Any uterine lacerations, independent of size, need to be sutured. Horizontal mattress suture or continuous everting suture should be used for this purpose. The uterine surface should be disinfected (e.g. diluted 0.5% *iodopovidone*, 1% chlorhexidine or 1:1000 potassium permanganate solution) before replacing the uterus. The replacement procedure can be done with the cow in standing up position, or, if it is a downer cow, hip-slings should be used to elevate the hindquarters or to help keep the cow standing [17]. These methods reduce the pressure being exerted by viscera on the opening of the pelvic canal into the abdominal cavity.

If lifting the cow is impossible, it should be placed in sternal decubitus with both legs extended caudally (New Zealand technique) or abducted (frog-leg position; ◘ Fig. 9.2). This position facilitates the replacement of the uterus.

> **Important**

Different techniques can be used to reduce the diameter of the uterus before or during its replacement. One to two kilogrammes of sugar or salt (NaCl) can be scattered on the endometrium, acting for approximately 5 min. This is an effective and practical method, allowing a significant reduction of the total mass by osmosis. Nevertheless, these substances can cause additional endometrial trauma and should be removed as soon as possible (within 5–10 min.) and the uterus washed and disinfected again. An alternative is to compress the uterus with a bandage from the end of the uterine horn towards the cervix. The bandage is gradually removed as the horn is replaced. The mass reduction is an important procedure when the uterine wall becomes more friable. In the latter case, the obstetrician's fingers can easily cause tears during the replacement procedure. To avoid tearing and other iatrogenic trauma, plenty of lubricant and "oven-mittens" or gloves should be used to handle and push the uterus through the vulva.

Tip

A clean moisty sheet should cover the prolapsed mass, and the dam restrained until veterinarian arrival to prevent environmental contamination and trauma of the uterine wall.

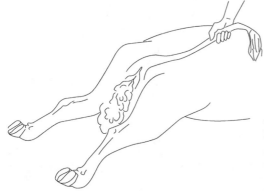

◘ Fig. 9.2 Optimal position for recumbent dams to replace the prolapsed uterus
Legend: Left drawing – The dam is in sternal decubitus with the hind legs pulled and restrained caudally. This position allows a downward 30°–45° inclination of the pelvis which facilitates the replacement of the uterus.

(Adapted from Weaver et al. [42] with permission from John Wiley & Sons). Right picture – Notice the metal rack that allows for washing and keeps the organ safe and clean during the reduction procedure. Note also the syringe and needle used for caudal epidural anaesthesia. (Courtesy of Margarida Batista da Costa and Rui Silva)

A caudal epidural anaesthesia is administered to prevent abdominal contractions of the dam. Dosage should be careful calculated (usually less than 10 mL of 2% lignocaine for a large cow) so as to prevent the cow from falling down during the replacement procedure. If the dam is standing, the uterus is elevated until its alignment with the birth canal. A towel can be used to hold the uterus in this position. The replacement of the uterus should be done as close as possible to the cervix, introducing the most cranial portion of the prolapse uterus, pushing it with the palm of the hands or closing the fingers into a fist, to prevent uterine tears. Use of oven-mittens is an excellent way of reducing the risk of trauma and is less slippery. If possible, the non-prolapsed (non-pregnant) uterine horn should be pushed firstly to obtain more space. A gentle massage around the uterus prevents uterine perforation. This initial procedure is crucial to perform the inversion of the uterine

wall, mainly if the birth canal is narrow. Simultaneously, an assistant should support the distal part, applying additional force to provoke pressure on the uterus towards the birth canal. When the distal portion of the uterine horn is close to the vulva, its end can be pushed directly into the inside the birth canal.

> **Tip**
>
> The lubrification and massage of the pro-lapsed mass can be helpful to replace it.

Once inside the abdominal cavity, inversion of the uterine horn should be made as deeply as possible. The complete reversal can be obtained, pumping approximately 10 L of warm water into the uterus. After a few minutes, the water should be siphoned out to prevent uterine rupture. Also, a 1.5 L bottle filled with warm water can be used as an arm extension for this purpose. The complete straightening of the uterine horns can prevent ischemic necrosis due to its invagination or even its relapse.

The use of retention sutures on the vulva is controversial. The re-prolapse of uterus is uncommon. The retention suture is unnecessary for those cases in which the replacement and reversion of both uterine horns are complete and myometrial contractility occurs. The retention sutures can provoke tenesmus, causing significant discomfort. Also, a partial re-invagination of the uterine horn can be imprisoned inside the vagina, causing undetected uterine necrosis. Nevertheless, when the uterine inertia occurs, a retention suture, such as the Bühner suture, can be used and should remain for 2–3 days (see ▶ Chap. 2).

> **Tip**
>
> An enlarged bladder should be emptied through massage, puncture or catheterization before replacing the uterus.

At the end, 30–50 UI of oxytocin should be administered. The subclinical hypocalcaemia is frequently associated with uterine prolapse, and the administration of 150–300 mL (at least) of 24% borogluconate is recommended. The aftercare consists of the administration of a nonsteroidal anti-inflammatory drug and a systemic broad-spectrum antimicrobial for two and four days, respectively. It is good practice to inject both drugs before replacing

the uterus for quick and pre-emptive protection. The vulvar suture, if used, should be removed three days after the prolapse reduction. The cow and the birth canal (vulva, vagina and cervix) should be re-evaluated within 12–24 h after treatment to check for potential complications (▶ Box 9.3).

> **Box 9.3   Complications of Uterine Prolapse**
> - Toxaemia.
> - Septicaemia.
> - Haemorrhage.
> - Uterine rupture with bladder or intestinal eventration.
> - Puerperal metritis.
> - Peritonitis.
> - Paresis.

9.2.3 **Prognosis**

The dam's survival rate is over 70% after treatment [11, 22] if timely and appropriate assistance is delivered. Multiparous dams giving birth to live foetuses without presenting a stage 3 hypocalcaemia and primiparous cows show the best survival prognostic two weeks after calving. Jubb et al. [22] observed that 84% of cows which suffered uterine prolapse conceived with a delay of approximately 10 days when compared to cows with normal calving. However, the delay of the interval between calving and conception can reach up to 50 days [29]. The recurrence of uterine prolapse in the next calving is uncommon.

Case Study 9.1: Uterine Prolapse

Two hours after calving, a significant prolapsed mass outside the vulva was observed in a Holstein dairy cow (600 kg weight live; third parturition). Dry placental membranes surrounding the uterus and placentomes were easily identified leading to the diagnosis of uterine prolapse. During physical examination, the dam was found standing up and alert, with clinical signs of hypovolaemia (sunken eyes

and skin tent >3 s). Subclinical hypocalcaemia was suspected. The cow also showed intense straining. Rectal temperature was 38.7 °C, and the cardiac frequency was 78 bpm. A caudal epidural anaesthesia (15 mL of 2% procaine) was performed to desensitize the perineal area, vaginal wall and rectum. The dam was restrained and placed in sloped floor elevating the hindquarters about 30°. The prolapsed mass was abundantly washed with warm water to remove faeces and to clean its surface. A damp towel was placed under the prolapsed mass, which was elevated until a horizontal plan with the vulva opening by two people, one in each side. Each foetal cotyledon was carefully separated from the maternal caruncles. The placentome should not be pulled, avoiding the caruncle tear and consequent haemorrhage. When the detachment of the foetal cotyledon was not possible, the chorioallantois membrane was cut around the placentome. The prolapsed mass was thoroughly washed and disinfected with a 0.5% iodopovidone solution. The endometrium was fully inspected, and interrupted horizontal mattress sutures were performed to close two partial lacerations of the uterine wall (approximately 5 cm of length each). The (previously) pregnant uterine horn was carefully introduced from the base to the end by successive pushes using both palms around the everted mass. At this time, the prolapsed mass was lubricated (liquid Vaseline) and massaged carefully. Once the uterus was relocated, a uterine lavage with 10 L of warm water was performed trying to ensure the complete inversion of the uterine horn. A 50 IU of oxytocin was injected intramuscularly to prevent uterine bleeding and promote uterine contraction and shrinking. Also, 150 mL of calcium borogluconate 24% were administered subcutaneously in three different locals. Without evidence of tenesmus, the retention suture was not applied. The aftercare involved the intramuscular administration of oxytetracycline (20 mg/kg/daily, i.m.) and flunixin meglumine (2 mg/kg/daily, i.m.) during five and three days, respectively. The dam was surveyed by the farmer during the next few days regarding the presence of tenesmus and vaginal secretions. Milk production increased in the following days and the cow was seen eating. A transrectal examination to evaluate the presence of potential adhesions of the uterus with viscera or abdominal wall and the uterine involution was performed five days after calving.

9.3 Birth Canal Bruises and Lacerations

Postpartum contusions and lacerations of the birth canal result mainly from the pressure exerted by the foetus or caused by the obstetrician (iatrogenic) during foetal manipulation.

Sometimes the foetus (the hooves are usually involved) damage the internal wall of the obstetrical canal causing rupture of the pudendal artery, forming a large haematoma in the lateral wall of the vagina. Usually a 3-day interval before draining the blood through a vaginal incision is required, to allow vessel haemostasis. More rarely, the artery will bleed into the vagina and could lead to death by acute blood loss. In these cases, a clamp should be applied to the artery, closing it completely. Clamps should only be removed after 3–5 days.

In any of these cases, the dam should be restrained to prevent abrupt movements. Systemic antimicrobials should be administered in order to avoid infections caused by *Clostridium* spp. or other bacteria.

The perineal lacerations are classified in three degrees according to the extension of the injury and the involved tissues (▶ Box 9.4). Vulvovaginal tears of the vulvar or vaginal walls larger than 2 cm are a risk factor for vaginitis, puerperal metritis and endometritis [41].

All these lesions should be immediately diagnosed by always doing a thorough examination of the birth canal after calving finishes.

Box 9.4   Classification of Perineal Lacerations
- First degree: The junction of vulvar labia with the skin, vestibule and mucosa of the vagina are involved.
- Second degree: The fibromuscular tissues that separate the rectum and vagina are torn, but the anal sphincter remains intact.
- Third degree: A complete rectovestibular rupture is observed, including the tear of the anal sphincter, the caudal portion of the rectum and the constrictor vestibuli muscle.

First-degree perineal lacerations and small lacerations near the vulva, usually less than 2 cm long, do not require suturing. The vaginal wall and vulvar labia remain undeformed after healing. Nevertheless, for more deep lac-erations, a Caslick suture is recommended (❒ Fig. 9.3a). In the second-degree lesions, the functional vulvar and vestibular confor-mations are probably lost. The reconstruction of the tissues is compulsory to re-establish the

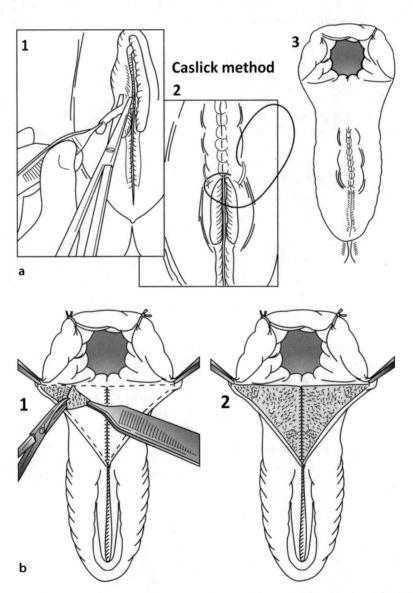

❒ **Fig. 9.3** Illustrated vaginoplasty for **a** Caslick's pro-cedure and **b** second-degree perineal lacerations
Legend: **a** Caslick's procedure. On the labia's dorsal third segment, a strip of tissue is excised at the mucocu-taneous junction on both sides to remove necrotic tis-sues and induce cicatricial adhesion (1). The length of the strip to be sectioned can vary according to the tear location. However, the vulva should remain free ven-trally, at least 3 cm, to allow urination. A continuous suture pattern is made (2). At the end, juxtaposition of the labia is ensured (3). A cut in the dorsal region may be required in the next parturition to ensure vulva fully dilatation. **b** Reconstruction of a second-degree lacera-tion. A triangular-shaped piece of the mucosa is excised on both sides of the vaginal vestibule and vulva. A Met-zenbaum scissor is used to debride lacerated tissues (1). The excision is made parallel to the rectum, and its extension depends on the length of the tear. Absorbable (USP 2-0 or 3-0) sutures are made from the inside out, juxtaposing the exposed mucosa with submucosa (2). Finally, a Caslick's suture is applied. (Modified from Fubini [9] with permission from Elsevier)

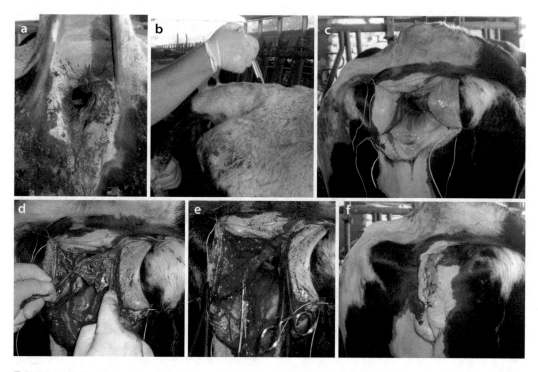

◘ Fig. 9.4 Vaginoplasty of second-degree perineal lacerations
Legend: **a** Initial aspect. The anal sphincter as well as the rectum remain intact. **b** Caudal epidural anaesthesia. **c** The perineal, anal sphincter and external genitalia were cleaned, and vulvar lips are temporally fixed to the skin improving the accessibility to the surgery local before disinfection. **d** A strip of tissue is excised on both sides of the vaginal vestibule and vulva. **e** A suture is made to juxtapose the exposed mucosa with submucosa; the surgical area was cleaned with sterile saline solution before closing the surgical wound. **f** A simple continuous (five stitches) pattern is dorsally applied to the vulvar lips leaving free ventral space for urination. (Courtesy of António Carlos Ribeiro)

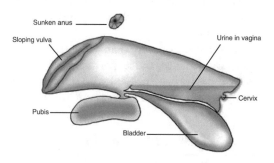

◘ Fig. 9.5 Schematic representation of urovagina (lateral view). (Jackson and Cockcroft [19] with permission from John Wiley & Sons)

juxtaposition of the vulvar lips and to close the vestibule (◘ Figs. 9.3b and 9.4). This will prevent urovagina (◘ Fig. 9.5), pneumovagina and vaginal or uterine infections, as well as their consequences on the dam's fertility. In the third-degree, untreated cases will lead

almost certainly to a rectovaginal fistula (◘ Fig. 9.6).

Large tears and especially the third-degree ones should not be reconstructed immediately after calving as tissues are fragile and difficult to identify. Good second intention healing should be promoted by the area washing frequently. Reconstruction should be only attempted when cicatrization is finished – usually 15–21 days after calving. After incising the lateral and cranial walls along the border between rectum and vaginal tissues, individual sutures are made juxtaposing the vestibular and rectal flaps (◘ Fig. 9.7). The sutures start cranially and finish where the skin is evident. Separate sutures of the perineal skin close the wound. Recently, Sato et al. [36] suggested a modified Aanes technique for cows, which consists of closing the vaginal roof with a continuous horizontal mattress suture followed

9

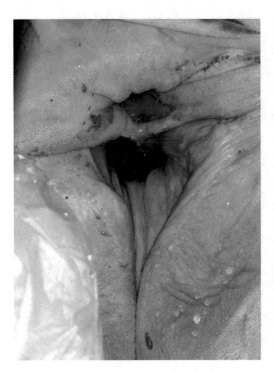

◨ **Fig. 9.6** Rectovaginal fistula (dorsal to the vaginal vestibule) consecutive to a third-degree perineal laceration, after second intention healing. (Courtesy of António Carlos Ribeiro)

by a purse-string pattern suture between the vaginal and rectal shelves. These perineal areas' rebuilding and suture are done under caudal epidural anaesthesia, and, preferably, the cow remains in the stand-up position during all of the described surgeries.

9.4 Calving Paralysis

Maternal obstetrical paralysis is caused by unilateral or bilateral compression of the obturator and sciatic nerves during Stage II of labour or forced extraction of the foetus. In the bilateral compression of these nerves, the cow cannot rise and stand up and will eventually develop the downer cow syndrome. Although damage to only one nerve is possible, most cases involve the obturator nerve(s) and the sciatic nerve(s). This is the reason why the designation of calving paralysis is favoured.

9.4.1 Obturator Nerve Paralysis

The obturator nerve has its spinal roots in L5 and L6. This nerve then runs along the medial aspect of the ilium shaft before exiting the pelvic cavity through the obturator foramen (◨ Fig. 9.8). The obturator nerve innervates the muscles responsible for the adduction of the respective hindlimb. Pressure on the nerve(s) from an oversized foetus (see, e.g. hiplock in ▶ Chap. 4) can cause damage, ischaemia or even transection of one or both nerves with subsequent temporary or long-term loss of nerve function, causing unilateral or bilateral inability to adduct the hindlimbs. In unilateral neuropathies, the cow will usually be able to rise and support weight, if doing so on a non-slippery flooring, but may show slight abduction of the affected limb and hind leg weakness (◨ Fig. 9.9). In bilateral neuropathies, the cow may still rise if the obturator nerve is the only nerve affected. However, even an exclusive damage of the obturator nerve will predispose cows to slipping and uncontrolled abduction on slippery floors, leading to more severe lesions (e.g. adductor muscle rupture). It is frequent to observe the dam alert, eating and in a dog-sitting position, since all the other spinal nerve roots remain unaffected. The obturator nerve does not innervate the skin and has no sensory function.

It is now proposed that obturator nerve paralysis will not cause permanent recumbency unless it is associated with sciatic nerve paralysis (which occurs very frequently).

9.4.2 Sciatic Nerve Paralysis

The sciatic nerve has its spinal roots in L6, S1 and S2 and then runs ventral to the sacrum. Two important branches of this nerve are the tibial nerve and the peroneal nerve. Usually, it is the L6 root of the sciatic nerve that suffers damage from compression of foetus against the sacrum during delivery of large foetus (◨ Fig. 9.8). In unilateral single sciatic neu-

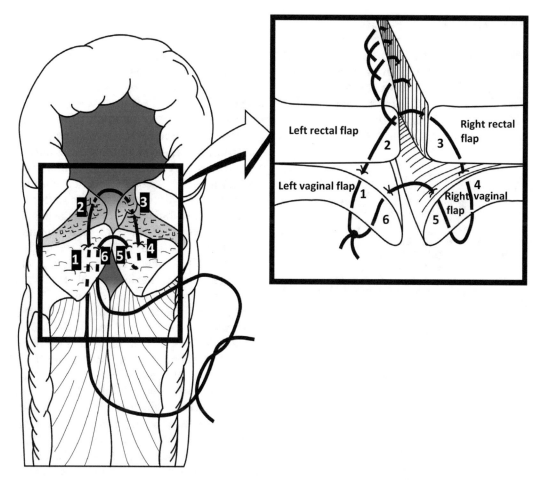

Fig. 9.7 Illustrative vaginoplasty of the third-degree perineal lacerations
Legend: One-stage repair modified from Goetz. An individual suture involving a six-bite vertical pattern is used after tissue debridement; USP 1 or 2 synthetic non-absorbable (e.g. monofilament polyamide) or nonreactive mono- or multifilament absorbable (e.g. polydioxanone and polyglactin 910) sutures can be used for this purpose. In each suture, the bites were made juxtaposing the left vestibular flap with the rectal flap without drilling the rectal mucosa (1 and 2). An inverse trajectory pattern is created on the right side (3 and 4). A dorsal to ventral bite is made on the right vaginal flap (5) followed by the ventral to dorsal bite on the left vaginal flap (6). Each bite is placed around 2 cm away from the edges, except near the rectal mucosa. A reinforced knot is required. Each individual suture pattern should be placed 1–1.5 cm apart. The non-absorbable suture, if used, should be removed around 14 days after the surgery. (Modified from Fubini [9] with permission from Elsevier)

ropathies, the cow may rise (most times needing some help) and will bear weight on the dorsum of the fetlocks and digits of the affected limb. Cattle with bilateral damage of both the sciatic and obturator nerves will not get up. Typically, it will lie in a froglike position with the hindlimbs stretched and abducted on both sides of the body or even, in the worst cases, having both hindlimbs splayed and extended perpendicular to the body's long axis.

More active/nervous animals will repeatedly try to rise prompting severe complications – e.g. fractures, coxofemoral luxation and rupture of the gastrocnemius muscle or tendon.

Prognosis is good for unilateral cases but poor for bilateral calving paralysis associated with recumbency, especially in heavy cows or those remaining for long in hard flooring. These animals will usually develop downer cow syndrome (see below).

9.4.3 Tibial Nerve Paralysis

Compression of the fibres that will branch onto the tibial nerve by an oversized calf will result in flexion of the hock and slight knuckling of the fetlock joint and typically involves both hind legs (■ Fig. 9.10).

Recovery may take several months.

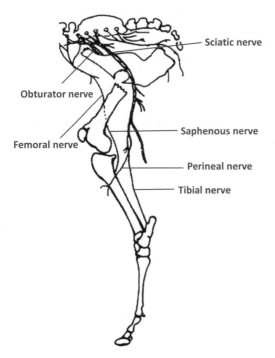

■ **Fig. 9.8** Trajectory of the obturator, sciatic and other hindlimbs nerves. (Modified from Ames [1] with permission from John Wiley & Sons)

9.4.4 Downer Cow Syndrome

The downer cow syndrome is a pathological complication of prolonged recumbency (over 6–12 h) that results from muscle and nerve damage. Usually, these cows are alert and show no other signs of disease. In adult cattle, prolonged recumbency causes a high and constant pressure on the hindlimb muscles and on peripheral nerves (peroneal and tibial nerves) originating a compartment syndrome. Compartment syndrome starts with resistance to vascular flow and oedema inside the muscle and nerve fascia on the down-side limb and from which areas of ischemia eventually develop. This muscle oedema/ischemia will result in swelling of the limb distal to the stifle joint and combined neurologic-muscular damage including pressure necrosis of muscle fibres and nerves confined within the fascial borders. Additionally, the sciatic and the peroneal nerves will be subjected to pressure at the caudal aspect of the femur and at the proximal fibula, respectively. Maternal obstetrical paralysis, other calving trauma (e.g. musculoskeletal injuries, pelvic fractures and coxofemoral or sacroiliac luxation), postpartum metabolic diseases (e.g. hypocalcaemia or severe hepatic lipidosis) and toxaemia are the primary causes of sternal decubitus which can lead to this syndrome.

The development of myopathies in the hindquarters will start around 6 h after decubitus if the cow is kept on the same side, espe-

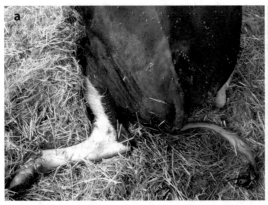

■ **Fig. 9.9** Clinical presentations of obturator nerve injury after difficult calving. **a** Unilateral abduction of the right leg consecutive to obturator nerve lesion of the ipsilateral side. **b** Typical frog-sitting or dog-sitting posture involving bilateral nerve lesions. (Originals from George Stilwell)

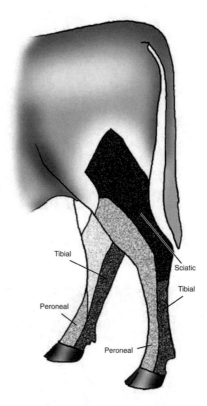

Fig. 9.10 Bilateral tibial nerve paralysis. (Original from George Stilwell)
Legend: Both fetlock articulations remain slightly flexed due to the lack of motor function of the tibial nerve

Fig. 9.11 Sensory innervation of the skin by the sciatic, peroneal and tibial nerves of the hindlimb. (Jackson and Cockcroft [18] with permission from John Wiley & Sons)
The sensory innervation (peripheral sensory nerve and spinal cord integrity) should be tested by skin clamping. The cow shows a conscious perception of skin sensitivity. The desensitization area can evidence the respective lesioned nerve

cially when good bedding is absent. After a few days of decubitus, other complicating lesions will start to develop – skeletal and cardiac muscle damage, sciatic nerve degeneration spinal cord lesions and skin bruises and sores [30]. Because of damage to the peroneal nerve after a few hours of recumbency on the same side, the fetlock joint will be permanently flexed, and the dorsal surface of the hoof may contact the ground. Skin below the hock becomes insensitive (Fig. 9.11).

Pharmacological therapeutics involves the administration of nonsteroidal analgesic anti-inflammatory drugs (e.g. flunixin meglumine) or even steroids (e.g. prednisolone or dexamethasone), vitamins B1 (neuroprotector) and E, selenium (antioxidants) compounds and solutions of calcium borogluconate, magnesium, organic phosphorus and glucose. Glucocorticoids have the advantage of promoting neoglucogenesis. Rubefacient sub-

stances (e.g. *Terebinthinae aetheroleum rectificatum* 5–6% and methyl salicylate 3–9% ointments; TID) are still topically used in the hindquarter, improving local hyperaemia and thereby accelerating the elimination of metabolites resulting from inflammation and ischemia. Antibiotherapy may be required to treat decubitus (pressure) ulcers and to prevent metritis, mastitis and systemic infection.

Probably more important than the medical treatments is the daily management of the down cow. Conservative therapeutics mainly consist of physiotherapy, which aims to reduce the pressure on the hindquarters and to improve blood flow. The side of decubitus should be alternated every 4–6 h, and a soft bed is crucial to decrease the pressure. The

Fig. 9.12 Elevation of the hindquarter using a Bagshaw hoist. (Original from George Stilwell)

hindquarters should be elevated two or three times a day using a Bagshaw hoist or hip-slings (■ Fig. 9.12), a girdle or an air mattress. These devices are associated with pressure necrosis of the *tuber coxae* or can increase pressure on the abdominal organs and the diaphragm, compromising respiration. Higher recovery rates are achieved with water tanks where the cow is put repeatedly to "float" for a few hours a day. Water tanks are available commercially, but they can also be built on farm when there is a downer cow (■ Fig. 9.13).

If the obturator nerve is affected bilaterally, the cow's pelvic limbs should be hobble

Fig. 9.13 Hydrotherapy as physiotherapy of downer cow syndrome
Legend: **a** A provisional water tank was built to allows for cow's flotation. (Courtesy of Pedro Alves (farmer)). **b** A transportable water tank providing flotation and feed to a downer cow. **c** The placement of the cow inside

the water tank. Great care should be taken to avoid trauma to the hip bones. Pulling the down cow into the tank before filling it is a preferable method. **d** It should be ensured that the cow is able to fluctuate (being able to put some weight on all limbs). (**b–d** photographs are courtesy of Ana Paula Peixoto)

Fig. 9.14 Prevention of cow's pelvic limb abduction. (Courtesy of João Fagundes)

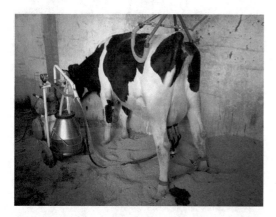

Fig. 9.15 Twice-daily milking a downer cow is very important to prevent discomfort and mastitis. (Original from George Stilwell)

just above the fetlock joints, with 50–60 cm distance between limbs. A soft rope should be used and applied to prevent tightening of the limb (**Fig. 9.14**).

Care of the downer cow is very important, and stockpersons should be aware of their responsibilities. Additionally, to the essential physiotherapy described above, water and palatable feed should always be available, twice a day milking should be done (**Fig. 9.15**), wet bedding should be removed, skin sores should be washed and disinfected, and pain should be constantly assessed and treated when necessary.

Several surveys (see Radostitis et al. [34]) have shown that approximately one third will recover, one third will die and one third will have to be put down. The outcome, of course,

varies with severity of the lesions but also with the conditions and management of the downer cow. The supreme importance of welfare in downer cows makes an accurate prognosis extremely important.

A list of signs in relation to prognosis is presented in Harwood [14]. Serum creatine kinase (CK; specific muscle damage), aspartate aminotransferase (AST) and lactate dehydrogenase (LDH) activities quickly increase, and can be used as prognosis indicators, but should be viewed critically. It was observed that recumbent dams presenting high serum CK levels during the first week (18,600 U/L on day 1, 16,300 U/L on day 2, 14,000 U/L on day 3, 10,900 U/L on day 4, 8500 U/L on day 5, 6200 U/L on day 6 and 3900 U/L on day 7; [5]; U = unit of enzyme activity) or AST > 171 U/L [39] during the first 4 days have a lower chance of recovering. More recently, Bilodeau et al. [3] observed that the total nucleated cell count (>4.5 cells/μL) and protein concentration (>0.39 g/L) in cerebrospinal fluid could be an indicator of poor prognosis of downer dairy cows. Both threshold values present an excellent specificity (>90%), but the sensibility remains low (<30%). Also, downer cows exhibiting serum cardiac troponin I (a biomarker of cardiac muscle damage) above 0.7 I ng/mL (sensibility = 54% and specificity = 78) is an indicator of poor prognosis [23].

At our knowledge, there is no absolute rule regarding length of time a cow may remain down before the prognosis becomes hopeless.

9.5 Retained Foetal Membranes

Usually, in cows, the placental expulsion occurs up to 6–8 h after calving. After this period, the foetal membrane progressively becomes a suitable bacterial growing medium. It is technically considered as retained placenta when the foetal membranes are not completely expelled within 12–24 h after parturition. In dairy farms, retained placenta should not occur in more than 5% of calvings. Retained placenta greatly increases the risk of metritis, endometritis, ketosis, abomasal displacement and mastitis. It also increases the "open days" in approximately 3 weeks, increasing in 1 week the number of days to the first artificial insemination.

9.5.1 Aetiopathogenesis and Diagnosis

The aetiology of retained placenta is not fully elucidated. It is a multifactorial condition (◻ Box 9.5), although some factors can cause this condition alone. During the periparturient period, a decrease in feed intake, in immune function and in serum calcium levels and an increase in negative energy balance, insulin resistance and bacterial contamination of the uterus occur mainly in high-producing dairy cows. High serum calcium levels (≥ 9.7 mg/dL) have been associated with fewer retained placentas and clinical endometritis [21] demonstrating the importance of the uterine contraction in the development of both processes. In fact, subclinical hypocalcaemia remains a significant risk factor for infectious and metabolic disorders during the postpartum period (◻ Table 9.1).

Box 9.5   Risk Factors of Retained Placenta (Odds Ratio According to Ghavi Hossein-Zadeh and Ardalan [12])
- Shortened gestation (less than 270 days; odds ratio: 3.82; 95% CI: 2.16–5.48; $P < 0.001$).
- Abortion (odds ratio: 8.46; 95% CI: 6.72–10.19; $P < 0.001$).
- Hormonal induced parturition (dexamethasone with or without $PGF_{2\alpha}$).
- Twinning (odds ratio: 2.76; 95% CI: 2.16–5.48; $P < 0.001$).
- Gender of offspring (male).
- Parity: multiparous cows (odds ratio: 2.69; 95% CI: 2.24–3.13; $P < 0.001$).
- Dystocia (odds ratio: 3.17; 95% CI: 2.18–4.16; $P < 0.001$).
- Stillbirth (odds ratio: 3.18; 95% CI: 2.94–3.42; $P < 0.001$).
- Uterine injuries.

- Milk fever (odds ratio: 3.66; 95% CI: 2.85–4.47; $P < 0.001$).
- Uterine atonia.
- Premature obstetrical assistance (less than 1 h after the appearance of foetal hooves).
- Cyclooxygenase inhibitors (e.g. flunixin meglumine).
- Foetotomy.
- Caesarean section.
- Immunosuppression (e.g. decreased interleukin-8 level, leukocyte chemotaxis and phagocytic).
- Infectious microorganisms (e.g. bovine viral diarrhoea virus).
- Nutritional deficiencies (e.g. vitamins A, D and E, selenium, carotene, selenium, iodine and zinc).

Tip

The single administration of a long-acting corticosteroid (e.g. 20 mg of dexamethasone trimethylacetate or betamethasone suspension; 80–90% effective) to induce parturition can have a positive impact on the maturation of placenta, reducing the incidence of retained placenta (less than 25% of the cases). Some long-acting corticosteroids are banned in the EU and USA for growth-promoting purposes.

Dehiscence of the placenta is originated from the structural tissue changes with the breakdown of collagen in the cotyledon-caruncle interface, due to the enzymatic activity of collagenase (collagenolysis and proteolysis) and other proteases. The maturation of placenta starts 3–5 days before parturition, when the different enzymatic activities induced by placental cortisol and other steroids play an essential role to accomplish the maturation of the placenta. The number and height of epithelial cells in maternal caruncles crypts decrease. Simultaneously, apoptosis rates of both maternal and foetal epithelial cells increase. One of the major immunological mechanisms involves the destruction of tro-

□ Table 9.1 Effects of subclinical hypocalcaemia and parity (primiparous vs. multiparous) on the main puerperal diseases and oestrus expression in dairy cows

Item	SCHC[a]		Multiparous[b]		P-value[c]		
	Odds ratio	95% CI	Odds ratio	95% CI	C	Par	C × Par
Displaced abomasum	3.71	1.20–11.45	2.12	0.68–6.66	<0.01	0.03	0.11
IMI	1.05	0.65–1.68	1.82	1.14–2.91	0.83	<0.01	0.84
Ketosis	5.47	1.80–16.65	1.68	0.55–5.12	<0.001	0.36	0.06
Metritis	4.25	2.62–6.89	1.72	1.05–2.78	<0.01	<0.01	0.02
Retained placenta	3.43	0.50–2.11	1.03	0.50–2.11	<0.001	0.05	0.03
Oestrus before 60 DIM	0.32	0.17–0.58	0.35	0.19–0.65	<0.001	<0.001	0.43

Modified from Rodríguez et al. [35] with permission from Elsevier

IMI intramammary infections, *DIM* days in milk, *95% CI* 95% confidence interval

[a]Cows were classified as subclinical hypocalcaemia (SCHC) when serum Ca ≤2.14 mmol/L (8.58 mg/dL); absence of SCHC is the reference point; samples were obtained between 24 and 28 h after calving

[b]Primiparous is the reference point

[c]*C* = effect of serum Ca concentration; Par = effect of parity; *C* × Par = interaction between calcaemia and parity

phoblastic cells during calving or abortion, leading to the recognition of major histocompatibility complex (MHC) class I peptides (antigens) by helper (h1) T lymphocyte cytokines (IL-2), which activate apoptosis via cytokines (IL-2). Also, the helper (h2) T lymphocytes recognize trophoblastic antigens, stimulating the activation of B lymphocytes. Both antigen-presenting cells and B lymphocytes produce interleukin-8 (IL-8), which is a chemoattractant cytokine attracting neutrophils, and activate phagocytosis in the cotyledon-caruncle interface. These structural tissue changes associated with myometrial contractions seem to be critical factors allowing for the separation of maternal caruncles from foetal cotyledons. The alternating uterine pressure allows for the occurrence of ischemia and hyperaemia in both caruncle and cotyledon tissues. As a consequence of these processes, the placenta is separated and then expelled through uterine contractions (see ▶ Chap. 3). In high-yielding dairy cows, a 20–30% decrease in immune activity, i.e. immune suppression, is commonly observed

and seems to be the major factor causing placental retention. Vitamin E, which serum levels decrease during peripartum [25], and selenium (trace element) are involved in immunomodulation and oxidative stress [28]. Supplementation with vitamin E and selenium around 2 weeks before calving tends to reduce the prevalence of retained placenta [2, 25].

9.5.2 Treatment and Prevention

The treatment of retained placenta is not consensual and is still open to discussion. During many years, manual removal of the foetal membranes, followed by local or systemic antimicrobial administration, was a common practice. This way bacterial growing medium is removed, the putrid odour is avoided, hygiene in the milking parlour improves, and flies are not attracted. Nevertheless, this practice is associated with the occurrence of haematomas, vascular thrombi and haemorrhages, due to the forced mechanical separation of foetal cotyledons. Uterine leukocyte phagocytosis is also suppressed, often resulting in bacterial overload and severe toxic metritis or endometritis. Also, insufficient cotyledon dehiscence of some placentomes may prevent complete removal of the placenta. The risk of uterine haemorrhages is also high. Some macroscopic and microscopic foetal cotyledon tissues remain attached to caruncles, even when removal was thought to be complete (► Box 9.6).

> **Box 9.6   Complications (Sequelae) due to Retained Placenta**
> - Endometritis.
> - Metritis.
> - Delay of uterine involution (expulsion of lochia and the regeneration of the endometrium).
> - Calving to first service interval increases.
> - Services per conception increases.
> - Pregnancy rate decreases.
> - Days open increases.
> - Ketosis.
> - Mastitis.

Currently, manual removal of the foetal membranes is not recommended. Placental membranes should stay inside the uterus until the complete cotyledons' dehiscence and natural expulsion. Gentle traction of the placenta can be performed during the days after calving, trying to remove an eventual free placenta. Possible treatment protocols involve intrauterine administration of antimicrobials (e.g. 1–2 g of tetracycline hydrochloride), although current evidence does not support its use. The intended objective is to prevent the occurrence of clinical metritis or endometritis as well as decrease reproductive indices, which is the major drawback of retained placenta. Nevertheless, intrauterine antimicrobials seem to interfere with the necrotizing process responsible for the dehiscence of the foetal membranes and so will not benefit recovery. Additionally, iatrogenic infection of the uterus is more likely.

The authors recommend no treatment of cows with retained placenta, unless signs of metritis or bacteraemia/toxaemia are evident

(e.g. rectal temperature >39.5 °C, foul-smelling uterine discharge, loss of appetite, reduction in milk yield). In these cases, systemic antimicrobial therapy is mandatory, due to bacterial infection of the uterus. Broad-spectrum antimicrobials should be used – oxytetracycline (20 mg/kg/day) or ceftiofur (1 mg/kg/day) should be administered for 3–5 days [13]. If necessary, this therapeutic protocol should be repeated until full recovery and/or expulsion of the placenta.

> **Important**
> The intrauterine antimicrobial therapeutics can lead to multiple chemical interactions between the drug and uterine contents. As a result, dehiscence of the foetal cotyledons can be delayed and the polymorphonuclear leukocytes inhibited. Also, the extra-label use of antimicrobials can increase the risk of residues in milk, due to different degrees of drug absorption.

The most commonly used hormone products in treating retained placenta are $PGF_{2\alpha}$ and oxytocin. These hormones play a role in uterine contraction, which persist during Stage III. However, their effectiveness to reduce the incidence of retained placenta is not fully demonstrated.

The prevention of the retained placenta should be addressed, mitigating the risk factors. Overall, appropriate nutritional management (positive energy balance and provision of adequate levels of antioxidants, such as vitamin E and selenium), hygiene and welfare/comfort should be provided, especially during the transient period.

9.6   Clinical Metritis

After calving, the cervix gradually closes, and the vulvar slacking disappears during the next 2–3 weeks. At this time, intrauterine contamination with environmental bacteria or fungi is likely to occur in most of the cows. In case of calving assistance, bacterial load and endometrial inflammation may be even higher. Usually, the contaminated uterus is cleared during the

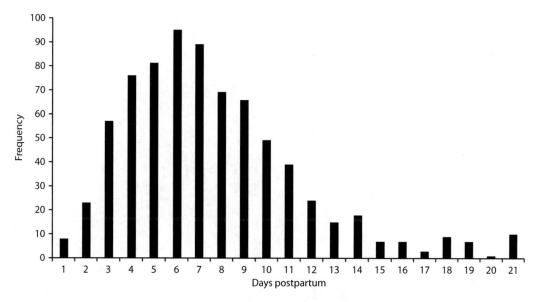

☐ Fig. 9.16 Frequency distribution of metritis incidence by days postpartum

Legend: The metritis cases (*n* = 753) occurred over a 1-year period in dairies in Ohio, New York and California. (Adapted from Galvão [10])

uterine involution. Metritis is an acute inflammatory disease in the uterus that only develops when microorganisms adhere to endometrial mucosa and colonize the epithelium. Approximately 95% of the puerperal or clinical metritis occurs within the first two weeks postpartum with a peak around the 5th–7th days (☐ Fig. 9.16) [10]. Heavy bacterial load, failure of the immune system and negative energy balance are the main factors that negatively affect the dam's defences, from which puerperal metritis may result. The concurrent severity of reduced feed intake, insulin resistance and weight loss contribute to increase the degree and duration of postpartum immunosuppression [26]. The incidence of postpartum uterine infections is, on average, around 20%, but it can affect more than 40% of the cows, mainly in problematic dairy farms.

Clinical metritis can cause severe toxaemia and septicaemia and even death of the dam. Reproductive and production impacts of metritis are related to a reduction in milk yield, the delay of the first oestrus and an increase in open days as well as the number of services per conception. Clinical metritis can predispose to other reproductive disturbance (e.g. endometritis and cystic ovarian disease), metabolic disease (e.g. ketosis and left dis-

placement abomasum), metastatic infections (e.g. hepatic and pulmonary abscesses, pyelonephritis, arthritis and endocarditis) and ovarian disorders.

9.6.1 Aetiopathogenesis and Diagnosis

Dubuc et al. [7] observed that the risk factors for metritis were associated with calving problems and metabolic status in the prepartum period (▸ Box 9.7). These researchers observed that dams suffering dystocia (odds ratio: 2.12; 95% confidence interval: 1.30–3.45; $P < 0.01$) and retained placenta (odds ratio: 6.25; 95% CI: 3.57–10.0; $P < 0.01$) at calving were more likely to develop metritis than dams presenting normal parturition. Retained placenta seems to be the major risk factor of metritis in dairy farms [12]. Cows presenting serum non-esterified fatty acids concentration ≥ 0.6 mmol/L (odds ratio: 1.59; 95% CI: 1.06–2.32; $P = 0.02$) and serum haptoglobin concentration ≥ 0.8 g/L (odds ratio: 2.17; 95% CI: 1.51–3.12; $P < 0.01$), evaluated between 1 and 7 days antepartum, were more likely to develop metritis [7]. The demand for energy to complete the immunological processes and the release of cytokines originate metabolic alterations (ketosis and hepatic lipidosis) and cause adverse consequences on reproductive indices. Moreover, bacterial endotoxins can induce significant inflammatory reactions, triggering pro-inflammatory immune responses during the periparturient period [8].

Box 9.7 Risk Factors for Metritis
- Dystocia.
- Stillbirth.
- Twins.
- Prolapsed uterus.
- Abortion.
- Retained placenta.
- Hypocalcaemia.
- Vulvovaginal lacerations (greater than 2 cm).
- Ketosis or increase in serum non-esterified fatty acid concentration.
- Negative energy balance in early postpartum.
- Parity (damage to the uterus is more common in primiparous).

In clinical metritis, inflammation involves the endometrium, the myometrium and even the serosa (perimetritis) and is classified in three grades (▶ Box 9.8). The main histopathological findings are prominent leukocyte infiltration combined with myometrial degeneration and oedema. The inflammatory reaction to bacteria infiltration causes an intrauterine accumulation and vulvar discharge of fetid watery red-brownish fluids. System reaction, including pyrexia (rectal temperature >39.5 °C), is present in only 50% of the cases. The puerperal or toxic metritis differs from other forms of clinical metritis due to a severe systemic illness. After three weeks postpartum, the uterine infection is classified as endometritis, i.e. only endometrium inflammation. As in clinical metritis, the uterine and vulvar discharges in endometritis can be scored according to colour and odour parameters (◻ Table 9.2). Pyometra is a particular condition of chronic endometritis. A progressive accumulation of intrauterine purulent content occurs caused by the closure

of the cervix in the presence of a persistent corpus luteum (CL).

Box 9.8 Classification of Postpartum Uterine Infections According to Sheldon et al. [37]

Clinical metritis (up to 21 days postpartum)
- Grade 1: Enlarged slightly tonic uterus, vaginal discharge presenting fetid watery red-brown without signs of systemic illness, within 21 days after calving.
- Grade 2 or puerperal metritis: Enlarged non-tonic uterus, vaginal discharge presenting fetid watery red-brown, fever (rectal temperature >39.5 °C) and other clinical signs of systemic illness, within 21 days after calving.
- Grade 3 or toxaemic metritis: Additional signs of toxaemia (such as cold extremities, depression and/or collapse). Uterus enlarged (most times filled with gas), with thin and tense walls.

Clinical endometritis (more than 21 days postpartum)
- Vaginal discharge presenting pus without systemic signs of illness. The cervix is open, reaching >7.5 cm in cervical diameter. The vaginal discharge can be purulent (>50% pus) or mucopurulent (approximately 50% pus, 50% mucus).

Pyometra
- Pyometra is a particular condition of chronic endometritis. A progressive accumulation of intrauterine purulent content occurs caused by the functional closure of the cervix in the presence of a CL. It may remain hidden for a few weeks although a small decrease of milk yield may be observed. If pyometra is not diagnosed during routine reproductive examination, septicaemia or toxaemia may eventually occur, associated to rectal temperature >39.5 °C and anorexia (systemic con-

---

dition). Commonly, vaginal discharge is not observed, or only small amount of pus is seen outside the vulva.

Modified by Szenci [40] and by the authors

▢ **Table 9.2** Vaginal or uterine discharge scores

Score	Clinical metritis [4]	Endometritis [24]
1	No discharge.	Clear mucus or translucent mucus.
2	Viscous discharge; not fetid; normal lochia; red, brown or clear coloured.	Cloudy mucus with flecks of pus.
3	Thick mucus discharge; not fetid; cloudy, clearing or clear.	Mucopurulent (approximately 50% mucus and 50% pus present).
4	Purulent or mucopurulent discharge; not fetid; chocolate brown coloured.	Purulent (>50% pus present).
5	Serous or watery discharge; fetid; thin; red or pink to chocolate brown coloured; necrotic tissue can be present.	Purulent or red-brown and foul-smelling.

Traditionally the most prevalent pathogens said to be involved in clinical metritis are *Escherichia coli*, *Trueperella pyogenes*, *Prevotella melaninogenica* and *Fusobacterium necrophorum* [20]. However, recent research has suggested that metritis is associated with a dysbiosis of the uterine microbiota, which is characterized by high abundance of *Bacteroides*, *Porphyromonas* and *Fusobacterium* [21]. Other Gram-positive (e.g. *Staphylococcus* spp. and *Streptococcus* spp.) and Gram-negative bacteria (e.g. *Enterobacter* spp., *Acinetobacter* spp., *Mannheimia haemolytica*, *Proteus mira-*

bilis, *Proteus vulgaris*, *Providencia rettgeri* and *Aeromonas hydrophila*) have also been found in cattle infected uterus. According to Williams et al. [43], the odour of vaginal mucus may reflect the number of bacteria in the uterus. Mucopurulent or purulent vaginal mucus is more related to *T. pyogenes*, *Proteus* spp. and *F. necrophorum* higher densities; fetid mucus odour is related to *Escherichia coli*, *Mannheimia haemolytica* and non-haemolytic streptococci. An association between bacteria species and uterine lesions is reported in ▢ Table 9.3.

▢ **Table 9.3** Categorization of bacteria and fungi according to the expected intrauterine pathogenic potential

Pathogens associated with uterine endometrial lesions

> *Trueperella pyogenes*
> *Escherichia coli*
> *Prevotella melaninogenica*
> *Fusobacterium necrophorum*

Potential pathogens frequently isolated from the bovine uterine lumen, but not commonly associated with uterine lesions

> *Bacillus licheniformis*
> *Enterococcus faecalis*
> *Mannheimia haemolytica*
> *Pasteurella multocida*
> *Peptostreptococcus* species
> *Staphylococcus aureus*
> Non-haemolytic streptococci

Opportunist contaminants transiently isolated from the uterine lumen and not associated with uterine lesions

> *Micrococcus* spp.
> *Clostridium perfringens*
> *Proteus* spp.
> *Klebsiella pneumoniae*
> *Providencia stuartii*
> Coagulase-negative staphylococci
> Alfa-haemolytic streptococci
> *Streptococcus acidominimus*
> *Aspergillus* spp.

Modified from Williams et al. [43] with permission from Elsevier
The bacterial and fungi isolates were collected from the uterine body on day 21 ± 1 and day 28 ± 1 postpartum. A total of 328 uterine swabs were obtained from 210 Holstein-Friesian of a dairy herd between 2000 and 2003

The diagnosis and classification of metritis depend on clinical signs (e.g. attitude, hydration status and rectal temperature), transrectal palpation or transrectal ultrasonographic examination and by vaginoscopy. The odour and macroscopic characteristics of vaginal discharges have diagnostic and monitoring values. Uterine or vaginal fluids should be sampled for microbiological cultures and antimicrobial resistance evaluation. The samples can be obtained using a sterilized vaginal speculum, using an arm-length glove or with the Metricheck® device (50-cm-long stainless-steel rod with a 4 cm hemisphere of silicon; [33]).

9.6.2 Treatment, Prognosis and Prevention

The treatment for clinical metritis should be done as soon as possible to prevent or mitigate complications. Several antimicrobials, e.g. ceftiofur, penicillin (21,000 IU/kg i.m. twice in day) ampicillin, amoxicillin or oxytetracycline, should be used for at least 3–5 days. For grade 2 or grade 3 clinical metritis, a systemic antibiotherapy associated with uterine lavage is recommended (▶ Box 9.9). Nevertheless, the use of intrauterine antimicrobials and antiseptics is still controversial mainly due to the endometrium irritation that they can cause. In toxic puerperal metritis (clinical mastitis grade 3), the septic shock should also be controlled with aggressive fluid therapy (20 L isotonic saline or 3–4 L hypertonic saline followed by oral fluids) and analgesics. Pinedo et al. [32] suggested the administration intravenously of 500 mL of hypertonic 7.2% saline solution and 500 mL of 50% dextrose (IV), as well as the administration of five oral boluses of aspirin (15.6 g of acetylsalicylic acid/bolus). Flunixin meglumine (2.2 mg/kg) is the nonsteroidal anti-inflammatory drug choice for endotoxaemia. The administration of calcium solution should be avoided in dams presenting toxaemia. The repeated use of oxytocin, e.g. 20–40 UI, three times in a day for 2–3 days, can be useful to improve the uterine contractibility. The use of $PGF_{2\alpha}$ also has a similar function, but its usefulness remains controversial as the inflamed myometrium produces large amounts of this hormone.

> **Box 9.9   Protocols to Treat Clinical Metritis**
> — Protocol 1: 1–2 mg/kg/day of ceftiofur subcutaneously or 20 mg/kg/day of oxytetracycline i.v. (or eventually i.m.) for 3–5 days. A uterine lavage with 0.1% chlorhexidine or 0.5% povidone-iodine may be attempted if the cervix is open, to fully remove fluids and gas from the uterus. This protocol is the preferred by the authors.
> — Protocol 2: One intrauterine infusion with 5 g of oxytetracycline followed by another infusion with 0.5% iodopovidone [27].
> — Protocol 3: Infusion of 200 mL of povidone-iodine diluted in 2 L of distilled water once a day for 3 days [32].

The prevention is based on two principles: decreasing cow's antepartum and postpartum susceptibility to bacteria by ensuring adequate immune response and mitigating the bacterial contamination during calving. Adequate nutritional support and a comfortable environment are practical measures that prevent severe immunity disturbs (◻ Table 9.4; ◻ Fig. 9.17). Szenci [40] suggests as a preventive measure the intrauterine application of 500 mL of 2% Lugol's iodine immediately after calving and again 6 h later. In dairy farms, the cows should be daily monitored for the first 2 weeks by clinical examination and by looking for milk production variations. As an alternative, a periodical examination of the cows, e.g. at 4th-, 7th-, and 12th-day postpartum, can be suitable. Also, biomarkers such as plasma levels of beta-hydroxybutyrate and haptoglobin are useful tools to diagnose clinical metritis timely.

◻ **Table 9.4** Management practices and monitoring targets to reduce the risks of metritis, purulent vaginal discharge and cytological endometritis in dairy cows

Management practices

Prevent consumption of dietary energy above requirement in the "far-off" dry period, 3–8 weeks before calving.

Provide for unrestricted feed bunk access: 75 cm of linear bunk space per cow or no more than four cows per five headlocks. All cows should have access to fresh feed delivery.

Provide space to allow for lying approximately 12 h/day or one free stall/cow or 10 m^2 of bedded pack/cow. Minimize pen moves and social group changes.

Build dry cow and early lactation pens for approximately 140% of the expected average number of calvings per month.

Provide heat abatement (fans and sprinklers) when the temperature humidity index exceeds 72.

Manage nutrition so that cows calve at a body condition score of 3.0 or 3.25 (on the 5-point scale), and maintain a minimum body condition of 2.5.

Monitoring targets

Lipid mobilization: Non-esterified fatty acids 0.5 mmol/L in the week before expected calving.

Subclinical and clinical ketosis: Beta-hydroxybutyrate 1.1 mmol/L in the first week and 1.4 mmol/L in the second week after calving.

Acute inflammatory response: Haptoglobin 0.8 g/L in the first week after calving. The normal serum level is 0.5–0.7 g/L. It is a useful market to estimate acute inflammatory processes.

Modified from LeBlanc et al. [26] with permission from Elsevier

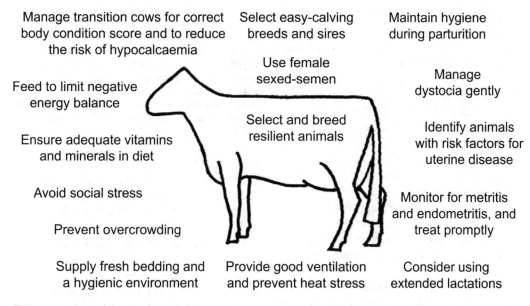

◻ **Fig. 9.17** Potential strategies to help prevent postpartum uterine infection. (Adapted from Sheldon et al. [38] with permission from Elsevier)

9.7 Endometritis

Endometritis is defined as an inflammation of the endometrium, occurring after 3 weeks post-partum and commonly caused by environmental bacteria (uterine infection). Clinical endometritis can be diagnosed by transrectal palpation of cervical diameter >7.5 cm from day 20 postpartum or by vaginoscopy detecting the presence of mucopurulent or purulent vaginal discharge from day 26 postpartum [24], when lochia disappears. Nevertheless, a disagreement between vaginal discharge (clinical endometritis) and cytological (cytological endometritis) findings can sometimes be observed [6].

In subclinical endometritis, the purulent uterine discharge is absent, and the diagnosis is based on cytology evaluation using conventional cytobrush, cytotape, biopsy or low-volume uterine lavage techniques [6, 31]. Sheldon et al. [37] proposed a threshold of 18% neutrophils in uterine cytology samples collected between the 21st and the 33rd day after calving or, as an alternative, >10% neutrophils between the 34th and the 47th day postpartum.

Pyometra is a chronic endometritis consisting of an enlargement of the uterine horns by the intrauterine accumulation of purulent or mucopurulent fluids (Fig. 9.18 Video Box) due to the presence of an active CL and a functional cervical closure. Nevertheless, some purulent fluid can pass through the cervical canal and can be observed in the vagina. Its diagnosis is easily made by ultrasonography detecting the flow of uterine contents, twin uterine walls and the presence of a persisting CL. The diagnosis and treatment of pyometra should be made early to prevent severe endometrial lesions and to restore endometrium function.

The treatment of endometritis aims to (1) remove pathogenic and environmental bacteria, (2) ensure local immunity efficacy, and (3) regenerate endometrium to prepare the uterus for a new conception.

The treatment of endometritis includes intrauterine (e.g. 0.5 g cephapirin) or systemic (e.g. 1 mg/kg cefquinome) administration of antimicrobials, intrauterine administration of antiseptic solutions (e.g. Lugol's iodine or poly-vinylpyrrolidone-iodine polycondensated) and/

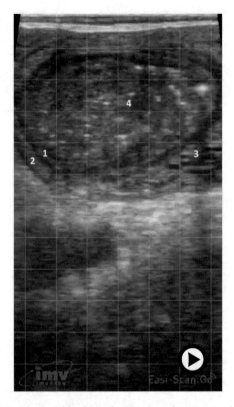

⬛ Fig. 9.18 Video box of pyometra: ultrasonographic appearance of uterine content and uterine wall (transversal cut) (► https://doi.org/10.1007/000-2qk)
Legend: (1) Endometrium. (2) Myometrium. (3) Small blood vessels of the uterine broad ligament. (4) Muco-purulent intraluminal content. The distinction between the endometrium and myometrium of the uterine wall is evident due to the inflammatory process. At transrectal palpation, both uterine horns were found enlarged in presence of a corpus luteum. Commonly, the uterine wall is found thin consecutive to this distension. 30 days postpartum. This sonogram is obtaining using Easi-Scan (7.5 MHz); each square = 1 cm^2. (Courtesy of Rui d'Orey)

or the use of PGF$_{2\alpha}$. Nevertheless, the treatment of clinical and subclinical endometritis using PGF$_{2\alpha}$ when an active corpus luteum is absent is rarely effective [40]. According to the authors (Stilwell, data not published), an optimal management of subclinical endometritis consists in waiting for the development of an active corpus luteum and subsequent administration of PGF$_{2\alpha}$. This clinical approach avoids the use of antimicrobials.

The treatment of pyometra is based on the systemic administration of PGF$_{2\alpha}$, which causes luteolysis and induces oestrus with cervix opening and uterine clearance. This treatment is commonly repeated 12–14 days later. It can be complemented by the intrauterine administration of antimicrobials at the time of the second injection of $PGF_{2\alpha}$.

Key Points

- Mechanical pressure of the foetus on the soft tissue is the main cause of lacerations of the birth canal during assisted or unassisted calving. Perforation by foetus hooves is the main cause of deeper tears of the uterus, cervix or vagina.
- The delay in uterine involution is highly correlated with poor fertility outcomes.
- Puerperal complications involving vascular compromise and soft tissue degeneration or necrosis can threaten the dam's life and should be considered a reproductive emergency.
- Vaginal and vulvar soundness and outlines should be careful evaluated, and corrected, if subjected to large injuries during calving.
- Clinical metritis occurs within 21 days after calving and is classified into three grades according to the severity of systemic clinical signs.
- Retained placenta is an important disturb of the Stage III of labour and a risk factor for uterine infection and metabolic diseases. Medical intervention is only needed if systemic signs of metritis are evident.
- Use of intrauterine antimicrobials should be avoided. Treatment should be based on systemic antimicrobials and uterine drainage/lavage.
- Treatment of endometritis can be done by intrauterine antimicrobials or prostaglandins when there is a corpus luteum present.

? **Questions**

1. Describe the relationships between hypocalcaemia and puerperal diseases.
2. What is the surgical treatment for the different degrees of perineal lacerations?
3. How do you make the clinical management of retained placenta and reduce its incidence in dairy farms?

✓ **Answers**

1. In cows, serum calcium levels above 8 mg/dL usually define normocalcaemia. Values below this threshold are indicative of subclinical hypocalcaemia. Clinical signs (milk fever) are evident for values lower than approximately 5.0 mg/dL. In periparturient and early postpartum high-yielding dairy cows, the deficit of serum calcium levels is frequent due to high calcium demand for colostrum or milk production. In a small percentage of the cows, sometimes more than 10%, a clinical hypocalcaemia (stage 1, ataxia; stage 2, sternal recumbency; and stage 3, lateral recumbency) is diagnosed and should be immediately treated. The progressive decrease of serum calcium levels reduces the availability of calcium ions (Ca2+) in the neuromuscular junction of skeletal, cardiac and smooth muscles. As a consequence, the contraction of the muscles becomes progressively scarcer and less intense than normal. The lower uterine contractile activity pattern induces uterine inertia, which is a significant predisposing risk of uterine prolapse, retained placenta, clinical metritis and endometritis. As an example, uterine flaccidity can be a significant predisposing factor of uterine re-prolapse. In this case, as conservative therapeutics, a solution of approximately 300 mL of calcium borogluconate 24% should be administered to improve uterine contractions. Also, for these cases, a retention suture is recommended. Another paradigmatic example concerns the retention of placental membranes due to clinical hypocalcaemia. In a small percentage of cows, the lack of mechanical expulsive forces in the endometrium may cause the retention of the placental membranes inside the uterus. The placenta is usually expelled a short time (sometimes within 1 h) after the treatment of the primary cause.

2. Perineal lacerations are usually caused by the pressure of foetal parts in the

vulvar labia and vaginal vestibule. Due to the convex shape of the frontal bones, the pressure of the foetal head on the dorsal vulvar commissure is common. Most times, vulvar tears occur due to intense abdominal contractions or during the traction of the foetus. These tears are mainly located dorsally, towards the anal sphincter and rectum. Depending on the affected tissues and length of the wounds, the perineal lacerations are classified as first, second and third degree. In first-degree perineal lacerations less than 2 cm, a suture of vulvar labia is not required since the vulvar coaptation is not affected. For those greater than 2 cm, two or three interrupted stitches using a USP 2-0 non-absorbable suture (e.g. monofilament polyamide) are suitable to suture the tears. The second-degree perineal lacerations may include the opening of the fibromuscular tissues that separate the rectum and the vagina. In this case, a surgery to restore the perineal structure is required. A necrotic or granulation/scar tissue debridement is made excising a triangular-shaped piece of the mucosa in both sides of the vaginal vestibule and vulva. A simple continuous suture pattern (e.g. a USP 2-0 synthetic multifilament absorbable polyglycolic acid) is performed from the inside out. At the end, a Caslick's suture is made to close the skin. In the third-degree perineal lacerations, both vaginal vestibule and rectal walls, including anal sphincter, need to be rebuilt. The necrotic or granulation tissue is debrided. A one-stage repair modified from Goetz, involving a six-bite vertical suture pattern, is a useful procedure to restore the integrity and functionality of all affected structures. Also, at the end, a Caslick's suture is required to ensure the coaptation of vulvar labia. The postoperative care consists of the administration of systemic antimicrobial for 4–5 days and analgesic for two days.

3. Overall, the treatment of retained placenta is not recommended in cows without systemic clinical signs of toxaemia or sepsis. The manual removal of foetal membranes can provoke local trauma, haemorrhages and uterine leukocyte phagocytosis depression, increasing the risk of clinical metritis or endometritis. Also, the intrauterine administration of antimicrobials in cows presenting retained placenta seems to not modify the incidence of clinical metritis and endometritis, as well as other reproductive outcomes, such as the days open and the interval between calving and conception. Additionally, there is evidence that some antimicrobials (e.g. oxytetracycline) can also reduce the local immunological activity. Nevertheless, gentle traction of the foetal membrane can be periodically attempted, expecting the timeline placentome dehiscence to be enough to cause the separation of the foetal cotyledons and the maternal caruncles. In cows presenting systemic illness, fever (rectal temperature >39.5 °C) or even a weak or obese general condition (body condition less than 2.5 or higher than 4; 5-point scale), a systemic administration of antimicrobials (e.g. 1 mg/kg/day of ceftiofur) is required until placenta expulsion and for at least 3–4 days. The evaluation of hyperketonaemia, (i.e. beta-hydroxybutyrate above 1.1 mmol/L in serum) and acute inflammation (i.e. serum haptoglobin above 0.8 g/L) within the first week after calving is suitable to identify cows at risk of developing uterine infections. This clinical approach using biomarkers can reduce the incidence of clinical metritis and endometritis, one of the most common complications of retained placenta.

References

1. Ames NK. Epidural anesthesia. In: Noordsy's food animal surgery. 5th ed. Ames: Wiley Blackwell; 2013. p. 39–50. https://doi.org/10.1002/9781118770344.ch4.

2. Aréchiga CF, Ortíz O, Hansen PJ. Effect of prepartum injection of vitamin E and selenium on postpartum reproductive function of dairy cattle. Theriogenology. 1994;41(6):1251–8. https://doi.org/10.1016/0093-691x(94)90482-x.

3. Bilodeau MÈ, Achard D, Francoz D, Grimes C, Desrochers A, Nichols S, Babkine M, Fecteau G. Survival associated with cerebrospinal fluid analysis in downer adult dairy cows: a retrospective study (2006-2014). J Vet Intern Med. 2018;32(5):1780–6. https://doi.org/10.1111/jvim.15305.

4. Chenault JR, McAllister JF, Chester ST Jr, Dame KJ, Kausche FM, Robb EJ. Efficacy of ceftiofur hydrochloride sterile suspension administered parenterally for the treatment of acute postpartum metritis in dairy cows. J Am Vet Med Assoc. 2004;224(10):1634–9. https://doi.org/10.2460/javma.2004.224.1634.

5. Clark RG, Henderson HV, Hoggard GK, Ellison RS, Young BJ. The ability of biochemical and haematological tests to predict recovery in periparturient recumbent cows. N Z Vet J. 1987;35(8):126–33. https://doi.org/10.1080/00480169.1987.35410.

6. Dubuc J, Duffield TF, Leslie KE, Walton JS, LeBlanc SJ. Definitions and diagnosis of postpartum endometritis in dairy cows. J Dairy Sci. 2010;93(11):5225–33. https://doi.org/10.3168/jds.2010-3428.

7. Dubuc J, Duffield TF, Leslie KE, Walton JS, LeBlanc SJ. Risk factors for postpartum uterine diseases in dairy cows. J Dairy Sci. 2010;93(12):5764–71. https://doi.org/10.3168/jds.2010-3429.

8. Eckel EF, Ametaj BN. Invited review: role of bacterial endotoxins in the etiopathogenesis of periparturient diseases of transition dairy cows. J Dairy Sci. 2016;99(8):5967–90. https://doi.org/10.3168/jds.2015-10727.

9. Fubini SL. Surgery of the perineum. Surgery of the bovine reproductive system and urinary tract (Chapter 12). In: Fubini SL, Ducharme NG, editors. Farm animal surgery. St. Louis: Saunders; 2004. p. 351–428. https://doi.org/10.1016/B0-72-169062-9/50016-2.

10. Galvão KN. Postpartum uterine diseases in dairy cows. Anim Reprod. 2012;9(3):290–6.

11. Gardner IA, Reynolds JP, Risco CA, Hird DW. Patterns of uterine prolapse in dairy cows and prognosis after treatment. J Am Vet Med Assoc. 1990;197(8):1021–4.

12. Ghavi Hossein-Zadeh N, Ardalan M. Cow-specific risk factors for retained placenta, metritis and clinical mastitis in Holstein cows. Vet Res Commun. 2011;35(6):345–54. https://doi.org/10.1007/s11259-011-9479-5.

13. Haimerl P, Arlt S, Borchardt S, Heuwieser W. Antibiotic treatment of metritis in dairy cows-A meta-analysis. J Dairy Sci. 2017;100(5):3783–95. https://doi.org/10.3168/jds.2016-11834.

14. Harwood JPP. Tackling the problem of the Downer Cow: cause, diagnosis and prognosis. Cattle Pract. 2003;2(2):89–92.

15. Heppelmann M, Krach K, Krueger L, Benz P, Herzog K, Piechotta M, Hoedemaker M, Bollwein H. The effect of metritis and subclinical hypocalcemia on uterine involution in dairy cows evaluated by sonomicrometry. J Reprod Dev. 2015;61(6):565–9. https://doi.org/10.1262/jrd.2015-015.

16. Hopper RM. Surgical correction of abnormalities of genital organs of cows: management of uterine prolapse. In: Youngquist RS, Threlfall WR, editors. Current therapy in large animal theriogenology. 2nd ed. Philadelphia: Saunders, Elsevier; 2007. p. 470–1. Radostitis OM, Gay CC, Blood DC, Hinchcliff KW. Metabolic diseases. In: Radostitis OM, editor. Veterinary medicine. 9th ed. London: WB Saunders; 2000. p. 1435–42.

17. Ishi M, Aoki T, Yamakawa T, Uyama T, El-Khodery S, Matsui M, Miyake Y. Uterine prolapse in cows: effect of raising the rear end on the clinical outcomes and reproductive performance. Vet Med. 2010;55(3):113–8.

18. Jackson PGG, Cockcroft PD. Clinical examination of the female genital system (Chapter 10). In: Clinical examination of farm animals. Oxford/Malden, MA: Blackwell Science Ltd; 2002. p. 125–40. https://doi.org/10.1002/9780470752425.ch10.

19. Jackson PGG, Cockcroft PD. Clinical examination of the nervous system (Chapter 14). In: Clinical examination of farm animals. Oxford/Malden, MA: Blackwell Science Ltd; 2002. p. 198–216. https://doi.org/10.1002/9780470752425.ch14.

20. Jeon SJ, Galvão KN. An advanced understanding of uterine microbial ecology associated with metritis in dairy cows. Genomics Inform. 2018;16(4):e21. https://doi.org/10.5808/GI.2018.16.4.e21.

21. Jeong JK, Kang HG, Kim IH. Associations between serum calcium concentration and postpartum health and reproductive performance in dairy cows. Anim Reprod Sci. 2018;196:184–92. https://doi.org/10.1016/j.anireprosci.2018.08.006.

22. Jubb TF, Malmo J, Brightling P, Davis GM. Survival and fertility after uterine prolapse in dairy cows. Aust Vet J. 1990;67(1):22–4.

23. Labonté J, Dubuc J, Roy JP, Buczinski S. Prognostic value of cardiac troponin I and L-lactate in blood of dairy cows affected by downer cow syndrome. J Vet Intern Med. 2018;32(1):484–90. https://doi.org/10.1111/jvim.14874.

24. LeBlanc SJ, Duffield TF, Leslie KE, Bateman KG, Keefe GP, Walton JS, Johnson WH. Defining and diagnosing postpartum clinical endometritis and its impact on reproductive performance in dairy

cows. J Dairy Sci. 2002;85(9):2223–36. https://doi.org/10.3168/jds.S0022-0302(02)74302-6.

25. LeBlanc SJ, Herdt TH, Seymour WM, Duffield TF, Leslie KE. Peripartum serum vitamin E, retinol, and beta-carotene in dairy cattle and their associations with disease. J Dairy Sci. 2004;87(3):609–19. https://doi.org/10.3168/jds.S0022-0302(04)73203-8.

26. LeBlanc SJ, Osawa T, Dubuc J. Reproductive tract defense and disease in postpartum dairy cows. Theriogenology. 2011;76(9):1610–8. https://doi.org/10.1016/j.theriogenology.2011.07.017.

27. Liu WB, Chuang ST, Shyu CL, Chang CC, Jack A, Peh HC, Chan JP. Strategy for the treatment of puerperal metritis and improvement of reproductive efficiency in cows with retained placenta. Acta Vet Hung. 2011;59(2):247–56. https://doi.org/10.1556/AVet.2011.004.

28. Mikulková K, Kadek R, Filípek J, Illek J. Evaluation of oxidant/antioxidant status, metabolic profile and milk production in cows with metritis. Ir Vet J. 2020;73:8. https://doi.org/10.1186/s13620-020-00161-3.

29. Murphy AM, Dobson H. Predisposition, subsequent fertility, and mortality of cows with uterine prolapse. Vet Rec. 2002;151(24):733–5.

30. Ohfuji S. Pathological evaluation of thigh muscle, sciatic nerve, and spinal cord in downer cow syndrome with emphasis on the prognostic significance. Comp Clin Pathol. 2019;28:117–27. https://doi.org/10.1007/s00580-018-2804-4.

31. Pascottini OB, Dini P, Hostens M, Ducatelle R, Opsomer G. A novel cytologic sampling technique to diagnose subclinical endometritis and comparison of staining methods for endometrial cytology samples in dairy cows. Theriogenology. 2015;84(8):1438–46. https://doi.org/10.1016/j.theriogenology.2015.07.032.

32. Pinedo PJ, Velez JS, Bothe H, Merchan D, Piñeiro JM, Risco CA. Effect of intrauterine infusion of an organic-certified product on uterine health, survival, and fertility of dairy cows with toxic puerperal metritis. J Dairy Sci. 2015;98(5):3120–32. https://doi.org/10.3168/jds.2014-8944.

33. Pleticha S, Drillich M, Heuwieser W. Evaluation of the Metricheck device and the gloved hand for the diagnosis of clinical endometritis in dairy cows. J Dairy Sci. 2009;92(11):5429–35. https://doi.org/10.3168/jds.2009-2117.

34. Radostits OM, Gay CC, Blood DC, Hinchcliff KW. Metabolic diseases. In: Radostits OM, editor. Veterinary medicine. 9th ed. London: WB Saunders; 2000. p. 1435–42.

35. Rodríguez EM, Arís A, Bach A. Associations between subclinical hypocalcemia and post-parturient diseases in dairy cows. J Dairy Sci. 2017;100(9):7427–34. https://doi.org/10.3168/jds.2016-12210.

36. Sato R, Kamimura N, Kaneko K. Surgical repair of third-degree perineal lacerations with rectovestibular fistulae in dairy cattle: a series of four cases (2010-2018). J Vet Med Sci. 2019;81(5):703–6. https://doi.org/10.1292/jvms.19-0004.

37. Sheldon IM, Lewis GS, LeBlanc S, Gilbert RO. Defining postpartum uterine disease in cattle. Theriogenology. 2006;65(8):1516–30. https://doi.org/10.1016/j.theriogenology.2005.08.021.

38. Sheldon IM, Molinari PCC, Ormsby TJR, Bromfield JJ. Preventing postpartum uterine disease in dairy cattle depends on avoiding, tolerating and resisting pathogenic bacteria. Theriogenology. 2020;150:158–65. https://doi.org/10.1016/j.theriogenology.2020.01.017.

39. Shpigel NY, Avidar Y, Bogin E. Value of measurements of the serum activities of creatine phosphokinase, aspartate aminotransferase and lactate dehydrogenase for predicting whether recumbent dairy cows will recover. Vet Rec. 2003;152(25):773–6. https://doi.org/10.1136/vr.152.25.773.

40. Szenci O. Recent possibilities for diagnosis and treatment of post parturient uterine diseases in dairy cow. JFIV Reprod Med Genet. 2016;4:170. https://doi.org/10.4172/2375-4508.1000170.

41. Vieira-Neto A, Lima FS, Santos JEP, Mingoti RD, Vasconcellos GS, Risco CA, Galvao KN. Vulvovaginal laceration as a risk factor for uterine disease in postpartum dairy cows. J Dairy Sci. 2016;99(6):4629–37. https://doi.org/10.3168/jds.2016-10872.

42. Weaver AD, Atkinson O, St. Jean G, Steiner A. Female urinogenital surgery (Chapter 6). In: Bovine surgery and lameness. 3rd ed. Oxford: John Wiley & Sons Ltd; 2018. p. 187–210.

43. Williams EJ, Fischer DP, Pfeiffer DU, England GC, Noakes DE, Dobson H, Sheldon IM. Clinical evaluation of postpartum vaginal mucus reflects uterine bacterial infection and the immune response in cattle. Theriogenology. 2005;63(1):102–17. https://doi.org/10.1016/j.theriogenology.2004.03.017.

Postpartum Guidelines

Anna Heaton A, Plate P, Weirich W, Shingleton GD, Thomsett A, Child J, D'Alterio GL. Post-partum conditions in cattle. Farm Health Online, Duchy College Rural Business School, UK; 2018. From: https://www.farmhealthonline.com/disease-management/cattle-diseases/post-partum-conditions/. Accessed on 11 May 2020.

Eilts BE. Bovine postpartum problems. Comparative theriogenology. Louisiana State University (Department of Veterinary Clinical Sciences), USA; 2012. From: http://therio.vetmed.lsu.edu/bovine_post_partum.htm. Accessed on 18 Sept 2020.

Suggested Reading

Crowe MA, Hostens M, Opsomer G. Reproductive management in dairy cows – the future. Ir Vet J. 2018;71:1. https://doi.org/10.1186/s13620-017-0112-y.

Földi J, Kulcsár M, Pécsi A, Huyghe B, de Sa C, Lohuis JA, Cox P, Huszenicza G. Bacterial complications of postpartum uterine involution in cattle. Anim

Reprod Sci. 2006;96(3–4):265–81. https://doi.org/10.1016/j.anireprosci.2006.08.006.

Mordak R, Stewart PA. Periparturient stress and immune suppression as a potential cause of retained placenta in highly productive dairy cows: examples of prevention. Acta Vet Scand. 2015;57:84. https://doi.org/10.1186/s13028-015-0175-2.

Prado TM, Schumacher J, Dawson LJ. Surgical procedures of the genital organs of cows. Vet Clin North Am Food Anim Pract. 2016;32(3):727–52. https://doi.org/10.1016/j.cvfa.2016.05.016.

Reichel MP, Wahl LC, Hill FI. Review of diagnostic procedures and approaches to infectious causes of reproductive failures of cattle in Australia and New Zealand. Front Vet Sci. 2018;5:222. https://doi.org/10.3389/fvets.2018.00222.

Cows' Obstetrical Case Studies in Images

Contents

Electronic Supplementary Material The online version of this chapter (https://doi.org/10.1007/978-3-030-68168-5_10) contains supplementary material, which is available to authorized users. The videos can be accessed by scanning the related images with the SN More Media App.

10.1 Removal of a Mummified Foetus

Originals from Dália Castro and António Carlos Ribeiro (🅾 Fig. 10.1).

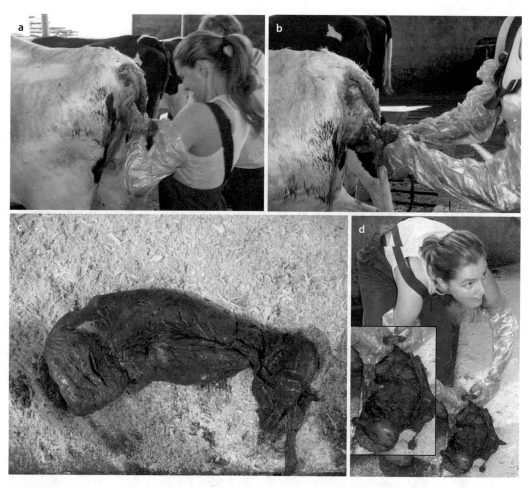

🅾 **Fig. 10.1** **a** Detection of a mummified foetus. The mummified foetus was found inside the vagina (natural luteolysis). Plenty lubrification was used to extract the foetus, because of foetal and placental desiccation. The orbital surface of the lacrimal bone was easily palpable and was used to fix the obstetrician's fingers for forced traction. The use of a double-jointed Krey-Schottler hook attached to an obstetrical snare would also be a suitable alternative to fix a foetal part. **b** Vulvar dilatation. Usually the vulva is only slightly dilated, even if luteolysis of the corpus luteum body is induced by PGF$_{2\alpha}$. Careful manipulation of the foetus is required to prevent vestibular and vulvar lacerations. **c** Mummified mass. The placental membranes are completed dehydrate (leather appearance) and cover the foetus. **d** Aseptic condition. Foetal mummification is an aseptic condition in which the placenta covers the foetus. Nevertheless, at expulsion time, the uterus can be contaminated by environmental bacteria and fungi due to the opened cervix and the handling to remove the foetus. A uterine lavage is recommended in the more difficult cases

10.2 Suture Retention (Vertical Mattress) on the Vulva After Cervicovaginal Prolapse Replacement

Originals from António Carlos Ribeiro (▣ Fig. 10.2).

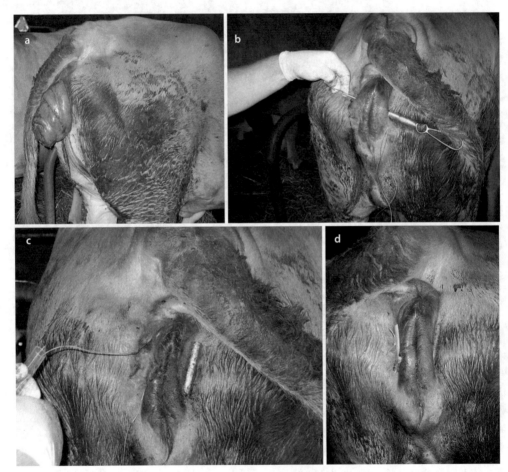

▣ **Fig. 10.2** **a** Cervicovaginal prolapse (degree IV vaginal prolapse) in a beef cow. **b** Use of a Bühner suture needle, after caudal epidural anaesthesia and washing and disinfection of the area. The needle is introduced deeply into the vulvar labia preventing vulvar lacerations mainly during tenesmus. A USP (United States Pharmacopeia) 4 or 5 nylon suture line is used. **c** Vertical rods (in this case two syringes) are placed in each side. **d** A surgeon's knot (clockwise double knot + single anticlockwise knot + single clockwise knot) is done provoking a strong tension in both vulvar labia. The diameter and tension of the suture thread are important factors to prevent vulvar tears. Also, 3–4 cm next to the ventral commissure of the vulva should be kept unsutured to prevent dysuria and urovagina. The suture can be removed between 1 and 2 weeks after prolapse replacement, preventing local infections or even fibrosis on the trajectory of the suture thread. It is expected that at this time the primary cause of vaginal prolapse has been solved. **e** Basic, effective and easy-to-perform retention sutures. **A**: Deep vertical mattress suture. **B**: Deep horizontal mattress suture (the buttons can be replaced by gauze). **C**: Vaginal pins (can replace the horizontal mattress suture). **D**: Bootlace suture technique. Two or three equidistant small eyelets (umbilical tape or hog nose rings) are inserted on each side of the vulvar mucocutaneous junction. These eyelets are X laced using an umbilical tape and both ends are tied ventrally to provoke juxtaposition of the vulvar lips. (**A**, **B** and **D**: Adapted from Prado et al. [4] with permission from Elsevier. **C**: Modified (FotoSketcher 3.60) from Jorgensen Laboratories, Inc. (USA) with permission)

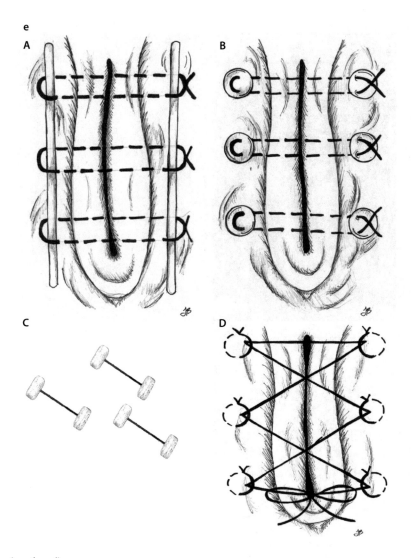

■ **Fig. 10.2** (continued)

10.3 Using Metallic Vaginal Pins (◘ Fig. 10.3)

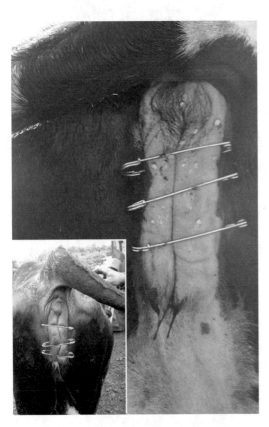

◘ **Fig. 10.3** Suture retention using metallic vaginal pins after uterus prolapse replacement. The perineum was washed and disinfected (e.g, chlorhexidine gluconate vaginal solution 1%), and three equidistant vaginal pins were used to close the vulva. Note the deep vulvar insertion of the pins to prevent vulvar laceration due to potential abdominal muscle contractions. Oxytocin, i.v., and carprofen and penicillin, i.m., were administered, with the latter continued for three days. Pins were removed after 48 h. (Courtesy of João Fagundes)

10.4 Perosomus Elumbis (■ Fig. 10.4)

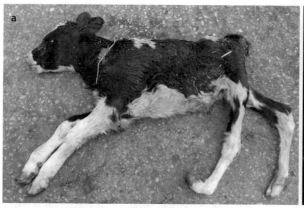
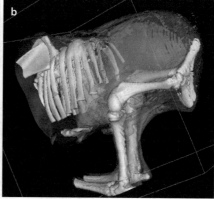

■ **Fig. 10.4** Perosomus elumbis. **a** Perosomus elumbis is a lethal congenital malformation from unknown origin. The atrophy of the hindlimbs with arthrogryposis, shortness of the hindquarter and normal forequarter are the most evident findings at calving time. These abnormalities are due to partial agenesis of the spinal cord and absence of lumbar, sacral and coccygeal spine. (Original from Ana Paula Peixoto). **b** Bone and soft tissue surfaces rendered computed tomography images of a perosomus elumbis. (Adapted from Agerholm et al. [1] with permission from Springer Nature)

10.5 Spina Bifida (■ Fig. 10.5)

■ **Fig. 10.5** Spina bifida of a dead newborn calf. No other external macroscopic malformations were observed in this clinical case. A medial groove can be observed in the magnified image. Spina bifida is a defective closure of dorsal vertebral laminae (central nervous system) resulting from abnormal neural tube fusion or closure. Its association with dicephalus, Arnold-Chiari malformation (foramen magnum herniation of cerebellar tonsils), duplication of thoracic organs or other malformations has been described in the literature [2, 3]. (Courtesy of Ana Paula Peixoto)

10.6 Schistosomus Reflexus in a Twin Pregnancy (■ Fig. 10.6)

10.7 Visceral Presentation in a Schistosomus Reflexus Case (■ Fig. 10.7)

10

■ **Fig. 10.6** Unusual clinical case of dystocia due to the presence of one schistosomus reflexus in a twin pregnancy. The other foetus was normal. Both dead foetuses were removed by caesarean section. (Courtesy of Paulo Teixeira)

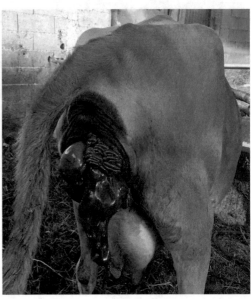

■ **Fig. 10.7** Visceral presentation of a schistosomus reflexus. The foetal viscera were detected in pelvic inlet and pulled outside the vulva. The abomasum (left), small intestine (right up) and amniotic membrane can be easily identified. The viscera should be torn or cut followed by foetotomy to reduce and remove foetal parts (extremities and trunk). (Original from Paulo Teixeira)

10.8 Vaginal Delivery of a Schistosomus Reflexus by Foetotomy (◐ Fig. 10.8)

◐ **Fig. 10.8** A schistosomus reflexus calf was
delivered by vaginal route
after partial foetotomy –
cuts were made at the
carpus and tarsus joint
and on the neck. (Original
from George Stilwell)

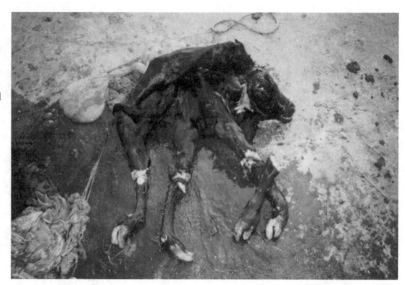

10.9 Cleft Palate (◐ Fig. 10.9)

◐ **Fig. 10.9** Absence of the soft and hard palate.
(Original from George Stilwell)

10.10 Kramer Curve Calf Snare: Technical Procedures (◫ Fig. 10.10)

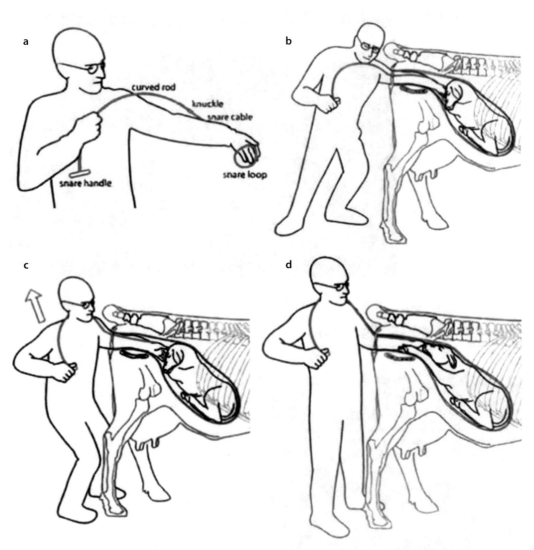

◫ **Fig. 10.10** Kramer curve calf snare use. **a** The loop of the snare is put on the obstetrician's hand and introduced into the uterus. **b** The loop is positioned over the lower jaw (deviation or retention of the head) or flexed foot of the foetus through the end of the respective extremity and must be tight. **c** The curved rod should be placed behind the obstetrician's neck and the snare handle is grasped. The snare handle is gentle pulled to adjust the tension on the curved rod. **d** The foetal head or limb is driven by the obstetrician's hand and extended by simultaneous gentle traction. (Modified from Jorgensen Laboratories, Inc. (USA) with permission)

10.11 Lifting the Downer Cow Using a Bagshaw Hoist (□ Fig. 10.11)

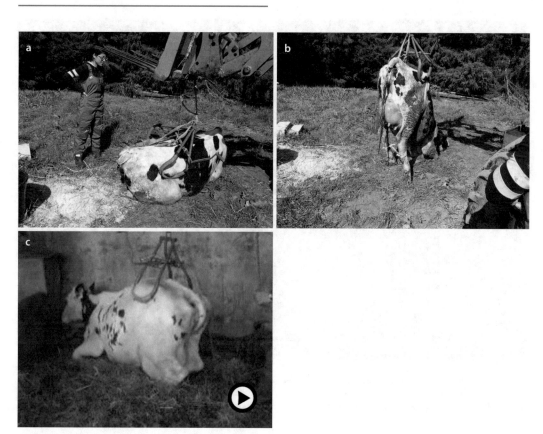

□ **Fig. 10.11** **a** The Bagshaw hoist is applied immediately below each tuber coxae which serve to support the hindquarter's weight of the cow. The Bagshaw hoist should be adequately adapted to both tuber coxae avoiding too much (risk of ilium fracture) or too little (Bagshaw hoist slippage) pressure. (Original from António Carlos Ribeiro). **b** The cow is raised smoothly. The hooves must remain in contact with the ground allowing for potential support and balance of the cow. The forequarter is elevated spontaneously by the cow or with the aid of two strong persons. Note that the hindlimbs are hoppled to prevent uncontrolled abduction. Usually, the cow obtains a self-equilibrium after a few minutes standing. The cow should not stay more than 15 min supported by the Bagshaw hoist to avoid lesions at tuber coxae level. **c** Video box: The Bagshaw hoist is pulled using a pulley fixed on the roof (► https://doi.org/10.1007/000-2qm)

10.12 Metallic Skin Staples Use
(■ Fig. 10.12)

■ **Fig. 10.12**   External aspect of a suture using metallic skin staples after caesarean section. Contrary to heifers, multiparous cows present a skin hard to pierce. An evaginated suture pattern can be easily made using individual stitches by application of metallic skin staples. (Original from António Carlos Ribeiro)

10

10.13 Uterine Prolapse in "Milk Fever" Cow (◻ Fig. 10.13)

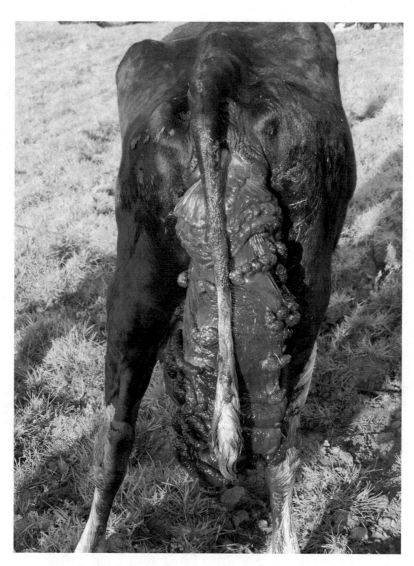

◻ **Fig. 10.13** Terceira, Azores. A cow was found down early in the morning with a prolapsed uterus and live calf lying beside. The cow was bloated and showed no pupilar or anal reflex. Milk fever was diagnosed and treated with 400 mL of 23% calcium borogluconate, by slow IV infusion. While the calcium was being given, the uterus was laid on a clean sheet and washed and disinfected with a warm iodine solution. When the infusion finished, the cow was stimulated to get up, and 5 mL of a local anaesthetic was injected in the epidural space between S5 and Co1. The uterus was lifted with the sheet by two persons (one on each side) keeping it as much as possible at vulva level. Oxytocin was given intra-myometrium. Using "hoven-mittens" to reduce trauma while increasing grasping ability, and plenty of lubricant, the uterus was slowly replaced into the abdominal cavity. Complete reversion was achieved by using a 1.5 L bottle filled with warm water, as an arm extension. (Courtesy of João Fagundes)

10.14 Umbilical Hernia
(■ Fig. 10.14)

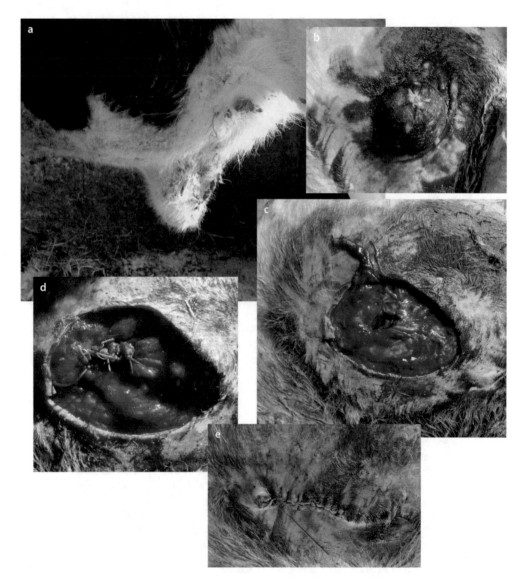

■ **Fig. 10.14** Top: **a** External aspect of an umbilical hernia in a 52-day-old Holstein-Frisian female. Umbilical hernias are a frequent congenital defect in calves. The females should be not used for replacement in herds due to the hereditability of this defect. Usually, the application of an umbilical hernia clamp is enough to treat umbilical hernia <5 cm in diameter, especially when detected in the first week of life. **b** No tissue adherences were observed at palpation and the hernia contents (simple hernia). **c** Umbilical ring (4–5 cm in diameter). The abdominal viscera are replaced and the umbilicus is removed. **d** The herniorrhaphy procedure involves a tension suture pattern (continuous or interrupted horizontal mattress) which should be used due to the pressure of abdominal viscera on this area. The closure of the umbilical ring (surgical repair) can be reinforced with an absorbable or non-absorbable suture USP 3 or USP 4 or umbilical tape. The overlay (if possible) of the surrounding tissue edges of the surgical wound can reinforce this area. Sometimes, when the umbilical ring is large (>8–10 cm), a mesh repair to support the surrounding soft tissue is recommended. **e** Ford interlocking suture pattern. (Original from João Simões). Bottom: Two cases from the same farm, in which an "outbreak" of very large umbilical hernias was detected. These calves were all born from cows sired by the same bull. Recovery after surgery was uneventful, but all progeny from that bull were sold for fattening. (Original from George Sitwell)

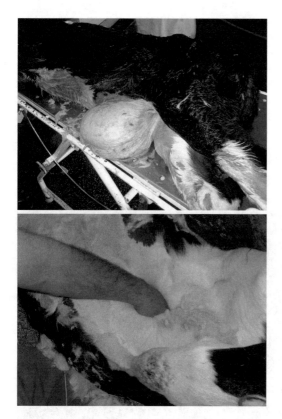

◘ Fig. 10.14 (continued)

10.15 Differential Diagnosis of Uterine Content Alterations by Transrectal Ultrasonography in Postpartum

A transrectal probe with a 7.5 MHz transductor (Aloka®) was used. Distance between bars = 1 cm. Originals from João Simões (◘ Fig. 10.15).

10

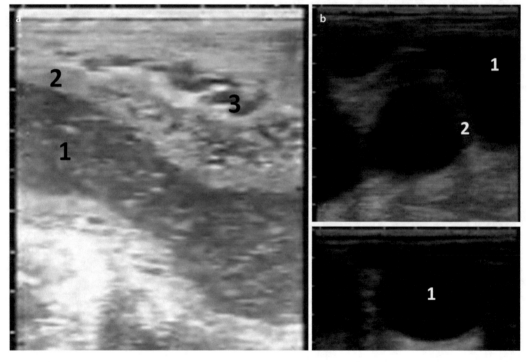

◘ **Fig. 10.15** **a** Endometritis at 56 days postpartum. Longitudinal ultrasonographic cut. (1) Mucopurulent uterine. (2) Uterine wall. (3) Blood vessel of the large ligament. **b** Mucometra in a Holstein-Frisian cow. Top: Longitudinal/oblique ultrasonographic cut. (1) Free uterine fluid. (2) Trabeculae (juxtaposition of two uterine wall sections) due to uterine horns spiral shape. Bottom: (1) Follicular cyst responsive for the mucometra occurrence. Ultrasonographic cut at the largest diameter. **c** Top left: Early foetal mortality detected at 52 days after artificial insemination in a Holstein-Friesian cow. All sonograms are represented using the same scale. (1) Transversal ultrasonographic cut of the foetus. The cardiac movements were not detected at real-time ultrasonography. (2) Thickness of the uterine wall is well evident. (3) Vessels of the broad ligament are evident. Like the thickness of the uterine wall, the larger diameter of the blood vessels is due to the inflammation pro- cess. (4) Small amount of foetal fluid. Similar to what may occur in late embryonic mortality (death between 28 and 42 days after artificial insemination), two outcomes can occur: the expulsion of the concept through the cervical canal or its complete reabsorption. Top right: A normal 48-day pregnancy in a Holstein-Friesian cow. By comparison to the previous sonogram, a larger foetus and a large amount of foetal fluids can be observed. The intact amniotic membrane is also evident. (1) Longitudinal (midsagittal) ultrasonographic cut of the foetus. (2) Amniotic membrane surrounding the foetus. Bottom left: The same foetus of the previous sonogram. (1) Transversal ultrasonographic cut of the foetus. (2) Amniotic membrane. (3) Umbilical cord. Observe its entrance into the foetal abdomen. Bottom right: Frontal ultrasonographic cut of a Holstein-Friesian 50-day-old foetus. Observe the head and abdomen as well as the four foetus limbs

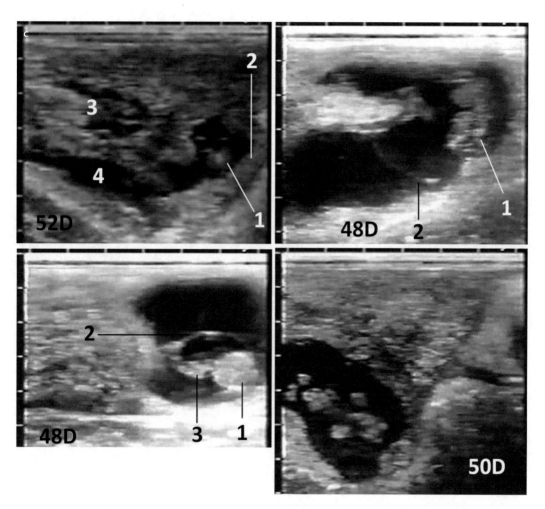

□ **Fig. 10.15** (continued)

References

1. Agerholm JS, Holm W, Schmidt M, Hyttel P, Fredholm M, McEvoy FJ. Perosomus elumbis in Danish Holstein cattle. BMC Vet Res. 2014;10:227. https://doi.org/10.1186/s12917-014-0227-2.
2. Gruys E. Dicephalus, spina bifida, Arnold-Chiari malformation and duplication of thoracic organs in a calf. Description of the case and critical discussion of the pathogenesis of the Arnold-Chiari malformation. Zentralbl Veterinarmed A. 1973;20(10): 789–800.
3. Madarame H, Ito N, Takai S. Dicephalus, Arnold-Chiari malformation and spina bifida in a Japanese black calf. Zentralbl Veterinarmed A. 1993;40(2): 155–60. https://doi.org/10.1111/j.1439-0442.1993. tb00611.x.
4. Prado TM, Schumacher J, Dawson LJ. Surgical procedures of the genital organs of cows. Vet Clin North Am Food Anim Pract. 2016;32(3):727–52. https://doi.org/10.1016/j.cvfa.2016.05.016.

Online Clinical Cases

Drost M, Samper J, Larkin PM, Gwen Cornwell D. Obstetrics: foetotomy. The visual guide to bovine reproduction. Gainesville: UF College of Veterinary Medicine; 2019. From: https://visgar.vetmed.ufl. edu/. Accessed on 18 Sept 2020.

Suggested Reading

Baird AN. Umbilical surgery in calves. Vet Clin North Am Food Anim Pract. 2008;24(3):467–vi. https:// doi.org/10.1016/j.cvfa.2008.06.005.
Górriz-Martín L, Neßler J, Voelker I, Reinartz S, Tipold A, Distl O, Beineke A, Rehage J, Heppelmann M. Split spinal cord malformations in 4 Holstein Friesian calves. BMC Vet Res. 2019;15(1):307. https://doi.org/10.1186/s12917-019-2055-x.
Quintela LA, Barrio M, Peña AI, Becerra JJ, Cainzos J, Herradón PG, Díaz C. Use of ultrasound in the reproductive management of dairy cattle. Reprod Domest Anim. 2012;47(Suppl 3):34–44. https://doi. org/10.1111/j.1439-0531.2012.02032.x.

Self-Evaluation in Calving, Obstetrical and Calf Management Subjects

Contents

© Springer Nature Switzerland AG 2021
J. Simões, G. Stilwell, *Calving Management and Newborn Calf Care*,
https://doi.org/10.1007/978-3-030-68168-5_11

11.1 **Instructions**

Multiple-choice questions (MCQ): only one of the four answers is correct. Please identify the correct choice.

True and false quiz questions: please assign the letters "T" for true or "F" for false to each answer option. The number of correct and incorrect answers may vary.

For other questions, the answers must be put in the correct order.

You will *find the solution* to each question at the end of this chapter.

❓ Questions

1. Please arrange the following terms numbered in ◗ Fig. 11.1:
   - Body of uterus.
   - Cervix.
   - Clitoris.
   - External urethral ostium.
   - Pubovesical pouch (peritoneal recess).
   - Rectogenital pouch (peritoneal recess).
   - Rectum.
   - Sacrorectal pouch (peritoneal recess).
   - Suburethral diverticulum.
   - Urinary bladder.
   - Uterine horn.
   - Vagina showing ridges.
   - Vesicogenital pouch (peritoneal recess).
   - Vestibulum.
   - Vulva.

2. Which suture pattern is more appropriate to close the uterine incision during a caesarean section in a cow (MCQ)?
   (a) Cushing's.
   (b) Halsted's.
   (c) Cushing's modified by the Utrecht method.
   (d) Lembert's.

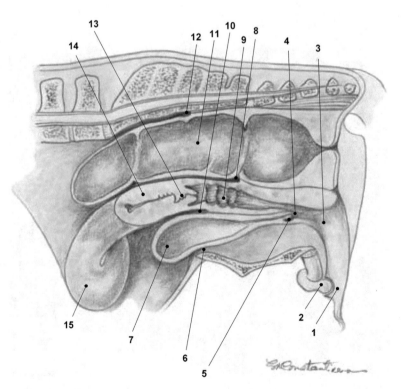

◗ **Fig. 11.1** Bovine external and internal genitalia. (Modified from Constantinescu [2] with permission from John Wiley & Sons)

3. In Stage III of labour (MCQ):
    (a) Uterine contractions immediately cease.
    (b) The cervix quickly closes.
    (c) Foetus is expelled.
    (d) Placenta is expelled.

4. Please assign the letters (T) for true and (F) for false to the following sentences:
    (a) T/F: The position of the foetus indicates the relationship between its longitudinal axis and the longitudinal axis of the birth canal.
    (b) T/F: To evaluate foetal viability in anterior presentation, we can assess the palpebral and sucking reflexes, as well as withdrawal reaction when pinching between the hooves.
    (c) T/F: Oxytocin is quite efficient in solving causes of obstruction, by increasing uterine contractions.
    (d) T/F: Caesarean sections performed with the dam in dorsal decubitus may lead to embolism and vena cava syndrome.

5. Which is the normal foetal position at calving (MCQ)?
    (a) Ventral or dorsopubic.
    (b) Dorsal or dorsosacral.
    (c) Left lateral or left dorso-iliac.
    (d) Right lateral or right dorso-iliac.

6. The pre-calving heifer in ◘ Fig. 11.2 shows signs of … (MCQ):
    (a) … naturally occurring ventral oedema.
    (b) … umbilical hernia.
    (c) … umbilical abscess.
    (d) … hydramnios.

7. A large (>5–7 cm) ventral rupture of the uterine wall (MCQ):
    (a) Is commonly related to uterine wall friability.
    (b) Only occurs in twin calving.
    (c) Is usually due to hormonal imbalance.
    (d) Only occurs during uterine torsion.

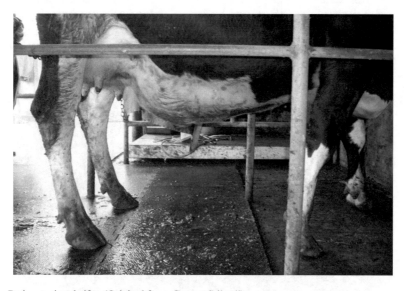

◘ **Fig. 11.2** Periparturient heifer. (Original from George Stilwell)

8. Please assign the letters (T) for true and (F) for false to the following sentences:
   (a) T/F: In caesarean sections, caudal epidural anaesthesia should always be performed to reduce myometrial contractions and facilitate surgery.
   (b) T/F: Utrecht suture pattern consists of an evaginated oblique suture that prevents adhesions between the wall of the uterus and the peritoneum.
   (c) T/F: Vaginal prolapse is a contraindication for performing a caesarean section.
   (d) T/F: The left flank approach is preferable in caesarean section to prevent the protrusion of intestinal viscera during surgery.

9. Look at ◘ Fig. 11.3 and choose the correct option (MCQ):
   (a) The "frog position" facilitates lung insufflation and regular breathing.
   (b) The calf naturally adopts this position immediately after birth.
   (c) This is a specific procedure to clear the low airways.
   (d) This is a specific procedure to improve thermogenesis.

10. At calving, the first intact foetal membrane which can be detected in the birth canal is ... (MCQ):
    (a) ... vitelline.
    (b) ... chorioallantois.
    (c) ... amnion.
    (d) ... all the three membranes.

11. What is the recommended procedure for schistosomus reflexus dystocia (MCQ)?
    (a) Waiting 24 h.
    (b) Pharmaceutical approach.
    (c) Caesarean section or foetotomy.
    (d) Pulling the calf with ropes.

12. During forced traction of the foetus (MCQ):
    (a) A halter head is always required.
    (b) A loop below the fetlock joint is contraindicated.
    (c) The obstetrical chain should be fixed above the fetlock joint.
    (d) Only one obstetrical chain or rope is used.

13. Look at ◘ Fig. 11.4 representing a maternal behaviour (MCQ):
    (a) The dam is rejecting the calf due to foetal fluid odours.
    (b) The dam intends to establish a cow-calf bond.

◘ **Fig. 11.3** Newborn calf in frog position. (Courtesy of Carlos Cabral)

Fig. 11.4 Dam and calf few minutes after calving. (Original from João Simões)

(c) This is an abnormal maternal behaviour.
(d) The dam rejects the calf due to the presence of blood.

14. Please assign the letters (T) for true and (F) for false to the following sentences:
   (a) T/F: Placental retention is the main postoperative complication of caesarean section, as a consequence of uterine mucosal trauma.
   (b) T/F: The foetus only acquires the birth posture in the Stage II of labour.
   (c) T/F: The contractions of the myometrium are enough to expel the foetus.
   (d) T/F: After the first bag is ruptured, we are facing a medical emergency, which should be solved in the following hour.

15. Look at ◻ Fig. 11.5 representing a prolonged Stage II (around 5 h) and choose the correct answer (MCQ):
   (a) The head dislocation occurred due to excessive traction.
   (b) Partial foetotomy was performed to prevent vulvar lacerations.

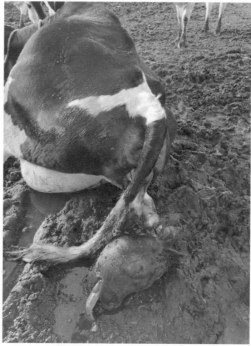

Fig. 11.5 Prolonged Stage II. (Courtesy of Carlos Cabral)

(c) There are two foetuses because the limbs are crossed.
(d) The foetus' head was oedematous and was removed after foetotomy to unlock the birth canal.

16. Please assign the letters (T) for true and (F) for false answers:
   (a) T/F: It is recommended to always use caudal epidural anaesthesia in dystocia.
   (b) T/F: To reduce a uterine torsion, we should roll the cow to the opposite side of the twist.
   (c) T/F: When extracting a foetus in a caesarean section, we should rotate and pull it in a caudolateral direction.
   (d) T/F: When the foetus is in equidistant ventro-transverse presentation, we should opt to use forced traction using the hindlimbs.

17. Please arrange the following terms numbered in ❏ Fig. 11.6:
    - Longitudinal cut.
    - Oblique caudal cut.
    - Oblique cranial cut.
    - Transversal cut.

18. In the easy calving indexes prediction, the direct calving index (MCQ):
    (a) Is not appropriate to improve immediate calving.
    (b) Refers to the ability of the dam to easily deliver a calf.
    (c) Refers to how easily a calf will be born.
    (d) Is more appropriate to heifers with more adequate pelvis.

19. Identify (assign the technical term) to the following obstetrical manoeuvres:
    (a) Force used to pull or extract the foetus or its parts.
    (b) The force exerted on foetal parts to push back the foetus from the birth canal into the uterine cavity.
    (c) Horizontal, vertical or oblique transverse presentation converted to an anterior or posterior longitudinal presentation.
    (d) Extension of a flexed joint of foetal extremities.
    (e) The force exerted upon the foetal longitudinal axis.

20. Look at ❏ Fig. 11.7 and identify the clinical condition (MCQ):
    (a) Vaginal prolapse Grade 3.
    (b) Vaginal prolapse Grade 4.
    (c) Bladder eversion.
    (d) Rectal prolapse.

21. Look at ❏ Fig. 11.8 representing calf's airway clearing (MCQ):
    (a) The calf should be kept in this position for 5 min.

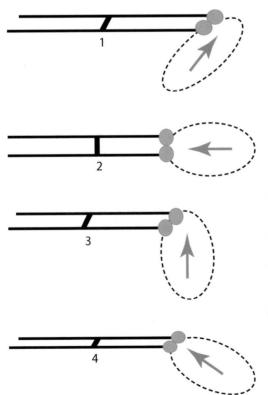

❏ **Fig. 11.6** Foetotomy cut types

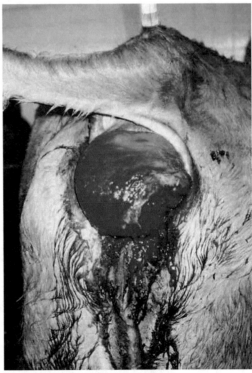

❏ **Fig. 11.7** Mass eversion. (Original from João Simões)

(b) Iodopovidone is used to stimulate breathing.

(c) The mucus comes out from the upper airways by gravity.

(d) The weight of abdominal viscera on the diaphragm helps mucus clearance.

22. Please assign the numbers (I) for Stage I, (II) for Stage II and (III) for Stage III.

(a) I/II/III: Chorioallantois sac enters the vagina.

(b) I/II/III: Loss of placental circulation due to rupture of the umbilical cord.

(c) I/II/III: Rupture of the umbilical cord.

(d) I/II/III: Onset of myometrial contractions.

(e) I/II/III: Restlessness behaviour.

(f) I/II/III: Persistence of mild myometrial and abdominal contractions.

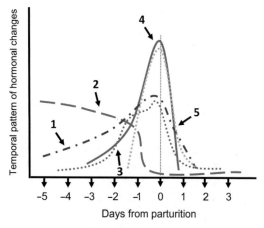

■ **Fig. 11.9** Timeline hormonal changes during the peripartum period

(g) I/II/III: Rupture of amniotic sac.

(h) I/II/III: Onset of abdominal contractions.

23. Please name the hormones numbered in ■ Fig. 11.9:
 − Foetal cortisol.
 − Oestrogens.
 − Oxytocin.
 − $PGF_{2\alpha}$.
 − Progesterone.

24. During the calving preparatory period, the cow usually reduces dry matter intake in the last… (MCQ):

(a) … 24 h.
(b) … 48 h.
(c) … 68 h.
(d) … 72 h.

25. The normal foetal disposition for calving is… (MCQ):

(a) anterior presentation, lateral position and extension of the forelimbs and head.

(b) posterior presentation, ventral position and extension of the hindlimbs.

(c) anterior presentation, dorsal position and extension of the forelimbs and head.

(d) posterior presentation, lateral position and extension of the hindlimbs.

26. During the calving preparatory period, the cow usually starts to increase number of lying/standing bouts in the last... (MCQ):
    (a) 1 h.
    (b) 2–4 h.
    (c) 4–6 h.
    (d) 6–8 h.

27. Simultaneously classify the dystocia according to foetal (F) or maternal (M) origin and as proximal (P), intermediate (I) or ultimate (U):
    (a) Abnormal foetal presentation, position and/or posture: F/M and P/I/U.
    (b) Birth canal undersize: F/M and P/I/U.
    (c) Uterine inertia: F/M and P/I/U.
    (d) Foetal oversize: F/M and P/I/U.
    (e) Hypocalcaemia and hypomagnesaemia: F/M and P/I/U.
    (f) Foetal congenital abnormalities: F/M and P/I/U.

28. The relationship between expected progeny differences (EPDs) and estimated breeding values (EBVs) is... (MCQ):
    (a) ... EPDs = ½ EBVs values.
    (b) ... EPDs = EBVs values.
    (c) ... EPDs = 2× EBVs values.
    (d) ... EPDs = ¼ EBVs values.

29. Please score (1–4) the calving difficulty to classify a dystocia:
    (a) Minor manual assistance.
    (b) No assistance.
    (c) Caesarean section or foetotomy.
    (d) One person + calf puller or >1 person.

30. Please assign the letters (T) for true and (F) for false to the following sentences.
    These factors predispose to uterine torsion:
    (a) T/F: Pregnancy in one uterine horn.
    (b) T/F: Twins.
    (c) T/F: Attachment of the broad ligaments along the great curvature of the uterus.

(d) T/F: Deep abdomen.
(e) T/F: The broad ligament is loose and quite large.
(f) T/F: Large foetuses.

31. A uterine torsion can be diagnosed by ... (MCQ):
    (a) ... vaginal palpation of both broad ligaments' displacement.
    (b) ... vaginal palpation of vaginal spirals.
    (c) ... ultrasonography detecting a thin uterine wall.
    (d) ... vaginal palpation of cervical folds.

32. In Schaffer's method used to solve uterine torsions (MCQ):
    (a) The cow is lying down for the opposite side of the uterine torsion.
    (b) The cow is kept in standing position.
    (c) The cow is always rotated only one time.
    (d) The cow is rolled 180°.

33. Cow's myometrial contractions can be blocked by using the following epinephrine solution and route (MCQ):
    (a) 1/10000; 5 mL; i.v.
    (b) 1/10000; 10 mL; i.v.
    (c) 1/10000; 5 mL; s.c.
    (d) 1/10000; 2 mL; i.m.

34. Please assign the letters (T) for true and (F) for false to the following sentences.
    Low caudal epidural anaesthesia:
    (a) T/F: Eliminates stimuli from a distended vaginal wall and cervix.
    (b) T/F: Local anaesthetics such as procaine are used to block nerve transmission to the spinal cord.
    (c) T/F: Xylazine cannot be administered to promote analgesia.
    (d) T/F: The effect mainly depends on the volume.
    (e) T/F: It is administered in lumbar epidural space.
    (f) T/F: Desensitize the perineum and vulva, but not the anal sphincter.
    (g) T/F: Usually only desensitize the sacral nerves S3, S4 and S5.

35. The break of the second water sac correspond to … (MCQ):
    (a) … chorioallantois membrane rupture.
    (b) … rupture of both foetal membranes.
    (c) … amniotic membrane rupture.
    (d) … rupture of both foetal membranes and umbilical cord.

36. The intrauterine fluid replacer can be obtained adding … (MCQ):
    (a) … 1 kg of methylcellulose to 45 L of warm water.
    (b) … 0.5 kg of methylcellulose to 45 L of warm water.
    (c) … 1 kg of methylcellulose to 45 L of cold water.
    (d) … 0.5 kg of methylcellulose to 45 L of cold water.

37. Identify and name the foetal origin dystocias illustrated in ◘ Fig. 11.10:
    (a) _____
    (b) _____
    (c) _____
    (d) _____
    (e) _____

38. Foetopelvic disproportion (MCQ):
    (a) Is more likely to occur in multiparous cows.
    (b) Can be prevented using direct calving indexes.
    (c) Can be predicted using the diameter of the pelvic outlet.
    (d) Is only related with foetal weight.

39. Look at ◘ Fig. 11.11 and choose the correct option (MCQ):
    (a) The intestinal viscera of the cow are pending from the vulva.
    (b) This is a posterior presentation of a perosomus elumbis.
    (c) This is a part of an emphysematous foetus.
    (d) This is a visceral presentation of a schistosomus reflexus.

40. Usually, the laparotomy incision length for a caesarean section is between … (MCQ):
    (a) … 10 and 20 cm.
    (b) … 20 and 30 cm.
    (c) … 30 and 50 cm.
    (d) … 50 and 70 cm.

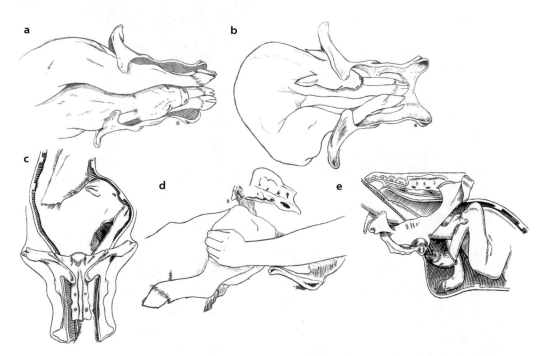

◘ **Fig. 11.10** Foetal dystocia due to maldisposition. (**a**, **b** and **d**: Original from Soraia Marques. **c** and **e**: Adapted from Parkinson et al. [3] with permission from Elsevier)

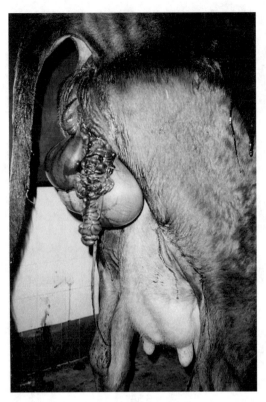

◘ **Fig. 11.11**   Stage II of the labour. (Original from João Simões)

◘ **Fig. 11.12**   Periparturient cow. (Courtesy of Miguel Saraiva Lima)

◘ **Fig. 11.13**   Foetal limbs after foetotomy. (Courtesy of Ana Paula Peixoto)

41. The best approach option to well exteriorize the uterus during a caesarean section involving an emphysematous foetus is … (MCQ):
    (a) … left paralumbar.
    (b) … right paralumbar.
    (c) … oblique.
    (d) … ventrolateral.

42. Gross contamination of the surgical site without active infection is a contaminated surgical site (MCQ):
    (a) Class I.
    (b) Class II.
    (c) Class III.
    (d) Class IV.

43. Look at ◘ Fig. 11.12 and choose the correct option (MCQ):
    This is …
    (a) … foetal membranes covering the foetus.
    (b) … an extensive necrosis of the uterus.
    (c) … a uterine prolapse covered by foetal membranes.
    (d) … viscera of a foetal monster.

44. To perform a laparotomy, the skin should be disinfected with … (MCQ):
    (a) … iodopovidone 1% or chlorhexidine 4% in ethanol 70%.
    (b) … iodopovidone 10% or chlorhexidine 4% in ethanol 70%.
    (c) … iodopovidone 10% or chlorhexidine 4% in ethanol 50%.
    (d) … iodopovidone 1% or chlorhexidine 10% in ethanol 70%.

45. Look at ◘ Fig. 11.13 and choose the correct option (MCQ):
    This is …

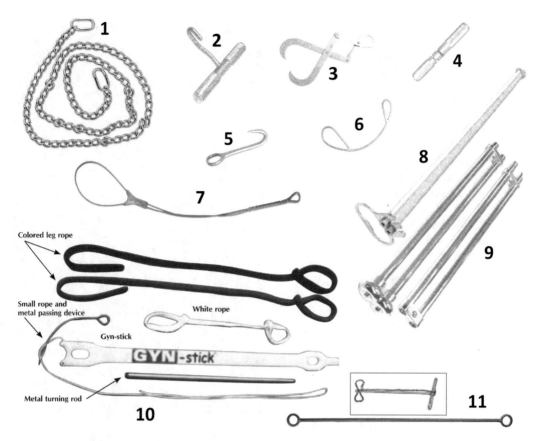

■ **Fig. 11.14** Obstetrical material. (Modified (FotoSketcher 3.60) from Jorgensen Laboratories, Inc. (USA) with permission)

(a) … both forelimbs of a foetus.
(b) … both hindlimbs of a foetus.
(c) … one forelimb and one hindlimb of a foetus.
(d) … one forelimb of two foetuses (twins).

46. In the proximal paravertebral anaesthesia (Farquharson method) to desensitize T13 nerves, the cannula needle is placed at cranial edge of the transverse process of … (MCQ):
    (a) … L1.
    (b) … T13.
    (c) … T12.
    (d) … L2.

47. Please assign the letters (T) for true and (F) for false to the following sentences:
    (a) T/F: Foetal mutation from posterior to anterior presentation is an appropriate and easy obstetrical manoeuver.
    (b) T/F: The placenta of cattle is synepitheliochorial.
    (c) T/F: One of the main signs of foetal stress or agony is the presence of meconium on the body surface of the foetus and in the amniotic fluid.
    (d) T/F: In foetal posterior presentation, we can evaluate the viability of the foetus by feeling the umbilical cord pulsation.

48. Please name the following terms numbered in ■ Fig. 11.14:
    − Calf snare.
    − Cornell detorsion rod.
    − Krey-Schöttler hook.
    − Kühn's obstetrical crutch-like instrument.
    − Obstetrical wire guide.

- Obstetrical chain handles.
- Obstetrical chains.
- Ostertag's blunt eye hooks.
- Saw wire handles.
- Thygesen's foetatome.
- Utrecht foetatome.

49. Obturator nerve paralysis ... (MCQ):
    (a) ... is a primary cause of the downer cow syndrome.
    (b) ... desensitizes the muscles responsible for the abduction of the hindlimb(s).
    (c) ... is always bilateral.
    (d) ... is always related with a poor prognosis.

50. To exteriorize the uterine wall (MCQ):
    (a) The large ligament should be sectioned.
    (b) A uterine incision should be made to grasp the foetus.
    (c) A rotation of the pregnant uterine horn, up to 180°, can be required.
    (d) The uterus is grasped near to the intercornual ligament and pulled.

51. Look at ◻ Fig. 11.15 representing a hobbled cow and choose the correct answer (MCQ):
    Hobbles are applied to prevent the ...
    (a) ... adduction of the hindlimbs.
    (b) ... rupture of suspensory ligament of the mammary gland.
    (c) ... cow kicking.
    (d) ... abduction of the hindlimbs.

52. During labour, if the second water bag is still intact inside the birth canal (MCQ):
    (a) This can be a reproductive urgency but not an emergency.
    (b) The foetus should be immediately extracted.
    (c) This means that a complete uterine torsion occurred.
    (d) This is always an abnormality.

53. During forced extraction of a foetus in anterior presentation, the alternate traction of the forelimbs (MCQ):

◻ **Fig. 11.15** Hobbling the cow. (Original from George Sitwell)

(a) Prevents hiplock.
(b) Reduces the foetal diameter at abdominal level.
(c) Does not have significant effect on calving ease.
(d) Reduces the foetal diameter at shoulder level.

54. In caesarean section, the uterine wall should be washed before replacing it into the abdominal cavity to ... (MCQ):
    (a) ... hydrate the uterus.
    (b) ... remove blood clots.
    (c) ... remove excessive suture lines.
    (d) ... look good.

55. Look at ◻ Fig. 11.16 and choose the correct answer (MCQ).
    (a) The obstetrical chains are inappropriately fixed to the limbs.
    (b) The obstetrical snare should be strongly pulled.

Fig. 11.16 Stage II of the labour. (Courtesy of João Fagundes)

Fig. 11.17 Obstetrical halter. (Courtesy of João Fagundes)

Fig. 11.18 Foetal death. (Courtesy of António Carlos Ribeiro)

(c) The foetus is trying to lick its feet.

(d) The dorsal region of the vulva is a point of resistance to the progression of the foetus' head.

56. Look at ◘ Fig. 11.17 and choose the correct answer (MCQ):

(a) A halter should be always used during forced traction.

(b) The halter is useful to prevent the foetus head retention in the birth canal.

(c) An obstetrical chain can be used as alternative to a snare.

(d) The halter is applied to the head inappropriately.

57. During caesarean section, the recommended USP size of the suture line to close the uterine incision is … (MCQ):

(a) … USP 2.

(b) … USP 2-0.

(c) … USP 1.

(d) … USP 1-0.

58. Look at ◘ Fig. 11.18 and choose the correct answer (MCQ):

This is …

(a) … a papyraceous mummified foetus, as the soft tissues are not intensively dried.

(b) … a haematic mummified foetus (brownish appearance due degeneration of red blood cells).

(c) … a macerated foetus.

(d) … an abortion occurring in early pregnancy.

59. Instead of using intraperitoneal antimicrobials in a caesarean section, the following concentration of iodopovidone intraperitoneal solution (1–4 L) can be used (MCQ):
    (a) 10%.
    (b) 5%.
    (c) 1%.
    (d) 0.5%.

60. First-degree perineal lacerations, less than 2 cm (MCQ):
    (a) Should always be sutured.
    (b) Cause significant vulvar labia deformations.
    (c) Cause urovagina.
    (d) Should be only disinfected.

61. Please arrange the following terms numbered in ◘ Fig. 11.19:
    — Abdomen.
    — Amputation of the head.
    — First forelimb.
    — Hindquarters/pelvis.
    — Lumbar area.
    — Thorax.

62. In retained foetal membranes (MCQ):
    (a) Manual removal of the foetal membranes is recommended.
    (b) The intrauterine administration of antimicrobials is always recommended.
    (c) Broad-spectrum antimicrobials should be used when evident signs of infection (e.g. hyperthermia) are observed.
    (d) The dehiscence of the placenta occurs quickly.

63. We classify a puerperal metritis as (MCQ):
    (a) Clinical metritis grade 2.
    (b) Clinical endometritis.
    (c) Clinical metritis occurring more than 21 days postpartum.
    (d) Subclinical metritis.

64. Please arrange the following terms numbered in ◘ Fig. 11.20:
    — External abdominal oblique muscle.
    — Internal abdominal oblique muscle.
    — Peritoneum.
    — Skin.
    — Transverse abdominal muscle.

65. Please assign the letters (T) for true and (F) for false to the following sentences.
    The anterior pituitary produces:
    (a) T/F: GnRH.
    (b) T/F: Follicle-stimulating hormone (FSH).

Anterior presentation

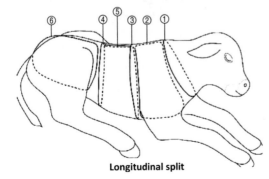

Longitudinal split

◘ **Fig. 11.19** Sequential cuts of foetotomy. (Modified from Vermunt and Parkinson [4] with permission from Elsevier)

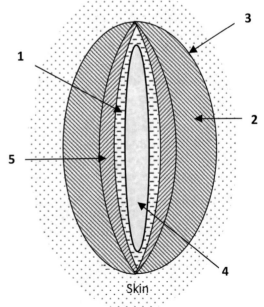

Skin

◘ **Fig. 11.20** Cutting sequence for laparotomy by flank approach

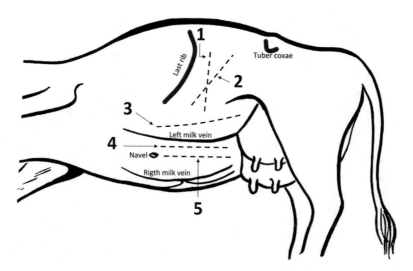

◘ Fig. 11.21 Surgical approaches for caesarean section. (Modified from Ames [1], with permission from John Wiley & Sons)

(c) T/F: Adrenocorticotropic hormone (ACTH).

(d) T/F: Luteinizing hormone (LH).

(e) T/F: Oxytocin.

(f) T/F: Prolactin hormone.

(g) T/F: Growth hormone (GH).

(h) T/F: $PGF_{2\alpha}$.

66. Please arrange the following terms numbered in ◘ Fig. 11.21:

   − Oblique.
   − Paralumbar.
   − Paramedian.
   − Ventral midline.
   − Ventrolateral.

67. The treatment of pyometra is based on (MCQ):

   (a) Topic administration of $PGF_{2\alpha}$.
   (b) Systemic administration of oxytocin.
   (c) Systemic administration of $PGF_{2\alpha}$.
   (d) Topic administration of oxytocin.

68. The recurrence of uterine prolapse after replacement (MCQ):

   (a) Is a common complication.
   (b) Can be caused by myometrial contractility.

◘ Fig. 11.22 Salers heifer presenting one limb at the vulva during labour. (Original from George Stilwell)

   (c) Can be caused ischemic necrosis of the uterus.
   (d) Can be caused by tenesmus.

69. Look at ◘ Fig. 11.22 and choose the correct option (MCQ).

   The most probable disposition of this foetus is…

   (a) … anterior presentation and dorso-pubic position.
   (b) … posterior presentation and dorso-pubic position.

(c) … anterior presentation and dorso-sacral position.

(d) … posterior presentation and dorso-sacral position.

70. The cow in ◘ Fig. 11.23 shows signs of a nerve paralysis after being down for a few days (MCQ).
    What nerve is most likely affected?
    (a) Tibial nerve.
    (b) Sciatic nerve.
    (c) Obturator nerve.
    (d) Peroneal nerve.

71. Look at ◘ Fig. 11.24 and identify (a–f) the sonograms where the following structures or uterine contents can be observed:
    ▬ Allantoic fluid.
    ▬ Amniotic membrane.

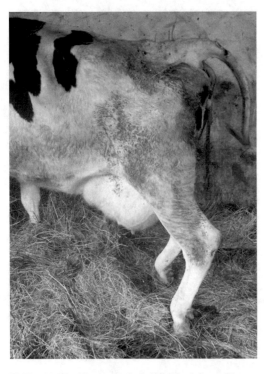

◘ **Fig. 11.23** Nerve paralysis. (Original from George Stilwell)

▬ Foetal fluids.
▬ Foetus.
▬ Free uterine fluid.
▬ Mucopurulent fluid.
▬ Ovarian cyst.

72. According to Score 1 of the Apgar score for newborn calves, the respiratory rate and effort is irregular when less than … (MCQ):
    (a) … 25 mpm.
    (b) … 15 mpm.
    (c) … 45 mpm.
    (d) … 35 mpm.

73. The meconium staining of the newborn calf is a sign of … (MCQ):
    (a) … low foetal vitality.
    (b) … foetal stress.
    (c) … eminent death.
    (d) … normality.

74. The calving jack can manage a maximum pull strength up to … (MCQ):
    (a) … 100 kg.
    (b) … 200 kg.
    (c) … 300 kg.
    (d) … 400 kg.

75. Look at ◘ Fig. 11.25 and choose the correct answer (MCQ):
    (a) The two orifices correspond to the double cervix.
    (b) It is a repair of a vulvar laceration degree 2.
    (c) It is a repair of a vulvar laceration degree 3.
    (d) The two orifices correspond to the inlet of each uterine horn.

76. Look at ◘ Fig. 11.26. The disinfection procedure of the surgical site using gauze pads with disinfectant should be performed from … (MCQ):
    (a) … A to B, C to D and C to E.
    (b) … C to A, C to B and D to E.
    (c) … B to A and D to E.
    (d) … D to E and A to B.

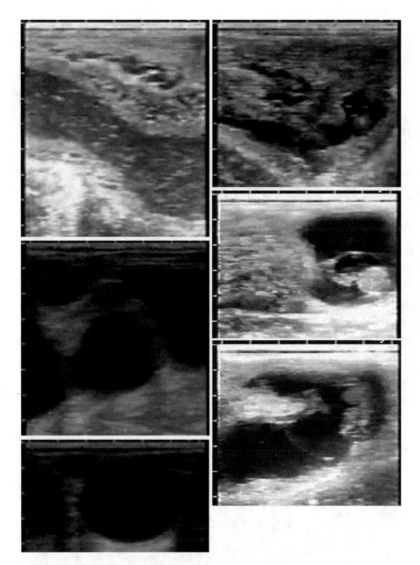

Fig. 11.24 Sonograms obtained from the uterine content of cows. Legend: 7.5 MHz transductor (Aloka®). Distance between bars = 1 cm. (Original from João Simões)

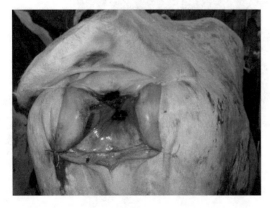

Fig. 11.25 Surgical repair of external genital. (Courtesy of António Carlos Ribeiro)

77. Look at ■ Fig. 11.27 and choose the correct option (MCQ):
    (a) This is the early Stage I of the labour.
    (b) The first water bag remains unruptured.
    (c) The second water bag remains unruptured.
    (d) This is the late Stage I of the labour.

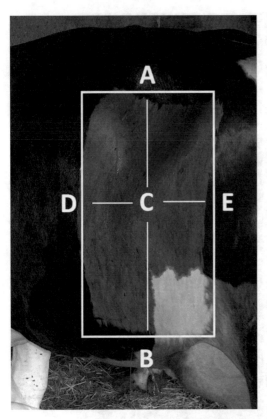

☐ **Fig. 11.26** Preparation of the surgical site for a cae-sarean section. (Courtesy of António Carlos Ribeiro)

✔ **Answers**

1. (1) Vulva; (2) Clitoris; (3) Vestibulum; (4) External urethral ostium; (5) Suburethral diverticulum; (6) Pubovesical pouch (peritoneal recess); (7) Urinary bladder; (8) Rectogenital pouch (peritoneal recess); (9) Vagina showing ridges; (10) Vesicogenital pouch (peritoneal recess); (11) Rectum; (12) Sacrorectal pouch (peritoneal recess); (13) Cervix; (14) Body of uterus; (15) Uterine horn
2. (c)
3. (d)
4. (a) F; (b) T; (c) F; (d) T
5. (b)
6. (a)
7. (a)
8. (a) F; (b) F; (c) F; (d) T
9. (a)
10. (b)
11. (c)
12. (c)
13. (b)
14. (a) T; (b) F; (c) F; (d) F
15. (d)
16. (a) F; (b) F; (c) T; (d) T

☐ **Fig. 11.27**   Labour of a heifer. (Original from João Simões)

17. (1) Oblique cranial cut; (2) Longitudinal cut; (3) Transversal cut; (4) Oblique caudal cut
18. (c)
19. (a) Traction; (b) Retropulsion or repulsion; (c) Version; (d) Extension; (e) Rotation
20. (d)
21. (c)
22. (a) I; (b) III; (c) II; (d) I; (e) I; (f) III; (g) II; (h) II
23. (1) Oestrogens; (2) Progesterone; (3) PGF2α; (4) Foetal cortisol; (5) Oxytocin
24. (a)
25. (c)
26. (c)
27. (a) F and P; (b) M and I; (c) M and P; (d) F and I; (e) M and I; (f) F and U
28. (a)
29. (a) 2; (b) 1; (c) 4; (d) 3
30. (a) T; (b) F; (c) F; (d) T; (e) T; (f) T
31. (b)
32. (d)
33. (b)
34. (a) T; (b) T; (c) F; (d) T; (e) F; (f) F; (g) T
35. (c)
36. (a)
37. (a) (Dystocia caused by) Twins; (b) Ventro-transverse presentation; (c) Lateral deviation of the head; (d) Hip flexion; (e) Hock flexion
38. (b)
39. (d)
40. (c)
41. (d)
42. (c)
43. (c)
44. (b)
45. (b)
46. (a)
47. (a) F; (b) T; (c) T; (d) T
48. (1) Obstetrical chains; (2) Obstetrical chain handles; (3) Krey Schöttler hook; (4) Saw wire handles; (5) Ostertag's blunt eye hooks; (6) Obstetrical wire guide; (7) Calf snare; (8) Utrecht foetatome; (9) Thygesen's foetatome; (10) Kühn's obstetrical crutch-like instrument; (11) Cornell detorsion rod
49. (a)
50. (c)
51. (d)
52. (a)
53. (d)
54. (b)
55. (d)
56. (b)
57. (a)
58. (b)
59. (c)
60. (d)
61. Cut 1: Amputation of the head; Cut 2: First forelimb; Cut 3: Thorax; Cut 4: Abdomen; Cut 5: Lumbar area; Cut 6: Hindquarters/pelvis
62. (c)
63. (a)
64. (1) Transverse abdominal muscle; (2) External abdominal oblique muscle; (3) Skin; (4) Peritoneum; (5) Internal abdominal oblique muscle
65. (a) F; (b) T; (c) F; (d) T; (e) F; (f) T; (g) T; (h) F
66. (1) Paralumbar; (2) Oblique; (3) Ventrolateral; (4) Paramedian; (5) Ventral midline
67. (c)
68. (d)
69. (d)
70. (a)
71. Allantoic fluid: (e, f); Amniotic membrane: (e, f); Foetal fluids: (d–f); Foetus: (d–f); Free uterine fluid: (b); Mucopurulent fluid: (a); Ovarian cyst: (c)
72. (d)
73. (b)
74. (d)
75. (c)
76. (a)
77. (c)

References

1. Ames NK. Noordsy's food animal surgery. 5th ed. Ames: John Wiley & Sons; 2014.
2. Constantinescu GM. Female genital organs. In: Schatten H, Constantinescu GM, editors. Comparative reproductive biology. Ames: Blackwell Publishing; 2007. p. 33–48. https://doi.org/10.1002/9780h470390290.ch2b.
3. Parkinson TJ, Vermunt JJ, et al. Dystocia in livestock: delivery per vaginam. In: Noakes DE, Parkinson TJ, England GCW, editors. Veterinary reproduction and obstetrics. Edinburgh: Elsevier; 2019. p. 250–76. https://doi.org/10.1016/B978-0-7020-7233-8.00014-8.
4. Vermunt JJ, Parkinson TJ. Defects of presentation, position and posture in livestock: delivery by foetotomy. In: Noakes DE, Parkinson TJ, England GCW, editors. Veterinary reproduction and obstetrics. 10th ed. Edinburgh: Elsevier; 2019. p. 277–90. https://doi.org/10.1016/B978-0-7020-7233-8.00015-X.

11

Correction to: Calving Management and Newborn Calf Care

Correction to:
J. Simões, G. Stilwell, Calving Management and Newborn Calf Care,
▶ **https://doi.org/10.1007/978-3-030-68168-5**

This book was inadvertently published with the incorrect order of authors in Chapters 2 and 5.

The correct order of the authors has now been updated in the chapters as George Stilwell, João Simões

The updated versions of these chapters can be found at
https://doi.org/10.1007/978-3-030-68168-5_2
https://doi.org/10.1007/978-3-030-68168-5_5

Supplementary Information

© Springer Nature Switzerland AG 2021
J. Simões, G. Stilwell, *Calving Management and Newborn Calf Care*,
https://doi.org/10.1007/978-3-030-68168-5

Appendix: Drug Posology for Cows. (Modified from [1])

Amoxicillin trihydrate	11–22 mg/kg s.c.; SID or BID
Acepromazine maleate	0.01–0.02 mg/kg i.v.; single dose 0.03–0.1 mg/kg i.m.; single dose
Ampicillin sodium	22 mg/kg s.c., i.v.; BID
Ampicillin trihydrate	4–22 mg/kg i.m. or s.c.; SID or BID
Butorphanol	0.05–0.2 mg/kg i.v. or i.m.; Q1–3 h (loading dose: 0.02–0.05 mg/kg)
Ceftiofur crystalline free acid	1.1–2.2 mg/kg i.m. or s.c.; SID
Ceftiofur sodium	1.1–2.2 mg/kg i.m. or i.v.; SID
Chlortetracycline	6–10 mg/kg i.m. or i.v.; SID 10–20 mg/kg (pre-ruminants calves) p.o.; SID
Chorionic gonadotropin (HCG)	1000–5000 IU i.v. or 10,000 IU i.m.; single dose
Danofloxacin	8 mg/kg s.c.; single dose 6 mg/kg s.c.; a second dose 48 h later (only repeat once)
Detomidine	0.003–0.01 mg/kg i.m. or i.v. (Q2–4 h) Epidural: 0.015 mg/kg diluted to 5 mL with sterile saline
Dexamethasone	0.1 mg/kg i.m. or i.v. (single dose or repeated one or more times)
Dextrose	500 mL (50% solution) i.v. or s.c. (adult cows)
Diazepam	0.4 mg/kg (calves) i.v.; single dose
Dihydrostrepto-mycin	11 mg/kg i.m. or s.c.; TID
Dinoprost tromethamine	25 mg i.m. or s.c.; single dose
Dopamine hydrochloride	2–10 µg/kg/min infusion i.v.
Doxapram hydrochloride	5–10 mg/kg i.v.; single dose
Enrofloxacin	7.5–12.5 mg/kg s.c.; single dose 2.5–5.0 mg/kg s.c.; SID
Epinephrine (1 mg/mL)	0.01–0.02 mL i.m. or s.c.; single dose 0.1–0.2 mL i.v.; single dose
Erythromycin base	2.2–15 mg/kg i.m.; SID or BID
Florfenicol	20 mg/kg i.m.; a second dose 48 h later 40 mg/kg i.m; single dose
Flunixin meglumine	1.1–2.2 mg/kg i.v.; SID or BID
Framycetin sulphate	5–10 mg/kg (calves) i.m.; BID or p.o.; SID
Furosemide	0.5 or 1 mg/kg i.v. or i.m. (adult cattle); SID or BID
Gonadorelin	100 mcg i.m. or i.v.; single dose 100 µg i.m.; single dose
Ketoprofen	2–4 mg/kg i.m. or i.v.; BID or SID
Lasalocid	1 mg/kg p.o.; SID
Lincomycin	5–10 mg/kg i.m.; SID or BID
Magnesium sulphate	0.02 mg/kg (with calcium gluconate) i.v. (slow); single dose 0.1 mg/kg s.c.; single dose
Mannitol	1–3 g/kg i.v.; single dose (hypovolaemic shock)
Marbofloxacin	2 mg/kg i.m., i.v. or s.c.; SID

Meloxicam	0.5 mg/kg i.v. or s.c.; single dose 0.5–1.0 mg/kg p.o.; Q24–48 h
Oxytetracycline	5–10 mg/kg i.m.; SID 20 mg/kg i.m.; Q48–72 h (Long action 200)
Oxytocin	Retained placenta: 20 IU i.m.; TID or QID for 2–3 days Milk letdown: 10–20 IU i.v.; single dose Obstetric: 30–50 IU i.m.; can be repeated after 30 min
Penicillin G procaine	0.15–0.22 mg/kg i.m. or s.c.; SID
Phenylbutazone	4 mg/kg i.v.; SID 10–20 mg/kg (loading dose) and then 5–10 mg/kg p.o.; Q24–48 h
Prednisolone	1–4 mg/kg i.v.; single dose or SID
Ranitidine hydrochloride	50 (calves) p.o.; QID
Romifidine	0.003–0.02 i.v. or s.c.; Q2–4 h
Spectinomycin	10–15 mg/kg s.c.; SID
Streptomycin	11 mg/kg i.m. or s.c.; BID
Sulfachloropyridazine	88–110 mg/kg i.v.; SID or BID 30–50 mg/kg (calves) p.o.; QID
Sulfadimethoxine	55–110 mg/kg p.o.; SID
Sulfadoxine/trimethoprim	15 mg/kg i.m., s.c. or i.v.; SID or BID 15–30 mg/kg i.m. or i.v.; SID or BID 15–30 mg/kg (pre-ruminant calves) p.o.; SID or BID
Thiamine hydrochloride (vitamin B1)	5–50 mg/kg i.v. or i.m.; BID
Triamcinolone acetonide	0.02–0.04 mg/kg i.m.; single dose
Tylosin	18 mg/kg i.m.; SID

Xylazine hydrochloride	0.02–0.10 mg/kg i.m.; single dose 0.02–0.10 mg/kg i.m.; single dose Standing sedation with a low incidence of recumbency: Quiet dairy breeds – 0.0075–0.01 mg/kg, i.v.; 0.015–0.02 mg/kg, i.m. Tractable cattle 0.01–0.02 mg/kg, i.v.; 0.02–0.04 mg/kg, i.m. Anxious cattle 0.02–0.03 mg/kg, i.v.; 0.04–0.06 mg/kg, i.m. Extremely anxious or unruly cattle 0.025–0.05 mg/kg, i.v.; 0.05–0.1 mg/kg, i.m.

Recommendation of the standard dose, administration route and interval of the drugs for adult cows and calves. The manufacturer's recommendations and local country regulations should be applied, unless the veterinarian assumes the off-label use responsibility of pharmaceutical drugs. The withdrawal period for milk and meat, according to the manufacturer's information, is required.

Abbreviations for the route of administration: i.v., intravenous; i.m., intramuscular; s.c., subcutaneous; and p.o-., oral (*per os*).

Abbreviations for the interval of administration: SID (*semel in die*), once a day; BID (*bis in die*), twice a day; TID (*ter in die*), three times a day; and QID (*quater in die*), four times a day.

Reference

1. Plumb DC. Plumb's veterinary drug handbook. Veterinary medicine. 9th ed. St. Paul: John Wiley & Sons, Inc.; 2018; Constable PD, Hinchcliff KW, Done SH, Grünberg W. A textbook of the diseases of cattle, horses, sheep, pigs and goats. 11th ed. St. Louis: Elsevier; 2017; Other references.

Index

Printed in the United States
by Baker & Taylor Publisher Services